Chromosomal Mutagenesis

METHODS IN MOLECULAR BIOLOGY™

John M. Walker, SERIES EDITOR

439. **Genomics Protocols:** *Second Edition, edited by Mike Starkey and Ramnanth Elaswarapu, 2008*
438. **Neural Stem Cells:** *Methods and Protocols, Second Edition, edited by Le slie P. Weiner, 2008*
437. **Drug Delivery Systems,** *edited by Kewal K. Jain, 2008*
436. **Avian Influenza Virus,** *edited by Erica Spackman, 2008*
435. **Chromosomal Mutagenesis,** *edited by Gregory D. Davis and Kevin J. Kayser, 2008*
434. **Gene Therapy Protocols:** *Volume 2, Design and Characterization of Gene Transfer Vectors, edited by Joseph M. LeDoux, 2008*
433. **Gene Therapy Protocols:** *Volume 1, Production and In Vivo Applications of Gene Transfer Vectors, edited by Joseph M. LeDoux, 2008*
432. **Organelle Proteomics,** *edited by Delphine Pflieger and Jean Rossier, 2008*
431. **Bacterial Pathogenesis:** *Methods and Protocols, edited by Frank DeLeo and Michael Otto, 2008*
430. **Hematopoietic Stem Cell Protocols,** *edited by Kevin D. Bunting, 2008*
429. **Molecular Beacons:** *Signalling Nucleic Acid Probes, Methods and Protocols, edited by Andreas Marx and Oliver Seitz, 2008*
428. **Clinical Proteomics:** *Methods and Protocols, edited by Antonio Vlahou, 2008*
427. **Plant Embryogenesis,** *edited by Maria Fernanda Suarez and Peter Bozhkov, 2008*
426. **Structural Proteomics:** *High-Throughput Methods, edited by Bostjan Kobe, Mitchell Guss, and Huber Thomas, 2008*
425. **2D PAGE:** *Volume 2, Applications and Protocols, edited by Anton Posch, 2008*
424. **2D PAGE:** *Volume 1, Sample Preparation and Pre-Fractionation, edited by Anton Posch, 2008*
423. **Electroporation Protocols,** *edited by Shulin Li, 2008*
422. **Phylogenomics,** *edited by William J. Murphy, 2008*
421. **Affinity Chromatography:** *Methods and Protocols, Second Edition, edited by Michael Zachariou, 2008*
420. **Drosophila:** *Methods and Protocols, edited by Christian Dahmann, 2008*
419. **Post-Transcriptional Gene Regulation,** edited by *Jeffrey Wilusz, 2008*
418. **Avidin–Biotin Interactions:** *Methods and Applications,* edited by *Robert J. McMahon, 2008*
417. **Tissue Engineering,** Second Edition, edited by *Hannsjörg Hauser and Martin Fussenegger, 2007*
416. **Gene Essentiality:** *Protocols and Bioinformatics,* edited by *Andrei L. Osterman, 2008*
415. **Innate Immunity,** edited by *Jonathan Ewbank and Eric Vivier, 2007*
414. **Apoptosis in Cancer:** *Methods and Protocols,* edited by *Gil Mor and Ayesha Alvero, 2008*
413. **Protein Structure Prediction,** Second Edition, edited by *Mohammed Zaki and Chris Bystroff, 2008*
412. **Neutrophil Methods and Protocols,** edited by *Mark T. Quinn, Frank R. DeLeo, and Gary M. Bokoch, 2007*
411. **Reporter Genes for Mammalian Systems,** edited by *Don Anson, 2007*
410. **Environmental Genomics,** edited by *Cristofre C. Martin, 2007*
409. **Immunoinformatics:** *Predicting Immunogenicity In Silico,* edited by *Darren R. Flower, 2007*
408. **Gene Function Analysis,** edited by *Michael Ochs, 2007*
407. **Stem Cell Assays,** edited by *Vemuri C. Mohan, 2007*
406. **Plant Bioinformatics:** *Methods and Protocols,* edited by *David Edwards, 2007*
405. **Telomerase Inhibition:** *Strategies and Protocols,* edited by *Lucy Andrews and Trygve O. Tollefsbol, 2007*
404. **Topics in Biostatistics,** edited by *Walter T. Ambrosius, 2007*
403. **Patch-Clamp Methods and Protocols,** edited by *Peter Molnar and James J. Hickman, 2007*
402. **PCR Primer Design,** edited by *Anton Yuryev, 2007*
401. **Neuroinformatics,** edited by *Chiquito J. Crasto, 2007*
400. **Methods in Lipid Membranes,** edited by *Alex Dopico, 2007*
399. **Neuroprotection Methods and Protocols,** edited by *Tiziana Borsello, 2007*
398. **Lipid Rafts,** edited by *Thomas J. McIntosh, 2007*
397. **Hedgehog Signaling Protocols,** edited by *Jamila I. Horabin, 2007*
396. **Comparative Genomics,** *Volume 2,* edited by *Nicholas H. Bergman, 2007*
395. **Comparative Genomics,** *Volume 1,* edited by *Nicholas H. Bergman, 2007*
394. **Salmonella:** *Methods and Protocols,* edited by *Heide Schatten and Abe Eisenstark, 2007*
393. **Plant Secondary Metabolites,** edited by *Harinder P. S. Makkar, P. Siddhuraju, and Klaus Becker, 2007*
392. **Molecular Motors:** *Methods and Protocols,* edited by *Ann O. Sperry, 2007*
391. **MRSA Protocols,** edited by *Yinduo Ji, 2007*
390. **Protein Targeting Protocols, Second Edition,** edited by *Mark van der Giezen, 2007*
389. **Pichia Protocols, Second Edition,** edited by *James M. Cregg, 2007*
388. **Baculovirus and Insect Cell Expression Protocols, Second Edition,** edited by *David W. Murhammer, 2007*
387. **Serial Analysis of Gene Expression (SAGE):** *Digital Gene Expression Profiling,* edited by *Kare Lehmann Nielsen, 2007*
386. **Peptide Characterization and Application Protocols,** edited by *Gregg B. Fields, 2007*
385. **Microchip-Based Assay Systems:** *Methods and Applications,* edited by *Pierre N. Floriano, 2007*
384. **Capillary Electrophoresis:** *Methods and Protocols,* edited by *Philippe Schmitt-Kopplin, 2007*
383. **Cancer Genomics and Proteomics:** *Methods and Protocols,* edited by *Paul B. Fisher, 2007*
382. **Microarrays, Second Edition:** *Volume 2, Applications and Data Analysis,* edited by *Jang B. Rampal, 2007*
381. **Microarrays, Second Edition:** *Volume 1, Synthesis Methods,* edited by *Jang B. Rampal, 2007*
380. **Immunological Tolerance:** *Methods and Protocols,* edited by *Paul J. Fairchild, 2007*
379. **Glycovirology Protocols,** edited by *Richard J. Sugrue, 2007*
378. **Monoclonal Antibodies:** *Methods and Protocols,* edited by *Maher Albitar, 2007*
377. **Microarray Data Analysis:** *Methods and Applications,* edited by *Michael J. Korenberg, 2007*
376. **Linkage Disequilibrium and Association Mapping:** *Analysis and Application,* edited by *Andrew R. Collins, 2007*
375. **In Vitro Transcription and Translation Protocols:** *Second Edition,* edited by *Guido Grandi, 2007*
374. **Quantum Dots:** *Applications in Biology,* edited by *Marcel Bruchez and Charles Z. Hotz, 2007*
373. **Pyrosequencing® Protocols,** edited by *Sharon Marsh, 2007*
372. **Mitochondria: Practical Protocols,** edited by *Dario Leister and Johannes Herrmann, 2007*
371. **Biological Aging:** *Methods and Protocols,* edited by *Trygve O. Tollefsbol, 2007*
370. **Adhesion Protein Protocols,** *Second Edition,* edited by *Amanda S. Coutts, 2007*
369. **Electron Microscopy:** *Methods and Protocols, Second Edition,* edited by *John Kuo, 2007*
368. **Cryopreservation and Freeze-Drying Protocols,** *Second Edition,* edited by *John G. Day and Glyn Stacey, 2007*
367. **Mass Spectrometry Data Analysis in Proteomics,** edited by *Rune Matthiesen, 2007*
366. **Cardiac Gene Expression:** *Methods and Protocols,* edited by *Jun Zhang and Gregg Rokosh, 2007*
365. **Protein Phosphatase Protocols:** edited by *Greg Moorhead, 2007*
364. **Macromolecular Crystallography Protocols:** *Volume 2, Structure Determination,* edited by *Sylvie Doublié, 2007*
363. **Macromolecular Crystallography Protocols:** *Volume 1, Preparation and Crystallization of Macromolecules,* edited by *Sylvie Doublié, 2007*

METHODS IN MOLECULAR BIOLOGY™

Chromosomal Mutagenesis

Edited by

Gregory D. Davis

Sigma Aldrich Biotechnology,
Sigma-Aldrich Corporation,
St. Louis, Missouri

Kevin J. Kayser

SAFC Biosciences,
Sigma-Aldrich Corporation,
St. Louis, Missouri

HUMANA PRESS TOTOWA, NEW JERSEY

999 Riverview Drive, Suite 208
Totowa, New Jersey 07512

www.humanapress.com

This publication is printed on acid-free paper. ∞
ANSI Z39.48-1984 (American Standards Institute)

Permanence of Paper for Printed Library Materials.
Cover illustration: Fig. 7 of Chapter 8 (Detection of promoter trap events using X-Gal staining) by Junji Takeda, Zsuzsanna Izsvák, and Zoltán Ivics

Production Editor: Rhukea J. Hussain
Cover design by Karen Schulz

For additional copies, pricing for bulk purchases, and/or information about other Humana titles, contact Humana at the above address or at any of the following numbers: Tel.: 973-256-1699; Fax: 973-256-8341; E-mail: orders@humanapr.com; or visit our Website: www.humanapress.com

Printed in the United States of America. 10 9 8 7 6 5 4 3 2 1

ISBN 13: 978-1-61737-833-1 e-ISBN 13: 978-1-59745-232-8

For Shannon, Jonah, Harrison, and Addison

Preface

Gene disruptions, knockouts, and stimulated homologous recombination are powerful site-specific mutagenesis tools that can be used for the elucidation of gene function, gene therapy, cell-line engineering, and target validation studies in drug discovery. In the past few years many international collaborations and scientific consortiums have formed whose directive is to disrupt every gene in the genome of given model organisms in an attempt to determine specific gene function. The enabling feature of most genome-wide gene knockout endeavors is an efficient method for chromosomal mutagenesis, such as those available for *Escherchia coli* and *Sacharomyces cerevisiae*. In particular, the ability of *E. coli* and yeast to uptake and incorporate PCR products guided to genomic sites by small primer-dictated sequences has been particularly useful for high throughput and economical chromosomal modification. While these model organisms serve well to delineate broad classifications of gene function, they still cannot directly address unique genetic features of certain organisms. Studying model systems or surrogates through systems biology approaches provides a mechanism to study complex systems; however, it is through the sum interaction of their parts that result in unique physiology, anatomy, disease states, and microbial virulence. Such cell-specific phenotypic qualities are a critical selective force imposed upon finding suitable target genes for drug and vaccine development.

Great disparities exist between organisms with regard to the relative ease of chromosomal mutagenesis and manipulation. Typical barriers to efficient homologous recombination include the difficulty of DNA delivery, particularly inefficient homologous recombination systems, chromatin interference, competing non-homologous integration pathways, and basic limitations in the ability of certain cells to survive cloning operations, as in the case of rat oocytes. Like yeast, some prokaryotes, such as *Streptococcus pneumoniae* and *Helicobacter pylori*, are highly transformable and efficient at incorporating donor DNA with minimal homologous ends into the genome. However, others such as *Mycobacterium tuberculosis* and *Clostridium perfringens*, are extremely difficult to mutate by conventional methods. Similarly for eukaryotic cells, it is profoundly unfortunate that human somatic cells remain generally resistant to efficient homologous recombination. Methods that attempt to overcome limitations in mammalian gene targeting, such as RNA interference (RNAi), have become widespread. While RNAi does not result in a complete disruption of gene activity, it has revolutionized eukaryotic reverse genetics by its procedural ease, rapid experimental turn-around, low expense, and, most significantly, application to diverse cell types. It is these qualities that permanent chromosomal manipulation methods should strive for in hopes of simplifying and improving gene targeting techniques. This volume will focus on a variety of chromosomal mutagenesis techniques for both prokaryotic and eukaryotic organisms. Methods covered include insertional gene disruptions, gene knockouts, stimulated homologous recombination techniques and novel tools based upon integrases, eukaryotic transposons, triplex forming oligonucleotides, group II introns, and engineered site-directed nucleases. In particular, we seek to highlight techniques that expand the genetic toolbox beyond model organisms into a wider variety of cell types and organisms.

Chinese hamster ovary (CHO) cells are an excellent example of a highly valuable cell type that lacks genetic tools and information. The chapter by Mitsuo Satoh and colleagues outlines an excellently detailed procedure which has been used to create the first biallelic gene knockouts in CHO cells, a hallmark achievement for industrial cell engineering. Since this method is based upon classical homologous recombination, it retains a high degree of flexibility in targeting virtually any region of the CHO genome. While government regulations restrict cell types, such as CHO, used for production of recombinant therapeutics and require the development of suitable mutagenesis methods, fundamental research allows consideration of many different cell-types, some of which are human-derived and have elevated rates of homologous recombination. Noritaka Adachi and colleagues describe protocols for gene targeting in one such cell line, Nalm-6, which has been used extensively as a model for leukemia research. In addition to having favorable recombination efficiencies, Nalm-6 retains a stable karyotype and normal p53 status making it more suitable for functional genomics studies.

One of the most useful and remarkable discoveries in recent years is that double-strand chromosomal breaks (DSBs) can stimulate homologous recombination (HR) by orders of magnitude. While the increase in HR efficiency is impressive, the ability to site-specifically apply this technique throughout the genome has been limited until very recently. The chapter by Jean-Pierre Cabaniols and Frédéric Pâques describes an efficient method using the I-*Sce*I meganuclease to create site specific mutations in CHO-K1 cells. These methods can also be applied to I-*Cre*I meganucleases which have been engineered to have flexible site-specificity. In another approach for targeting DSBs, zinc-finger binding domains have recently been combined with the *Fok*I nuclease to create zinc-finger nucleases (ZFNs). The combination of these two eukaryotic and prokaryotic modular protein domains has encountered great success. The chapter by Matthew Porteus describes the bioinformatic intricacies associated with ZFN and target site design and subsequent protocols for gene targeting in somatic mammalian cells. The chapter by Dana Carroll and colleagues describes excellent methods which expand ZFN gene targeting applications to *Drosophila melanogaster* and *Caenorhabditis elegans*. In many instances, as in several of the previously mentioned chapters, gene targeting methods require a gene-targeting vector to be constructed to introduce modified DNA into the chromosome, often a time consuming and expensive procedure. The chapter by Derrick Rancourt and colleagues outlines a very efficient and streamlined approach, termed Orpheus Recombination, using enhanced phage-based recombination to rapidly generate targeting vectors.

For years, *E. coli* and *Drosophila* genetics have benefited greatly from genome-wide transposon experiments to delineate gene function. Until recently, the same benefits were not realized in mammalian cells due to lack sufficiently active insertional elements. Two attractive features of insertional mutagenesis are that the mechanisms of insertion do not typically compete with or depend greatly on homologous recombination functions of the cell, and once an insertion is achieved at a specific site, a molecular tag is left in the genome that can easily be used to locate the genomic locus associated with a given mutant phenotype. The nuances of insertional element location are nicely described in the chapter by David Largaespada and Lara Collier where they detail the procedures required to identify and characterize insertional events in mouse tumor cells resulting from integration of the Sleeping Beauty (SB) transposon. The chapter by Zoltan Ivics and colleagues describes upstream methods of transformation and selection of the SB transposon to create germ-line mutations in mice. These applications of the SB transposon are an extremely important new development in cancer genetics since they can mutate a wider variety of tissue types than previous systems, often resulting in the creation of more relevant solid tumor phenotypes. Gene trapping is a novel method that allows for random insertional disruption of genes while also providing a means to report expression via gene-reporter splicing. Thomas Floss and Frank Schnütgen provide a chapter which describes protocols for characterizing conditional gene trap ES clones.

Transposon insertion is typically a random event, which is a very useful quality for relating observed phenotypes to a genomic locus. However, when researchers are focused on the function of a specific gene, finding an insertional clone from a random library can be labor intensive. The chapter by Craig Coates, Joseph Kaminski, and colleagues outlines protocols to target the *piggyBac* transposon by creating gene fusions to GAL4 binding domains. A significant aspect of this work is that the transposon retains sufficient activity when fused to the GAL4 binding domain for practical insertional mutagenesis. In a similar application, Dan Voytas and colleagues describe site-specific targeting of the Ty5 retrotransposon by a gene fusion approach exploiting the specific interaction of Ty5 with the Sir4 protein. Not only does the Ty5-Sir4 fusion method provide for site specific integration, but may also result in a new tool for determining DNA binding sites of transcription factors by leaving the inserted Ty5 insertional tag at the binding site. In addition to targeting specific genes for functional genomics applications, single site specificity is often desired when delivering transgenes to minimize off-target effects in gene therapy applications. In the chapter by Annahita

Keravala and Michele Calos they describe methods for site specific chromosomal integration mediated by ϕC31 integrase. Unlike similar applications with Cre and Flp recombinases, ϕC31 integrase results in an irreversible and stable integration of DNA. As mentioned previously, procedural ease is highly valued in gene targeting procedures, and the use of PCR-based donor DNA synthesis, or, if possible, small synthetic DNA, can greatly simplify work involved in assembling gene targeting materials. In the chapter by Peter Glazer and colleagues methods are provided for site specific gene modification using synthetic triplex-forming oligonucleotides (TFOs) and peptide nucleic acids (PNA).

Relative to mammalian eukaryotic genetics, the huge diversity of the single-cell microbial world can present very unique barriers for chromosomal mutagenesis. *Mycobacterium tuberculosis* is a leading world scourge, killing over 3 million people worldwide each year. Unfortunately, *M. tuberculosis* grows incredibly slow as typical prokaryotes go, and, to make matters worse, it is notoriously difficult to mutate. The chapter by Martin Pavelka outlines a suicide vector approach which enables the creation of point mutations, insertions, and deletions that delicately alter functionality with minimal impact on surrounding genes. As mentioned previously, *E. coli* and yeast can have minimal requirements for homologous sequence length to guide substrate DNA to specific genomic sites allowing site specificity to be efficiently changed through simple oligo synthesis. While this is natural phenomenon in yeast, the use of short homologous arms (~50 bp) in *E. coli* has required the expression of phage proteins to enhance homologous recombination. Lambda phage proteins have a limited enteric gram-negative host range for enhancing homologous recombination. Julia van Kessel and Graham Hatfull have contributed a chapter describing the discovery and implementation of similar phage proteins that enhance homologous recombination in Mycobacteria. This recombineering approach serves to minimize the length of homologous sequences required in donor DNA and make homologous recombination more competitive with the dominant illegitimate recombination mechanism in Mycobacteria. Clostridial species have also long suffered for lack of efficient genetic systems, and fatality rates for some species have recently surpassed that of methicillin-resistant *Staphylococcus aureus* (MRSA) in the UK. The chapter by Yue Chen and Phalguni Gupta presents a targeted insertional disruption method based on the Ll.LtrB group II intron which has greatly reduced the time required to create targeted gene disruptions in *Clostridium perfringens*.

In addition to providing more advanced detail on mutagenesis methods, we have intended to cover state-of-the-art techniques that are staged to expand, if not revolutionize, genetic analysis in long neglected and relevant cell types. The editors would like to extend their greatest thanks to all the participating authors for their willingness and enthusiasm to contribute. We thank David Casey and Patrick Marton at Humana Press and John Walker at the University of Hertfordshire for their support throughout this work. We would also like to thank Don Ennis at the

University of Louisiana, Douglas Berg at Washington University, Nigel Minton at the University of Nottingham, and Clive Svendsen at the University of Wisconsin for helpful discussions during the assembly of this volume.

Gregory D. Davis can be contacted at Greg.Davis@sial.com and Kevin Kayser can be contacted at Kevin.Kayser@sial.com.

Gregory D. Davis and Kevin J. Kayser

Contents

Contributors

NORITAKA ADACHI • *Kihara Institute for Biological Research, Yokohama City University, Yokohama, Japan*
KELLY J. BEUMER • *Department of Biochemistry, University of Utah School of Medicine, Salt Lake City, UT*
ANA BOZAS • *Department of Biochemistry, University of Utah School of Medicine, Salt Lake City, UT*
TROY L. BRADY • *Department of Genetics, Development and Cell Biology, Iowa State University, Ames, IA*
JEAN-PIERRE CABANIOLS • *CELLECTIS S.A., Romainville, France*
MICHELE P. CALOS • *Department of Genetics, Stanford University School of Medicine, Stanford, CA*
DANA CARROLL • *Department of Biochemistry, University of Utah School of Medicine, Salt Lake City, UT*
YUE CHEN • *Department of Infectious Diseases and Microbiology, Graduate School of Public Health, University of Pittsburgh, Pittsburgh, PA*
JOANNA Y. CHIN • *Department of Genetics and Department of Therapeutic Radiology, Yale University School of Medicine, New Haven, CT*
SAREINA CHIUNG-YUAN WU • *Department of Entomology, Texas A&M University, College Station, TX*
CRAIG J. COATES • *Department of Entomology, Texas A&M University, College Station, TX*
LARA S. COLLIER • *Department of Genetics, Cell Biology and Development, University of Minnesota, Minneapolis, MN*
THOMAS FLOSS • *Institute of Developmental Genetics, GSF-National Research Center for Environment and Health, Neuherberg, Germany*
PETER M. GLAZER • *Department of Genetics and Department of Therapeutic Radiology, Yale University School of Medicine, New Haven, CT*
PHALGUNI GUPTA • *Department of Infectious Diseases and Microbiology, Graduate School of Public Health, University of Pittsburgh, Pittsburgh, PA*
GRAHAM F. HATFULL • *Pittsburgh Bacteriophage Institute and Department of Biological Sciences, University of Pittsburgh, Pittsburgh, PA*
KENICHI ITO • *Department of Biochemistry and Molecular Biology, University of Calgary, Calgary, Alberta, Canada*
ZOLTÁN IVICS • *Max Delbrück Center for Molecular Medicine, Berlin, Germany*
ZSUZSANNA IZSVÁK • *Max Delbrück Center for Molecular Medicine, Berlin, Germany*
JOSEPH M. KAMINSKI • *Department of Entomology, Texas A&M University, College Station, TX*

ANNAHITA KERAVALA • *Department of Genetics, Stanford University School of Medicine, Stanford, CA*
JULIA C. VAN KESSEL • *Pittsburgh Bacteriophage Institute and Department of Biological Sciences, University of Pittsburgh, Pittsburgh, PA*
HIDEKI KOYAMA • *Kihara Institute for Biological Research, Yokohama City University, Yokohama, Japan*
AYA KUROSAWA • *Kihara Institute for Biological Research, Yokohama City University, Yokohama, Japan*
DAVID A. LARGAESPADA • *Department of Genetics, Cell Biology and Development, University of Minnesota, Minneapolis, MN*
KOMMINENI J. MARAGATHAVALLY • *Department of Entomology, Texas A&M University, College Station, TX*
J. JASON MORTON • *Department of Biochemistry, University of Utah School of Medicine, Salt Lake City, UT*
FRÉDÉRIC PÂQUES • *CELLECTIS S.A., Romainville, France*
MARTIN S. PAVELKA, JR. • *Department of Microbiology and Immunology, University of Rochester Medical Center, Rochester, NY*
MATTHEW PORTEUS • *Depts. of Pediatrics and Biochemistry, UT Southwestern Medical Center, Dallas, TX*
DERRICK E. RANCOURT • *Department of Biochemistry and Molecular Biology, University of Calgary, Calgary, Alberta, Canada*
MITSUO SATOH • *Tokyo Research Laboratories, Kyowa Hakko Kogyo Co., Ltd., Tokyo, Japan*
ERICA B. SCHLEIFMAN • *Department of Genetics, Yale University School of Medicine, New Haven, CT*
CLARICE L. SCHMIDT • *Department of Genetics, Development and Cell Biology, Iowa State University, Ames, IA*
FRANK SCHNÜTGEN • *Department of Molecular Hematology, University of Frankfurt Medical School, Frankfurt, Germany*
JUNJI TAKEDA • *Center for Advanced Science and Innovation & Department of Social and Environmental Medicine, Osaka University, Osaka, Japan*
JONATHAN K. TRAUTMAN • *Department of Biochemistry, University of Utah School of Medicine, Salt Lake City, UT*
TERUHISA TSUZUKI • *Department of Integrative Biomedical Sciences, Kyushu University, Fukuoka, Japan*
DANIEL F. VOYTAS • *Department of Genetics, Development and Cell Biology, Iowa State University, Ames, IA*
KNUT WOLTJEN • *Department of Biochemistry and Molecular Biology, University of Calgary, Calgary, Alberta, Canada*
NAOKO YAMANE-OHNUKI • *Tokyo Research Laboratories, Kyowa Hakko Kogyo Co., Ltd., Tokyo, Japan*
KAZUYA YAMANO • *Tokyo Research Laboratories, Kyowa Hakko Kogyo Co., Ltd., Tokyo, Japan*

1

Biallelic Gene Knockouts in Chinese Hamster Ovary Cells

Naoko Yamane-Ohnuki, Kazuya Yamano, and Mitsuo Satoh

Summary

Chinese hamster ovary (CHO) cells are the most common host cells and are widely used in the manufacture of approved recombinant therapeutics. They represent a major new class of universal hosts in biopharmaceutical production. However, there remains room for improvement to create more ideal host cells that can add greater value to therapeutic recombinant proteins at reduced production cost. A promising approach to this goal is biallelic gene knockout in CHO cells, as it is the most reliable and effective means to permanent phenotypic change, owing to the complete removal of gene function. In this chapter, we describe a biallelic gene knockout process in CHO cells, as exemplified by the successful targeted disruption of both *FUT8* alleles encoding α-1,6-fucosyltransferase gene in CHO/DG44 cells. Wild-type alleles are sequentially disrupted by homologous recombination using two targeting vectors to generate homozygous disruptants, and the drug-resistance gene cassettes remaining on the alleles are removed by a Cre/*loxP* recombination system so as not to leave the extraphenotype except for the functional loss of the gene of interest.

Key Words: Biallelic gene knockouts; Chinese hamster ovary (CHO) cells; homologous recombination; *FUT8*; targeted disruption; biopharmaceutical production.

1. Introduction

Recombinant protein expression technology in mammalian cell cultures is the principal means of the commercial production of glycobiopharmaceuticals. In fact, all approved therapeutic recombinant antibodies as well as erythropoietin have been manufactured in mammalian cells *(1,2)*, and include the majority of the recombinant therapeutic proteins currently used in clinics. One of the most common host cell lines used in the manufacture of these therapeutics is the Chinese hamster ovary (CHO) cell line, which represents the major new class of universal hosts in biopharmaceutical production. In pharmaceutical production, it is unacceptable for the quality of ingredients to vary depending on culture conditions and/or producer clone character, because the variations would affect therapeutic efficacy and could trigger severe clinical problems. The characteristics of CHO

From: *Methods in Molecular Biology, vol. 435: Chromosomal Mutagenesis*
Edited by: G. Davis and K. J. Kayser © Humana Press Inc., Totowa, NJ

cells—ease of genetic manipulation, high cloning efficiency, good proliferation in large-scale suspension culture, easy adaptability to serum- and protein-free media, and both high productivity and stability—are the characteristics necessary for industrial application, and thus robust processes have been developed to produce them. These processes are proven to produce safe and effective biopharmaceutical molecules equivalent to those observed in nature. However, improvements are eagerly awaited to create more ideal host cells by adding greater value to therapeutic recombinant proteins at reduced production cost by improving not only cell growth and viability but also posttranslational modification. Genetic modification of CHO cells is a promising approach to achieve these goals.

Biallelic gene disruption in CHO cells is the most reliable and effective way to completely remove gene function. Other loss-of-function technologies, for example, antisense and RNA interference, reduce target function but do not eliminate it. Recently, we succeeded in generating a mutant CHO cell line in which both *FUT8* alleles, encoding the only mammalian α-1,6-fucosyltransferase gene, are deleted by sequential gene targeting ***(3)***. *FUT8* knockout cells produce completely defucosylated therapeutic antibodies with significant enhancement of antibody-dependent cellular cytotoxicity (ADCC), which is one of the major immunological mechanisms responsible for the clinical efficacy of tumor cell eradication and which is controlled solely by α-1,6-fucosylation on N-linked oligosaccharides of antibodies ***(4)***. Hemizygous cell lines still retain enough α-1,6-fucosyltransferase activity for fully fucosylated antibody production, suggesting that biallelic gene disruption is required for the complete loss of α-1,6-fucosylation.

Gene targeting in mammalian somatic cells is difficult to achieve and requires exceedingly laborious and time-consuming processes. The difficulty arises from the fact that, in somatic cells, nonhomologous recombination events occur several orders of magnitude more frequently than homologous recombination events ***(5)***. In general, homologous recombination is estimated to occur with more than 100-fold lower frequency in somatic cells than in murine embryonic stem (ES) cells ***(6,7)***. In the case of CHO cells, two additional obstacles further complicate gene targeting. One is "targeted gene-templated extension;" the truncated end of a targeting vector sequence is frequently extended in CHO cells by replication of the target homologous region on the genome, and is randomly integrated elsewhere ***(8–12)***. Such recombinants not only still retain uncorrected target gene loci, but also carry corrected targeting vector sequences, which are the same as designed homologous recombinant loci, at ectopic sites. This event gives rise to many pseudo homologous recombinants that are indistinguishable from homologous recombinants by diagnostic genomic polymerase chain reaction (PCR) for the detection of the target fragment (mostly 1–2 Kb in length), because the vector is often extended for several kilobase beyond the target homologous region ***(11,12)***. The other complication to gene targeting is that chromosomal abnormalities, which significantly affect the copy number and chromosomal location of the target loci ***(13)***, accumulate in CHO cells ***(14,15)***. Most cultured somatic cell lines are aneuploid owing to chromosomal aberration in return for the acquisition of immortality during establishment and cultivation ***(16)***. CHO cells showed a pseudodiploid karyotype ($2n = 21$) just after establishment ***(17)***. To date, however, most of the sublines lose the native karyotype and chromosomal structure during long-term cultivation and/or artificial mutagenesis ***(14)***.

Considering these factors, it is desirable to design a "good" targeting vector to increase the chance of a targeted recombination event. An isogenic homologous gene sequence is thought to be necessary for efficient gene targeting in murine ES cells ***(18)***. Therefore, to construct *FUT8*-targeting vectors, we have used the CHO *FUT8* genomic fragment, including the first coding exon isolated from the CHO cell λ-genomic library, and succeeded in the gene targeting. In our experience, targeting vectors in which a homologous region is obtained by genomic PCR containing several possible mismatched base pairs have also worked well in CHO cells, which is consistent with the report that there is no obvious evidence of a strict isogenicity requirement for gene targeting in human cells ***(19)***. Target sequence homology length in the vector is another key factor in targeting frequency. In general, a target vector including a longer homologous region achieves a higher targeting frequency ***(20)***. In the case of CHO cells, several kilobase of a homologous gene sequence is required on each side of a selection marker on the target vector in order to improve the homologous recombination efficiency ***(12)***.

It is also critical that the target vector carries an efficient selection system to enrich the targeted recombinants from a huge number of transfectants. For this purpose, typically, a positive–negative selection (PNS) strategy is used to enrich the targeted recombinants by about 2- to 10-fold in mammalian somatic cells ***(21)***. We have used the PNS vectors carrying the modified *FUT8* exon by replacing the translation initiation site with a drug-resistance gene cassette, in which either the neomycin-resistance gene (*Neo*^{r}) or the puromycin-resistance gene (*Puro*^{r}) is used as a positive selection marker, and the diphtheria toxin gene (*DT*) is used to kill randomly integrated nontargeted recombinants as a negative selection marker (*see* **Fig. 1A** and **ref.** *3*). In CHO/DG44 cells, the enrichment efficiency of homologous recombinants by *DT* negative selection is estimated to be approximately twofold (unpublished data), which might be owing to the fact that DT-resistance mutation occurs easily in the CHO genome ***(22)***. To eliminate chromosomal position effects that hamper the selection marker gene expression on the genome, it is effective to place insulator sequences on each side of the *DT* expression cassette to improve the enrichment efficiency by more than threefold (unpublished data). Another selection strategy is the use of a promoter trap (PT) system, which is reported to yield up to 5000–10,000-fold enrichment of homologous recombinants ***(21)***. PT vectors are designed to drive a positive selection marker gene under the control of the endogenous promoter of

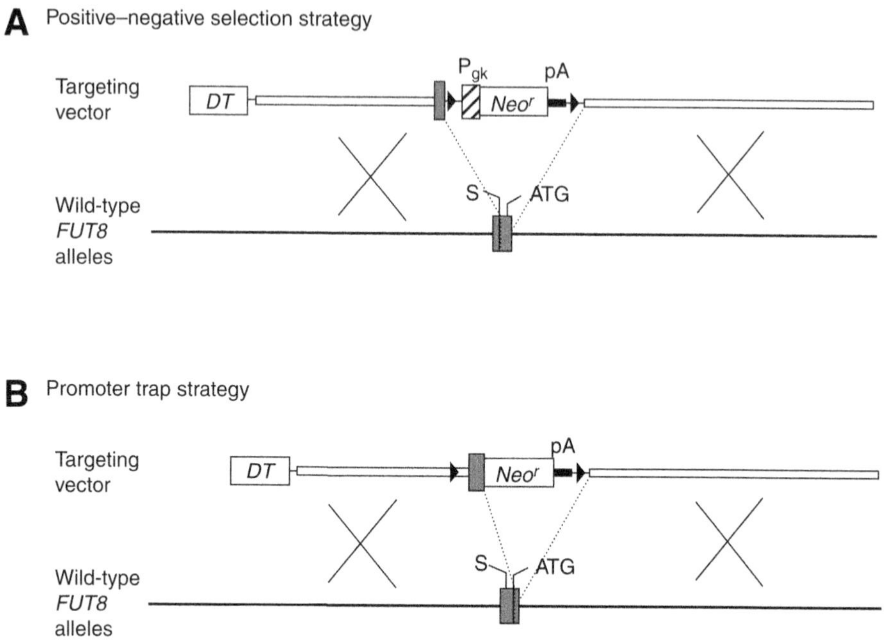

Fig. 1. Targeting vectors for *FUT8* gene knockout in CHO cells. Wild-type *FUT8* alleles of CHO cells and the targeting constructs. **(A)** PNS strategy: exon 2 (filled boxes) is modified by replacing a segment containing the translation initiation site with the drug-resistance gene cassette, flanked by two *loxP* sites (filled triangles). **(B)** PT strategy: the promoterless *Neo*^{r} is fused in-frame to the translation initiation site of *FUT8* gene on exon 2. Resulting hybrid selection marker cassette is floxed.

the target gene. In CHO/DG44 cells, the enrichment efficiency of homologous recombinants by the PT system in which the promoterless *Neo*^{r} is fused in-frame to the endogenous *FUT8* gene (*see* **Fig. 1B**), is estimated to be at least 10-fold. The PT strategy, irrespective of *DT* modification flanked by insulators, has achieved a targeting efficiency dozens of times greater than that of the PNS strategy (*see* **Table 1**).

Before gene targeting, it is necessary to estimate the copy number and chromosomal locations of the target loci by Southern blot analysis and fluorescence *in situ* hybridization. The chromosomal morphology of CHO cells is not as stable as that of ES cells and may change during long-term cultivation by our intervention, giving rise to variants derived even from a single clone. For example, in the case of CHO/DG44, the karyotypes vary from lab to lab; the chromosome number is reported to be $2n = 18$ ***(14)***, $2n = 20$ ***(23)***, or $2n = 34$ ***(24)***. Thus, we should note the possibility that the number of target loci is not always two copies. Assuming that several rounds of gene targeting would be necessary for the complete functional loss of more than three copies of target loci, it is desirable to design floxed

Table 1
Gene Targeting Efficiency in CHO/DG44 Cells

	Screened *Neo*^{r} clones	PCR-positive clones	Targeted clones
PNS vector	45,000	39	1
PT vector	3000	84	1
PT vector with modified *DT*	2500	29	1

targeting vectors in which the positive selection markers are removable by the Cre/*loxP* recombination system. On the other hand, CHO cells are known to have "functional hemizygosity" on many gene alleles ***(25)***, and only one round of gene targeting is enough to remove the full function of the target gene.

Screening scales to identify targeted clones depend on the targeting frequency in host cells as well as that in target gene loci. In the targeting of the *FUT8* loci in CHO/DG44 using PNS vectors, approx 45,000 transfectants were screened to find only one hemizygous target clone. On the other hand, in the second round of targeting, 7000 transfectants were enough to identify a homozygous disruptant ***(3)***. In addition, using the very same PNS vectors, the targeting event has occurred at much higher frequencies of 1/900–1/3200 in some clones derived from original CHO/DG44 cells as hosts (unpublished data). These discrepancies in targeting frequency among clones derived from the same parent cells have never been reported in ES cells. Considering that CHO cells consist of diverse clones with genetic heterogeneity owing to chromosomal rearrangement as described earlier, it is possible that targeting frequency varies by sublines and/or by labs. In CHO/DG44, phenotypic variations, for example, morphology, growth rate, and cell adhesiveness, have no relation to the gene-targeting frequency, although they do affect the cloning efficiency of the targeted recombinants. Therefore, it is recommended that a suitable clone for gene targeting is used as a host if several CHO clones are available.

Phenotypic selection is a powerful strategy for efficient enrichment for the targeted recombinants. Some groups have succeeded in gene targeting at only one locus in CHO cells by means of a selection strategy appropriate for the expected phenotype of targeted events. In these cases, single targeting events are detected by drug resistance either because hemizygous mutants are applied to gene targeting ***(8–12)*** or because single replacement is effective for phenotypic change ***(26)***. However, single targeting is not always sufficient to dramatically change the target gene function. In the case of glycosylation-related genes, lectins specific for the modified glycans are good candidates for positive phenotypic selection drugs. Although phenotypic selection with a lentil aggulutinin, which specifically binds to α-1,6-fucosylated trimannosyl-core structure of N-linked oligosaccharides and kills the cells expressing the structure, does not work well in the identification of hemizygous *FUT8* clones, it has great impact on the enrichment for the homozygous disruptants with greater than 3000-fold efficiency (unpublished data).

Here, we describe the details of the targeted disruption of both *FUT8* alleles in CHO/DG44 by the PNS strategy (*see* **Fig. 2**). Wild-type *FUT8* alleles are sequentially disrupted by homologous recombination using two independent vectors, both of which target the first coding exon of the *FUT8* gene. After homozygous disruptants are generated, the drug-resistance gene cassettes flanking the two *loxP* sites on *FUT8* alleles are removed by transient expression of Cre recombinase.

2. Materials

2.1. Cell Culture

1. CHO/DG44 cells (*see* **Note 1**).
2. T-175 flasks for adherent cell culture (Greiner Bio-One, Frickenhausen, Germany).
3. Dulbecco's phosphate-buffered saline (D-PBS): Ca^{2+}- and Mg^{2+}-free solution is available from Invitrogen (Carlsbad, CA).
4. Iscove's modified Dulbecco's medium (IMDM; Invitrogen)-10% fetal bovine serum (FBS, Invitrogen) medium: IMDM supplemented with 10% FBS, hypoxantin thymidine (HT) supplement (Invitrogen), and 50 μg/mL of gentamycin (tissue culture grade, Nacalai Tesque, Kyoto, Japan). Prepare gentamycin as a 10 mg/mL (200X) stock by dissolving gentamycin in D-PBS or distilled water, filter-sterilize it, and store at 4°C.
5. G418-selection medium: IMDM-10% FBS medium supplemented with 600 μg/mL of G418. Prepare G418 stock solution by dissolving G418 sulfate (Nacalai Tesque) 50 active mg/mL in D-PBS or distilled water, filter-sterilize it, and store at –20°C.
6. G418/puromycin selection medium: IMDM-10% FBS medium supplemented with 500 μg/mL of G418 sulfate and 15 μg/mL of puromycin. Prepare G418 stock solution as described earlier. Prepare puromycin stock solution by dissolving puromycin dihydrochloride (Sigma-Aldrich, St. Louis, MO) 5 mg/mL in D-PBS or distilled water, filter-sterilize it, and store at –20°C.
7. 0.05% (w/v) Trypsin-PBS: trypsin liquid (Invitrogen) is diluted at 0.05% (w/v) in D-PBS.
8. Freezing medium: IMDM-10% FBS medium, FBS, and dimethylsulfoxide Hybri-MAX® (Sigma-Aldrich) are mixed at a 2:2:1 ratio before preparation of master plates for screening. After mixing IMDM-10% FBS with FBS thoroughly, add Hybri-MAX. Take care not to mix Hybri-MAX and FBS.

2.2. Transfection for Homologous Recombination

1. *FUT8* targeting vector: pKOFUT8Neo, pKOFUT8Puro (*see* **Fig. 2** and **Note 2**). The vector plasmids are purified by CsCl density gradient centrifugation (*see* **Note 3**).
2. *Sal*I (New England BioLabs, Beverly, MA).
3. Phenol/chloroform/isoamylalcohol: a mixture of Tris-HCl-equilibrated phenol, chloroform, and isoamylalcohol at a 25:24:1 ratio is available from Invitrogen.
4. 5 *M* NaCl.
5. 70% (v/v) EtOH.
6. Tris-EDTA buffer (TE): 10 m*M* Tris-HCl, 1 m*M* EDTA, pH 8.0.

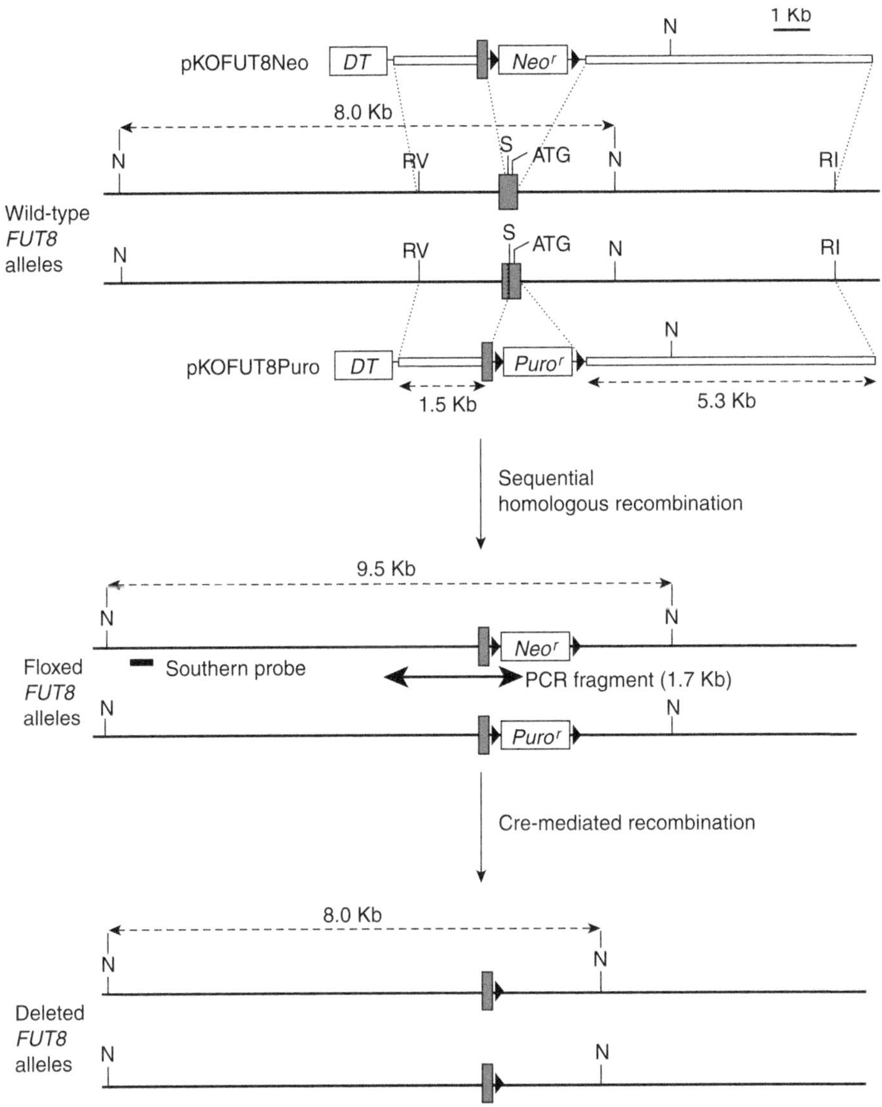

Fig. 2. Targeted disruption of *FUT8* alleles in CHO cells. *FUT8* alleles and PNS type targeting constructs. Exon 2 (filled boxes) is modified by replacing a segment containing the translation initiation site with the drug-resistance gene cassette, flanked by two *loxP* sites (filled triangles). After sequential gene targeting, the drug-resistance gene cassettes are removed from both *FUT8* alleles by transient expression of Cre recombinase. Relevant *Nhe*I (N), *Sac*I (S), *Eco*RV (RV), and *Eco*RI (RI) recognition sites are shown.

7. K-PBS buffer: 137 m*M* KCl, 2.7 m*M* NaCl, 8.1 m*M* Na_2HPO_4, 1.5 m*M* KH_2PO_4, and 4.0 m*M* $MgCl_2$. A 10X of Mg^{2+}-free K-PBS stock solution (1.37 *M* KCl, 27 m*M* NaCl, 81 m*M* Na_2HPO_4, and 15 m*M* KH_2PO_4) is prepared by dissolving 10.21 g of KCl, 158 mg of NaCl, 1.15 g of Na_2HPO_4, and 204 mg of KH_2PO_4 in 100 mL of

distilled water. A 10X of $MgCl_2$ stock solution (40 m*M* $MgCl_2$) is prepared by dissolving 812 mg of $MgCl_2 \cdot 6H_2O$ in 100 mL of distilled water. Both 10X stocks are filter-sterilized and stored at 4°C. Just before electroporation, a 1X working solution is prepared by mixing 10 mL of 10X Mg^{2+}-free K-PBS and 10 mL of 10X $MgCl_2$ with 80 mL of distilled water. The solution is then filter-sterilized and kept on ice.

8. Electroporation cuvets with an electrode gap of 0.2 cm (Bio-Rad Laboratories, Hercules, CA).
9. Gene Pulser II (Bio-Rad Laboratories).
10. 10-cm dishes for adherent cell culture (Greiner Bio-One).

2.3. Screening for Homologous Recombinants by Genomic PCR

1. Stereoscopic microscope.
2. A 96-well flat-bottomed plate for adherent cell culture (Greiner Bio-One or Asahi Techno Glass, Tokyo, Japan).
3. A 96-well round-bottomed plate for adherent cell culture (Asahi Techno Glass or Beckton Dickinson, Franklin Lakes, NJ).
4. Lysis buffer: 10 m*M* Tris-HCl, pH 7.5, 10 m*M* EDTA, pH 8.0, 10 m*M* NaCl, 0.5% (w/v) N-lauroylsarcosine, and 1 mg/mL proteinase K. A cell lysis solution (10 m*M* Tris-HCl, pH 7.5, 10 m*M* EDTA, pH 8.0, 10 m*M* NaCl, 0.5% [w/v] sarcosyl) is prepared by dissolving 0.29 g of NaCl and 2.5 g of N-lauroylsarcosine in 400 mL of distilled water, adding 5 mL of 1 *M* Tris-HCl, pH 7.5, and 10 mL of 0.5 *M* EDTA, pH 8.0. The final volume was adjusted to 500 mL with distilled water before filtration, and the solution was stored at room temperature. Proteinase K stock solution is prepared by dissolving proteinase K (Sigma-Aldrich) in distilled water to 50 mg/mL followed by storage at –20°C. Just before cell lysis, a working solution is prepared by mixing 500 μL of proteinase K stock solution with 24.5 mL of cell lysis solution.
5. NaCl/EtOH: just before preparation of genome DNA, a solution is prepared by mixing 600 μL of 5 *M* NaCl with 40 mL of EtOH and keeping it at room temperature. The solution is cloudy, but this is inconsequential. Before use, mix the solution again, because the salt will precipitate.
6. 70% (v/v) EtOH.
7. TE-RNase: TE supplemented with 200 μg/mL of ribonuclease A. Prepare RNase stock solution by dissolving ribonuclease A (type X-A, Sigma-Aldrich) in distilled water to 10 mg/mL and store at –20°C.
8. PCR primers: forward primer 5′-CTT GTG TGA CTC TTA ACT CTC AGA G-3′ is designed in an intron of *FUT8* not represented in the targeting vectors. Reverse primer 5′-GAG GCC ACT TGT GTA GCG CCA AGT G-3′ is specifically bound in the PGk promoter sequence of drug-resistance gene cassettes (*see* **Fig. 1**).
9. Ex *Taq* polymerase (Takara Bio, Shiga, Japan).

2.4. Diagnosis of Homologous Recombinants by Southern Blotting

1. A 24-well flat-bottomed plate for adherent cell culture (Greiner Bio-One or Asahi Techno Glass).

2. A 6-well flat-bottomed plate for adherent cell culture (Greiner Bio-One or Asahi Techno Glass).
3. Lysis buffer (*see* **Subheading 2.3.**).
4. NaCl/EtOH (*see* **Subheading 2.3.**).
5. 70% (v/v) EtOH.
6. TE-RNase (*see* **Subheading 2.3.**).
7. *Nhe*I (New England BioLabs).
8. PCR primers: forward primer 5′-GTG AGT CCA TGG CTG TCA CTG-3′ and reverse primer 5′-CCT GAC TTG GCT ATT CTC AG-3′ are designed in an intron region of *FUT8* not represented in the targeting vectors (*see* **Fig. 1**).
9. Ex *Taq* polymerase (Takara Bio).
10. Hybond-N+ (GE Healthcare Biosciences, Piscataway, NJ).
11. Ultraviolet crosslinker.
12. 0.4 *N* NaOH.
13. 0.2 *M* Tris-HCl, pH 7.5–2 × SSC.
14. Hybridization buffer: 5 × SSPE, 5 × Denhardt's solution, and 0.5% (w/v) sodium dodecyl sulfate (SDS). A 50 × Denhardt's solution is available from Nacalai Tesque.
15. 10 mg/mL of salmon sperm DNA.
16. (α-^{32}P)dCTP.
17. Megaprime DNA-labeling systems, dCTP (GE Healthcare Biosciences).
18. Probe Quant G-50 Micro Columns (GE Healthcare Biosciences).
19. 2 × SSPE-0.1% (v/w) SDS.
20. 0.2 × SSPE-0.1% (v/w) SDS.

2.5. Cre Expression

1. *Cre* expression vector: pBS185 (*see* **Note 4**).
2. TE: 10 m*M* Tris-HCl, 1 m*M* EDTA, pH 8.0.
3. K-PBS buffer (*see* **Subheading 2.3.**).
4. Electroporation cuvets with electrode gap of 0.2 cm (Bio-Rad Laboratories).
5. Gene Pulser II (Bio-Rad Laboratories).
6. 10-cm dishes for adherent cell culture (Greiner Bio-One).

3. Methods

3.1. Preparation of Linearized Targeting Vector

For electroporation, the targeting vectors are linearized at a unique restriction site such as *Sal*I in the plasmid backbone. It is recommended that transformation conditions, for example, maintenance media, cell culture, DNA per cell ratio, and applied voltage, should be optimized for the introduction of large constructs that must remain intact. We have found that maintaining CHO/DG44 in a low concentration of serum before transfection slightly improves transformation efficiency, which is preferable to the cells internalizing the intact vectors (unpublished data).

1. Cut the vector with *Sal*I and check for complete digestion by agarose gel electrophoresis. In a large-scale digest, 2–3 U of endonuclease per microgram of DNA and no more than 1 µg/µL of DNA are used.
2. Extract the DNA twice with an equal volume of phenol/chloroform/isoamylalcohol and once with chloroform.
3. Precipitate the DNA with 0.05 volumes of 5 *M* NaCl and 2.5 volumes of EtOH, and chill at –20°C (*see* **Note 5**).
4. Just before electroporation, centrifuge at 4°C, rinse the DNA with ice-cold 70% (v/v) EtOH, and allow it to air-dry for 20 min.
5. Resuspend the DNA at 1 µg/µL with 0.1X TE.

3.2. Transfection for Hemizygous Disruption

1. Passage CHO/DG44 cells in T-175 flasks for 2–3 d before electroporation. For transfection, the fifth to tenth passages of monolayer cultured cells (50–60% confluent) are used.
2. Rinse the cell monolayer twice with PBS and expose to 5 mL of 0.05% (w/v) trypsin-PBS at 37°C for 5 min. After tapping the flasks several times, resuspend cells gently in 15 mL of IMDM-10% FBS medium.
3. Centrifuge the cell suspension at 100 × *g* for 5 min and aspirate off the supernatant. Resuspend the cells gently in 10 mL of ice-cold K-PBS and put the cell suspension on ice. Determine the total cell number.
4. Centrifuge the cell suspension again at 100 × *g* for 5 min and aspirate off the supernatant, then resuspend at a density of 8×10^6 cells/mL.
5. Mix gently 4 µg of linearized plasmid pKOFUT8Neo and 200 µL of cell suspension, then transfer to a prechilled electroporation cuvet. Repeat the process for the required number of cuvets. Put the cuvet on ice for 10 min (*see* **Note 6**).
6. After wiping and tapping one of the cuvets, place it in the electroporation holder. Electroporation is carried out at 350 V, 250 µF using Gene Pulser II (*see* **Note 7**). Repeat the process for the required number of cuvets. Put the cuvet on ice again for 10 min.
7. Transfer the content of the cuvet into 10 mL of prewarmed IMDM-10% FBS medium and resuspend. Wash the inside of the cuvet once with the medium and combine with the cell suspension. Repeat the process for the required number of cuvets.
8. Divide 5 mL of the cell suspension evenly between two 10-cm dishes, each containing 5 mL of IMDM-10% FBS medium.
9. Place the dishes for 18–24 h at 37°C in a 5% CO_2 atmosphere and change the medium to G418 selection medium. Culture them for 10–14 d, renewing every 3–4 d with G418 selection medium.

3.3. Isolation of Drug-Resistant Colonies

The protocols described in this section and in **Subheading 3.4.** follow the Ramirez-Solis microextraction method in a 96-well format *(27)*.

1. Dispense 100 µL of IMDM-10% FBS medium into each well of the required number of 96-well flat-bottomed plates and store the wells at 37°C until ready for use. Dispense 20 µL of 0.05% (w/v) trypsin-PBS into each well of the required number of 96-well round-bottomed plates.

2. A 10-cm dish in which G418-resistant colonies appear is rinsed twice with 7 mL of PBS. Then add 7 mL of PBS to the plate, enough to completely cover the surface of the dish.
3. Using a sterile, disposable yellow microtip and a micropipet (e.g., Gilson's Pipetman P-100) set at 20 µL, pick an isolated G418-resistant colony by gentle scraping and place it into a well of the 96-well round-bottomed plate containing 0.05% (w/v) trypsin-PBS. Using a new microtip for each colony, repeat the process until a colony has been placed into each required well of the 96-well plate (*see* **Note 8**).
4. After incubating the plate at 37°C for 15–20 min, dispense 50 µL of IMDM-10% FBS into each well. Pipet up and down more than 10 times to disaggregate the cells and transfer the entire volume of cell suspension to the 96-well flat-bottomed plate containing IMDM-10% FBS medium. Repeat the process for the required number of plates.
5. Place the plates at 37°C in a 5% CO_2 atmosphere. The next day, change the medium to G418 selection medium. Culture for 3–5 d until the cells are more than 50% confluent.
6. Dispense 100 µL of IMDM-10% FBS medium into each well of the required number of new 96-well flat-bottomed plates and place them at 37°C until ready for use.
7. Each well of the plates growing isolated G418-resistant cells is rinsed two to three times with 100 µL of PBS and exposed to 25 µL of 0.05% (w/v) trypsin at 37°C for 15–20 min. After tapping the plates several times, dispense 25 µL of IMDM-10% FBS and pipet up and down more than 10 times to disaggregate the cells.
8. Dispense 50 µL of freezing medium to each well and pipet thoroughly. Then, 50 µL of each cell suspension is transferred to a new 96-well plate containing IMDM-10% FBS medium and suspended (replica plate). The plate containing the rest of the cell suspension (master plate) is sealed, wrapped with paper towels, and stored in a styrofoam box at –80°C.
9. Place the replica plates at 37°C in a 5% CO_2 atmosphere. The next day, change the medium for fresh IMDM-10% FBS medium. Culture for 3–5 d.

3.4. Screening for Hemizygous Clones by Genomic PCR

1. Each well of the replica plates growing cells is rinsed two to three times with 100 µL of PBS. Dispense 50 µL of lysis buffer into each well and incubate the plates overnight at 60°C in a humidified chamber such as a sealed plastic container with wet paper towels.
2. Cool the plates to room temperature. Dispense 100 µL of NaOH/EtOH to each well and place at room temperature for more than 30 min until the precipitated DNA is visible.
3. Invert the plate very slowly and discard the solution. Blot the plates gently on paper towels to discard the excess solution.
4. Add 150 µL of 70% (v/v) EtOH to each well, invert the plate very slowly to discard the solution, and blot the plates gently on paper towels. Repeat the washing process three times.
5. After the final wash, blot the plates gently on paper towels to discard the excess solution and allow them to air-dry for 20 min.
6. Dispense 30 µL of TE-RNase into each well, scrape the bottom of each well with a microtip to detach the DNA, and mix very well by pipeting. Incubate the plates overnight at 37°C in a humidified chamber.

7. For PCR analysis, 10 μL of the DNA solution is used in 25 μL of reaction mixture containing Ex *Taq* polymerase, 0.2 μ*M* dNTPs, 5% (v/v) dimethylsulfoxide, 0.5 μ*M* forward primer 5′-CTT GTG TGA CTC TTA ACT CTC AGA G-3′, and 0.5 μ*M* reverse primer 5′-GAG GCC ACT TGT GTA GCG CCA AGT G-3′ (*see* **Note 9**). PCR is carried out by heating at 94°C for 3 min and subsequent 30 cycles of 94°C for 1 min, 60°C for 1 min, and 72°C for 2 min. Positive clones are identified by 1.8 Kb fragments specific for the recombinant *FUT8* locus by 1.75% (w/v) agarose gel electrophoresis.

3.5. Diagnosis of Hemizygous Clones by Southern Blotting

1. Dispense 500 μL of IMDM-10% FBS medium into each well of the required number of 24-well flat-bottomed plates and place them at 37°C until ready for use.
2. A master plate including a well of the genomic PCR-positive clone described earlier is removed from the styrofoam box, stored at –80°C, and placed at 37°C in a 5% CO_2 atmosphere for 10 min.
3. Transfer the entire volume of the thawed cell suspension to a well containing IMDM-10% FBS medium and suspend. Place at 37°C in a 5% CO_2 atmosphere.
4. The next day, change the medium to fresh IMDM-10% FBS medium. Culture 3–5 d until the cells are more than 50% confluent.
5. Dispense 2 mL of IMDM-10% FBS medium into each well of the required number of 6-well flat-bottomed plates and store them at 37°C until ready for use.
6. Rinse each well with 200 μL of PBS and expose to 100 μL of 0.05% (w/v) trypsin-PBS at 37°C for 5 min. After tapping the plates several times, add 100 μL of IMDM-10% FBS medium to each well and pipet up and down.
7. Transfer the entire volume of the cell suspension to a well containing IMDM-10% FBS medium and suspend. Place at 37°C in a 5% CO_2 atmosphere.
8. The next day, change the medium to fresh IMDM-10% FBS medium. Culture 3–5 d until the cells are more than 70% confluent.
9. Each well of the replica plates growing cells is rinsed two to three times with 1.5 mL of PBS. Dispense 1.5 mL of lysis buffer into each well and incubate the plates overnight at 60°C in a humidified chamber.
10. Cool the plates to room temperature. Dispense 3 mL of NaOH/EtOH into each well, pipet up and down very gently using a sterile, wide-bore blue microtip (e.g., Rainin's Wide-pore tips HR-1000W). Then store the plates at room temperature for more than 30 min until the precipitated DNA is visible.
11. Using a sterile yellow microtip, scrape the bottom of a well gently to collect the DNA and hook the pellet. Rinse the pellet with ice-cold 70% (v/v) EtOH and allow it to air-dry for 20 min in a microtube.
12. Add 100–150 μL of TE-RNase to each microtube. Incubate the tubes overnight at 37°C in a humidified chamber. Store at 4°C.
13. Cut 12 μg of DNA with 20 U of *Nhe*I in 120 μL of reaction mixture.
14. Precipitate the DNA with 1.2 μL of 0.5 *M* EDTA, pH 8.0, and 300 μL of EtOH then chill at –20°C. Centrifuge at 4°C and discard the supernatant thoroughly. After air-drying for 10 min, resuspend the DNA with TE.
15. Electrophorese 10 μg of DNA on 0.6% (w/v) agarose gel at 20 V.
16. After nicking the DNA in the gel with an ultraviolet crosslinker, transfer it by capillary blotting to a nylon membrane (Hybond-N+) with 0.4 *N* NaOH.

17. Rinse the blotting membrane with 0.2 *M* Tris-HCI, pH 7.5–2 × SSC and allow it air-dry at room temperature. Bake the membrane at 80°C for 1–2 h.
18. Prehybridize the blot with hybridization buffer containing thermal denatured 100 μg/mL of salmon sperm DNA at 65°C for 3 h.
19. A Southern probe is prepared by PCR in 20 μL of reaction mixture containing Ex *Taq* polymerase, 0.2 μ*M* dNTPs, 0.5 μ*M* forward primer 5′-GTG AGT CCA TGG CTG TCA CTG-3′, and 0.5 μ*M* reverse primer 5′-CCT GAC TTG GCT ATT CTC AG-3′. PCR is carried out by heating at 94°C for 1 min followed by 30 cycles of 94°C for 30 s, 55°C for 30 s, and 74°C for 1 min. Then, 230 bp of amplified fragment is purified by 1.75% (w/v) agarose gel electrophoresis and labeled with (α-^{32}P)dCTP and Megaprime DNA labeling systems. The ^{32}P-labeled probe is purified with Probe Quant G-50 Micro Columns.
20. Hybridize the blot overnight with the ^{32}P-labeled probe at 65°C.
21. Wash the blot twice with 2 × SSPE-0.1% (v/w) SDS at 65°C for 15 min and once with 0.2 × SSPE-0.1% (v/w) SDS at 65°C for 15 min.
22. Expose the blot to X-ray film. A hemizygous clone gives 8.0 Kb of a wild-type allele specific fragment and 9.5 Kb of a recombinant allele-specific fragment at a ratio of 1:1.

3.6. Establishment of Homozygous Disruptants

1. Plasmid pKOFUT8Puro linearized with *Sal*I (*see* **Subheading 3.1.**) is introduced into the fifth to tenth passage of monolayer-cultured hemizygous disruptant cells (>50% confluent) as described in **steps 1–8** of **Subheading 3.2.** Place the dishes for 18–24 h at 37°C in a 5% CO_2 atmosphere and change the medium to G418/puromycin selection medium.
2. Culture the dishes for 10–14 d, renewing them every 3–4 d with G418/puromycin selection medium (*see* **Note 10**).
3. Drug-resistant colonies are isolated and replica plates are prepared as described in **Subheading 3.3.**
4. Isolate genomic DNAs from each clone and subject them to Southern blot analysis as described in **Subheading 3.5.** A homozygous disruptant clone gives only 9.5 Kb of a recombinant allele-specific fragment.

3.7. Cre Expression

1. Plasmid pBS185 is introduced into the fifth to tenth passage of monolayer-cultured homozygous disruptant cells (>50% confluent). After transfection with a cuvet as described in **steps 1–7** of **Subheading 3.2.**, the cell suspension is diluted 1:2000 with IMDM-10% FBS medium. Distribute 10 mL of the cell suspension into each 10-cm dish.
2. Place the dishes for 18–24 h at 37°C in a 5% CO_2 atmosphere and change the medium to fresh IMDM-10% FBS medium. Culture the dishes for 10–14 d, renewing them every 3–4 d with IMDM-10% FBS medium (*see* **Note 11**).
3. Colonies are isolated as described and replica plates are prepared as described in **steps 1–8** of **Subheading 3.3.**
4. Place the replica plates at 37°C in a 5% CO_2 atmosphere. The next day, change the medium to G418/puromycin selection medium. Culture the plates for 10–14 d, renewing them every 3–4 d with G418/puromycin selection medium.

5. Identify the positive clone wells in which no cells are viable.
6. Isolate genomic DNAs from each identified clone and subject them to Southern blot analysis as described in **Subheading 3.5.** A recombinant clone in which the drug-resistance gene cassettes are excluded from both *FUT8* alleles gives only 8.0 Kb of a recombinant allele-specific fragment.

4. Notes

1. The CHO/DG44 cell line in which dihydrofolate reductase (*DHFR*) gene loci are deleted *(28)*, is commonly used for mammalian recombinant protein production because the *DHFR* gene amplification system is available to yield high productivity. Several approved biopharmaceuticals have been manufactured by CHO/DG44. The cell line is a generous gift of Lawrence Chasin of Columbia University. Most CHO/DG44 clones have two *FUT8* loci (unpublished data).
2. In brief, the vectors are constructed as follows: first, a 9.0-Kb fragment of the *FUT8* gene including the first coding exon is isolated by screening the CHO-K1 cell λ-genomic library (Stratagene, La Jolla, CA) with the Chinese hamster *FUT8* cDNA as a probe. Second, a 234-bp segment containing the translation initiation site is replaced with either the Neo^r or the $Puro^r$ cassette, flanked by two *loxP* sites, from plasmid pKOSelectNeo or pKOSelectPuro (Lexicon, Woodlands, TX), respectively. Third, the *DT* cassette from plasmid pKOSelectDT (Lexicon) is inserted beyond the 5′ homologous region on the vectors. The resulting targeting vectors, pKOFUT8Neo and pKOFUT8Puro, include the 1.5 Kb 5′ homologous sequence, the 5.3 Kb 3′ homologous sequence, and a unique *Sal*I site for linearization of the vectors.
3. Do not use plasmid purification kits, even though the supplier's instructions say that kits are available for large plasmids. When the kits are used, the targeting vector plasmids are often easily denatured in the purification process.
4. The Cre expression vector pBS185 is the generous gift of Invitrogen (formerly, Life Technologies).
5. All steps subsequent to ethanol precipitation are carried out in a sterile hood. Ethanol rinsing, air-drying, and resuspension of DNA are performed by sterile techniques in parallel with cell preparation.
6. We routinely prepare approx 20 cuvets containing DNA and cells in 10 min, and electroporate them in turn.
7. Time constants are approx 3.6–4.2 under this condition.
8. It is important to pick colonies of various size and morphologies. Approximately 20–30 minutes will have elapsed by the end of the picking process. For a positive control, leave one well blank at the bottom-right corner of each plate (*see* **Note 9**).
9. We always use the blank well of each replica plate to hold the genome DNA of positive control cells, which have the extended 5′ homologous sequence beyond the target region (out of *Eco*RV site; *see* **Fig. 2**) on the genome (*see* **Note 8**).
10. Phenotypic selection is carried out in this step. Phenotypic selection drugs are added to cell culture after 4–5 d, because the phenotypic changes in targeted clones appear about 3–4 d after transfection.
11. We routinely obtain approx 200–400 colonies per dish.

Acknowledgments

The authors would like to thank Ms. Machi Kusunoki and Ms. Miho Inoue for technical assistance.

References

1. Chu, L. and Robinson, D. K. (2001) Industrial choices for protein production by large-scale cell culture. *Curr. Opin. Biotechnol.* **12,** 180–187.
2. Wurn, F. M. (2004) Production of recombinant protein therapeutics in cultivated mammalian cells. *Nat. Biotechnol.* **22,** 1393–1398.
3. Yamane-Ohnuki, N., Kinoshita, S., Inoue-Urakubo, M., et al. (2004) Establishment of *FUT8* knockout Chinese hamster ovary cells: an ideal host cell line for producing completely defucosylated antibodies with enhanced antibody-dependent cellular cytotoxicity. *Biotechnol. Bioeng.* **87,** 614–622.
4. Shinkawa, T., Nakamura, K., Yamane, N., et al. (2003) The absence of fucose but not the presence of galactose or bisecting *N*-acetylglucosamine of human IgG1 complex-type oligosaccharides shows the critical role of enhancing antibody-dependent cellular cytotoxicity. *J. Biol. Chem.* **278,** 3466–3473.
5. Sedivy, J. M. and Sharp, P. A. (1989) Positive genetic selection for gene disruption in mammalian cells by homologous recombination. *Proc. Natl. Acad. Sci. USA* **86,** 227–231.
6. Arbones, M. L., Austin, H. A., Capon, D. J., and Greenburg, G. (1994) Gene targeting in normal somatic cells: inactivation of the interferon-γ receptor in myoblasts. *Nat. Genet.* **6,** 90–97.
7. Hanson, K. D. and Sedivy, J. M. (1995) Analysis of biological selections for high-efficiency gene targeting. *Mol. Cell. Biol.* **15,** 45–51.
8. Adair, G. M., Nairn, R. S., Wilson, J. H., et al. (1989) Targeted homologous recombination at the endogenous adenine phosphoribosyltransferase locus in Chinese hamster cells. *Proc. Natl. Acad. Sci. USA* **86,** 4574–4578.
9. Pennington, S. L. and Wilson, J. H. (1991) Gene targeting in Chinese hamster ovary cell is conservative. *Proc. Natl. Acad. Sci. USA* **88,** 9498–9502.
10. Nairn, R. S., Adair, G. M., Porter, T., et al. (1993) Targeting vector configuration and method of gene transfer influence targeted correction of the *APRT* gene in Chinese hamster ovary cells. *Somat. Cell Mol. Genet.* **19,** 363–375.
11. Aratani, Y., Okazaki, R., and Koyama, H. (1992) End extension repair of introduced targeting vectors mediated by homologous recombination in mammalian cells. *Nucleic Acids Res.* **20,** 4795–4801.
12. Scheerer, J. B. and Adair, G. M. (1994) Homology dependence of targeted recombination at the Chinese hamster *APRT* locus. *Mol. Cell. Biol.* **14,** 6663–6673.
13. Prouty, S. M., Hanson, K. D., Boyle, A. L., et al. (1993) A cell culture model system for genetic analysis of the cell cycle by targeted homologous recombination. *Oncogene* **8,** 899–907.
14. Warner, T. G. (1999) Enhancing therapeutic glycoprotein production in Chinese hamster ovary cells by metabolic engineering endogenous gene control with antisense DNA and gene targeting. *Glycobiology* **9,** 841–850.
15. Siciliano, M. J., Stallings, R. L., and Adair, G. M. (1985) The genetic map of the Chinese hamster and the genetic consequences of chromosomal rearrangements in CHO cells, in *Molecular Cell Genetics*, (Gottesman M. M, ed.), Wiley, New York, pp. 95–135.

16. Mamaeva, S. E. (1998) Karyotypic evolution of cells in culture: a new concept. *Int. Rev. Cytol.* **178,** 1–40.
17. Deaven, L. L. and Petersen, D. F. (1973) The Chromosomes of CHO, an aneuploid Chinese hamster ovary cell line: G-band, C-band, and autoradiographic analyses. *Chromosoma* **41,** 129–144.
18. Riele, H., Maandag, E. R., and Beans, A. (1992) Highly efficient gene targeting in embryonic stem cells through homologous recombination with isogenic DNA constracts. *Proc. Natl. Acad. Sci. USA* **89,** 5128–5132.
19. Sedivy, J. M. and Dutriaux, A. (1999) Gene targeting and somatic cell genetics: a rebirth or a coming age? *Trend Genet.* **15,** 88–90.
20. Deng, C. and Capecchi, M. R. (1992) Reexamination of gene targeting frequency as a function of extent of homology between the targeting vector and target locus. *Mol. Cell. Biol.* **12,** 3365–3371.
21. Hanson, K. D. and Sedivy, J. M. (1995) Analysis of biological selection for high-efficiency gene targeting. *Mol. Cell. Biol.* **15,** 45–51.
22. Kohno, K. and Uchida, T. (1987) Highly frequent single amino acid substitution in mammalian elongation factor 2 (EF-2) results in expression of resistance to EF-2-ADP-ribosylating toxins. *J. Biol. Chem.* **262,** 12,298–12,305.
23. Derouazi, M., Martinet, D., Besuchet Schmutz, N., et al. (2006) Genetic characterization of CHO production host DG44 and derivative recombinant cell lines. *Biochem. Biophys. Res. Commun.* **340,** 1069–1077.
24. Davies, J. and Reff, M. (2001) Chromosome localization and gene-copy-number quantification of three random integrations in Chinese-hamster ovary cells and their amplified cell lines using fluorescence in situ hybridization. *Biotechnol. Appl. Biochem.* **33,** 99–105.
25. Siminovitch, L. (1985) Mechanisms of genetic variation in Chinese hamster ovary cells, in *Molecular Cell Genetics*, (Gottesman, M. M., ed.), Wiley, NY, pp. 869–879.
26. Kido, M., Miwatani, H., Kohno, K., Uchida, T., and Okada, Y. (1991) Targeted introduction of a diphtheria toxin-resistant point mutation into the chromosomal EF-2 locus by *in vivo* homologous recombination. *Cell Struct. Func.* **16,** 447–453.
27. Ramirez-Solis, R., Rivera-Perez, J., Wallace, J. D., Wims, M., Zheng, H., and Bradley, A. (1992) Genomic DNA microextraction: a method to screen numerous samples. *Anal. Biochem.* **201,** 331–335.
28. Urlaub, G., Mitchell, P. J., Kas, E., et al. (1986) Effect of gamma rays at the dihydroforate reductase locus: deletions and inversions. *Somat. Cell Mol. Genet.* **12,** 555–566.

2

Highly Proficient Gene Targeting by Homologous Recombination in the Human Pre-B Cell Line Nalm-6

Noritaka Adachi, Aya Kurosawa, and Hideki Koyama

Summary

Gene targeting provides a powerful means for studying gene function by a reverse genetic approach. Despite recent rapid progress in gene knockdown technologies, gene knockout studies using human somatic cells will be of greater importance for analyzing the functions of human genes in greater detail. Although the frequency of gene targeting is typically very low in human cultured cells, we have recently shown that a human precursor B cell line, Nalm-6, exceptionally allows for high-efficiency gene targeting by homologous recombination. In addition, we have developed a quick and simplified method to construct gene-targeting vectors, which is applicable to all sequenced organisms as well as embryonic stem cells. The combination of the simplified vector construction technology and the highly efficient gene-knockout system using Nalm-6 cells has enabled us to disrupt virtually any locus of the human genome within one month. Our system will greatly facilitate gene-knockout studies in human cells.

Key Words: Colony formation; electroporation; gene targeting; genomic PCR; homologous recombination; Ku86; MultiSite Gateway technology; Nalm-6; nonhomologous end-joining; targeting vector.

1. Introduction

Gene targeting provides a powerful means of studying gene function by a reverse genetic approach *(1,2)*. This technology relies on a homologous DNA recombination reaction that occurs between a targeting vector and the host genome. In mice, a great number of genes have been knocked out thus far using embryonic stem (ES) cells, and their physiological functions have been elucidated. However, given species difference and distinct genetic backgrounds between humans and model organisms, gene-knockout studies using human somatic cells should be of greater importance when attempting to reliably analyze the function of human genes, particularly for diagnostic and therapeutic purposes. However, in human cultured

From: *Methods in Molecular Biology, vol. 435: Chromosomal Mutagenesis*
Edited by: G. Davis and K. J. Kayser © Humana Press Inc., Totowa, NJ

cells, the frequency of gene targeting is typically too low for such systematic genetic analysis to be feasible *(2,3)*. Moreover, constructing targeting vectors involves time-consuming, complicated processes, being another rate-limiting step in gene-targeting experiments. To overcome these constraints in knocking out human genes, we have developed a novel system that enables rapid disruption of human genes of interest. Our system includes (1) a quick and simplified method for vector construction, which is based on the commercially available MultiSite Gateway® Technology *(4)*, and (2) the use of a human precursor B cell line, Nalm-6, which appears highly proficient in homologous recombination *(4–8)*. With this system, one can disrupt virtually any genomic locus within 1 month. Moreover, homozygous knockout mutants lacking a human gene of interest can be created within 2–3 months.

The Nalm-6 cell line was established from the peripheral blood of a 19-year-old man with acute lymphoblastic leukemia *(9)*. One of the premises of gene-targeting experiments would be the use of a cell line that is karyotypically stable. In this regard, Nalm-6 displays a stable diploid karyotype with a single reciprocal translocation *(10)*. In addition, Nalm-6 has a doubling time of 20–22 h and a high plating efficiency of approx 80% *(6)*. Furthermore, Nalm-6 cells express normal p53 with wild-type functions *(11)*. These advantageous properties of Nalm-6 further underscore its usefulness in gene-knockout studies of human genes.

2. Materials

2.1. Construction of Targeting Vectors

1. ExTaq™ DNA polymerase (Takara Bio Inc., Otsu, Japan).
2. Polymerase chain reaction (PCR) primers (*see* **Note 1**): kku86-1, 5′-GGGGACAACTTTGTATAGAAAAGTTGAGTGGTAGTTGTCTCTGAAGGGTC-3′; kku86-2, 5′-GGGGACTGCTTTTTTGTACAAACTTGCAGCTGCCTGGAAACAAAGTTCCA-3′; kku86-3, 5′-GGGGACAGCTTTCTTGTACAAAGTGGTAAGATGGATGCTTGTCTAGGCGG-3′; and kku86-4, 5′-GGGGACAACTTTGTATAATAAAGTTGTCCATGCTCACGATTAGTGCATCC-3′.
3. MultiSite Gateway Three Fragment Vector Construction Kit (Invitrogen, Carlsbad, CA; *see* **Note 2**): pDONR P4-P1R and pDONR P2R-P3, BP clonase II enzyme mix, 2 µg/µL proteinase K solution, 5X LR Clonase Plus reaction buffer, and LR Clonase Plus enzyme mix.
4. Entry clones for selectable marker genes (pENTR lox-Hyg and pENTR lox-Puro) *(4)*.
5. pDEST DTA-MLS (a modified version of destination vector pDEST R4-R3; *see* **Note 2**) *(4)*.
6. Luria Bertani (LB) agar plates containing either kanamycin (50 µg/mL) or ampicillin (50 µg/mL).
7. Restriction enzyme I-*Sce*I, 10X I-*Sce*I reaction buffer, and 10 mg/mL bovine serum albumin (New England Biolabs, Beverly, MA).
8. PCI: TE-saturated phenol:chloroform:isoamyl alcohol equal to 25:24:1. Store at 4°C.
9. CI: chloroform:isoamyl alcohol = 24:1. Store at 4°C.

10. 3 *M* Sodium acetate. Dissolve 40.81 g of sodium acetate in 80 mL of water. Adjust to pH 5.2 with acetic acid. Add water to 100 mL.
11. TE buffer: 10 m*M* Tris-HCl, 0.1 m*M* EDTA (pH 8.0). Store at 4°C.

2.2. Electroporation of Targeting Vector Into Nalm-6 Cells

1. Growth medium: ES medium (Nissui Seiyaku Co., Tokyo, Japan) supplemented with 10% calf serum (Hyclone, Logan, UT), and 50 μ*M* 2-mercaptoethanol.
2. Mg^{2+}, Ca^{2+}-free saline G: 130 m*M* NaCl, 5.3 m*M* KCl, 1.1 m*M* Na_2HPO_4, 1.1 m*M* KH_2PO_4, and 6.1 m*M* glucose. Store at 4°C after autoclaving.
3. 100 m*M* $MgCl_2$ (filter-sterilized). Store at 4°C.
4. 100 m*M* $CaCl_2$ (filter-sterilized). Store at 4°C.
5. Saline G *(12)*: 130 m*M* NaCl, 5.3 m*M* KCl, 1.1 m*M* Na_2HPO_4, 1.1 m*M* KH_2PO_4, 6.1 m*M* glucose, 0.49 m*M* $MgCl_2$, and 0.9 m*M* $CaCl_2$. Add 2.5 mL of 100 m*M* $MgCl_2$ and 4.5 mL of 100 m*M* $CaCl_2$ to 500 mL of Mg^{2+}, Ca^{2+}-free saline G. Store at 4°C.
6. Linearized targeting vector (I-*Sce*I-digested pKU86-Hyg).

2.3. Colony Formation

1. 2.25X ES medium. Dissolve 21.8 g of powdered ES medium (Nissui Seiyaku Co.), 4.7 g $NaHCO_3$, 0.68 g glutamine, and 8.1 μL 2-mercaptoethanol in 1 L of water, and stir at room temperature for 30–60 min. Sterilize with a 0.22-μm-pore-size membrane filter and store at 4°C (*see* **Note 3**).
2. Calf serum (Hyclone).
3. 0.33% (w/v) agarose solution. Dissolve SeaKem LE agarose (Cambrex Bio Science, Rockland, ME) in water. After autoclaving, keep at 60°C until immediately before use.
4. Hygromycin B (filter-sterilized, 100 mg/mL). Store at 4°C.

2.4. Colony Isolation and Selection of Targeted Clones

1. Growth medium containing hygromycin B (0.4 mg/mL).
2. Lysis buffer: 20 m*M* Tris-HCl (pH 8.0), 250 m*M* NaCl, and 1% (w/v) sodium dodecyl sulfate (SDS).
3. Proteinase K (10 mg/mL). Store in aliquots at –20°C.
4. Saturated NaCl solution.
5. *Taq* DNA polymerase.
6. PCR primers: kku86-5, 5′-ATCGCGGTCAAGACAAAGAATGGG-3′; kku86-6, 5′-CAGCCTCCACATAGGCAGAATGTA-3′; universal primer A, 5′-AATAATGGTTTCTTAGACGTGCG-3′; and universal primer B, 5′-AGGTTCACTAGTACTGGCCATTG-3′ *(4)*.

2.5. Cre-Mediated Excision of Selection Marker

1. Cre expression vector (pBS185, Invitrogen).
2. Growth medium.
3. Saline G.
4. 2.25X ES medium.
5. Calf serum (Hyclone).

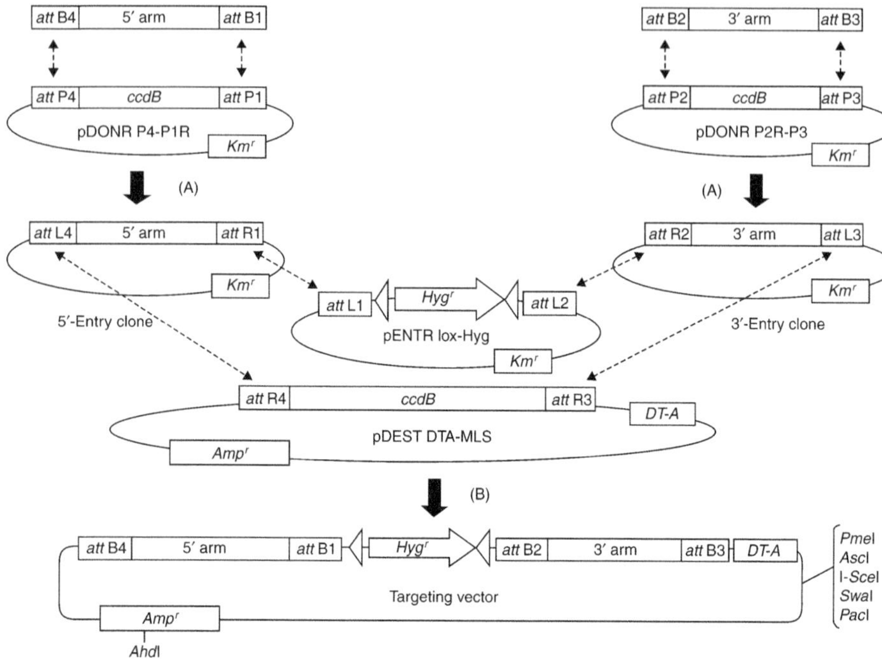

Fig. 1. Schematic representation of targeting vector construction. The method is based on the MultiSite Gateway system, and consists of three steps: (1) genomic PCR to amplify *att*B-flanked 5′- and 3′-arms, (2) BP recombination to generate 5′- and 3′-entry clones (**A**), and (3) LR recombination between four plasmids to generate targeting vector (**B**). Triangles represent *lox*P sequences. For simplicity, pENTR lox-Puro is not shown. *Hyg^r*, hygromycin-resistance gene; *DT-A*, a gene that codes for a diphtheria toxin A fragment; *Km^r*, kanamycin-resistance gene; *Amp^r*, ampicillin-resistance gene.

6. 0.33% (w/v) agarose solution.
7. Hygromycin B (filter-sterilized, 100 mg/mL).

3. Methods

To make the most of the MultiSite Gateway system for simplifying targeting-vector construction, we have generated a series of entry clones for floxed drug-resistance genes (such as pENTR lox-Hyg, pENTR lox-Puro, and pENTR lox-His) and a modified version of destination vector, pDEST DTA-MLS, which possesses a diphtheria toxin A fragment (DT-A) gene as well as multiple linearization sites (*Pac*I, *Swa*I, I-*Sce*I, *Asc*I, and *Pme*I) *(4)* (**Fig. 1**). Owing to the absence of any ligation steps and the presence of the multiple linearization sites, it is unnecessary to search for appropriate restriction sites for vector construction/linearization. Additionally, the floxed marker genes should be valuable when attempting to knockout two or more genes, as a floxed region can

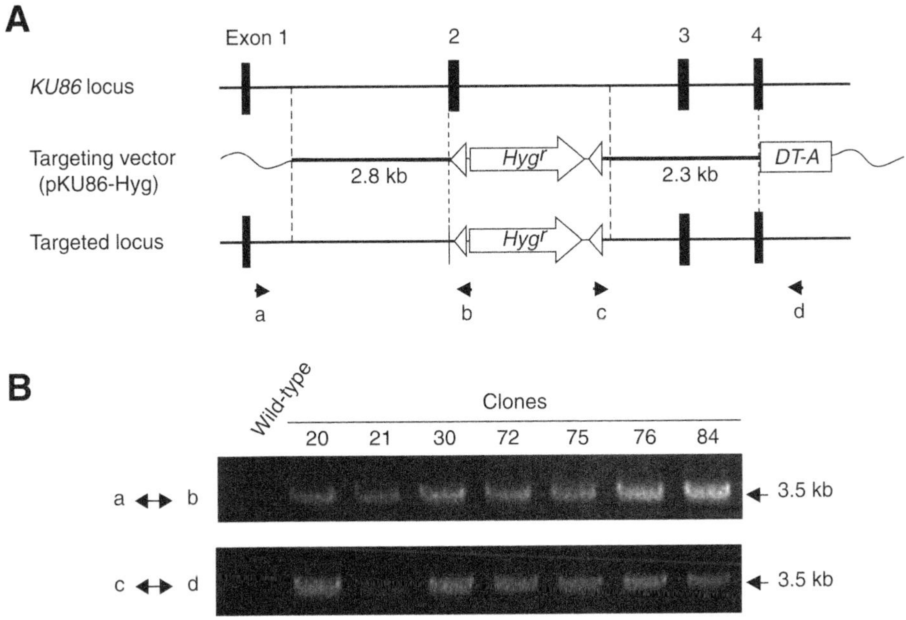

Fig. 2. Heterozygous disruption of the human *KU86* gene. (**A**) Scheme for *KU86* disruption in Nalm-6 cells. The human *KU86* gene consists of 21 exons, located on chromosome 2q35 (http://genome.ucsc.edu). The targeting vector pKU86-Hyg is expected to delete exon 2 (and most part of intron 2) after homologous recombination. Arrows a–d stand for PCR primers: a, kku86-5; b, universal primer B; c, universal primer A; and d, kku86-6. (**B**) PCR analysis of targeted clones. Eighty-four hygromycin-resistant clones were screened for heterozygous disruption of the *KU86* gene, seven of which were confirmed to be correctly targeted clones.

easily be removed from the genome by transient expression of Cre recombinase. In this chapter, we introduce the method to construct gene-targeting vectors for the human *KU86* gene, the product of which is critical for the nonhomologous end-joining pathway of DNA double-strand break repair ***(13–15)***.

We also describe the protocols for DNA transfection and colony formation/isolation using human Nalm-6 cells. As in mouse ES cells, transfection of targeting vector should be achieved by electroporation. Despite an initial report of successful Nalm-6 gene targeting with the Gene Pulser system (Bio-Rad, Hercules, CA) ***(5)***, at least in our hands, this system did not confer high transfection efficiencies (as low as 10^{-7}). We, therefore, describe the heterozygous disruption of the human *KU86* gene **(Fig. 2)** using our standard protocol with a Shimadzu machine that promises high transfection frequencies (~10^{-4}). The Amaxa Nucleofector technology is also recommended for transfection of Nalm-6 cells (Amaxa, Gaithersburg, MD; http://www.amaxa.com/nalm-60.html).

3.1. Construction of Targeting Vectors

1. Amplify *KU86* genomic fragments by PCR with ExTaq™ DNA polymerase (*see* **Note 4**) using Nalm-6 genomic DNA as a template and primers kku86-1 and kku86-2 for the 2.8-kb 5′-arm, and kku86-3 and kku86-4 for the 2.3-kb 3′-arm **(Fig. 2A)** (*see* **Note 1**). The PCR reaction should be performed in a more than 100-μL solution (e.g., 6 tubes of 20 μL solution) under the following condition: denaturation at 94°C for 2 min, followed by 40 cycles of 94°C for 30 s, 68°C for 1 min, and 72°C for 2 min; and the final extension at 72°C for 5 min.
2. Purify and quantitate the PCR products (*see* **Note 5**).
3. Perform a BP recombination reaction between each *att*B-flanked PCR fragment and the *att*P-containing donor vector (pDONR P4-P1R or pDONR P2R-P3), to generate entry clones. Add the following components to a 0.5-mL microcentrifuge tube at room temperature and mix well by vortexing:

pDONR P4-P1R or pDONR P2R-P3 (150 ng/μL)	1 μL (150 ng)
PCR product (*att*B-flanked 5′- or 3′-arm)	50 fmoles
TE buffer (pH 8.0)	8 μL

4. Vortex BP clonase II enzyme, mix briefly (*see* **Note 6**). Add 2 μL to the components above and mix well by vortexing briefly twice.
5. Incubate at 25°C for 4–5 h.
6. Add 1 μL of 2 μg/μL proteinase K solution and incubate at 37°C for 10 min (*see* **Note 7**).
7. Transform 5 μL of the reaction into 50 μL of competent *Escherichia coli*. Select for kanamycin-resistant clones on LB agar plates containing 50 μg/mL kanamycin (*see* **Note 8**).
8. Pick 10–20 colonies to isolate plasmids by the standard alkaline-SDS method, and electrophorese the plasmids (circular DNA) on agarose gels **(Fig. 3A)**. Check 2–3 candidate plasmids by digesting with appropriate restriction enzymes, followed by agarose gel electrophoresis (*see* **Note 9**).
9. Purify and quantitate the correct entry clones **(Fig. 3B)** (*see* **Note 10**).
10. Perform a MultiSite Gateway LR recombination reaction between the three entry clones (namely, 5′- and 3′-entry clones, plus pENTR lox-Hyg or pENTR lox-Puro) and the modified destination vector pDEST DTA-MLS, to generate targeting vectors (pKU86-Hyg and pKU86-Puro). Add the following components to a 0.5-mL microcentrifuge tube at room temperature and mix well by vortexing:

5X LR Clonase Plus reaction buffer	4 μL
pDEST DTA-MLS (60 ng/μL)	1 μL (60 ng)
5′-Entry clone	25 fmoles
3′-Entry clone	25 fmoles
pENTR lox-Hyg or –Puro	25 fmoles
TE buffer (pH 8.0)	16 μL

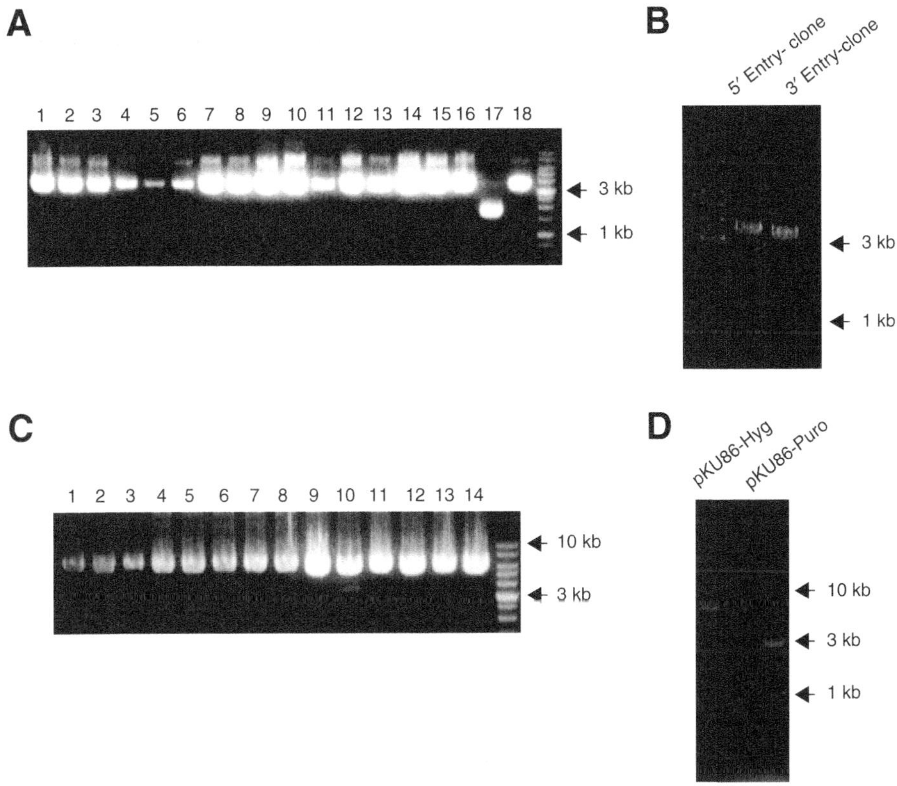

Fig. 3. Agarose gel electrophoresis of candidate plasmids. Note that all the plasmids are undigested and therefore circular. **(A)** Screening of 5′-entry clone candidates after BP recombination. Clone 17 appears to have lost the *ccdB* gene, whereas the other plasmids are all seemingly correct. **(B)** 5′- and 3′-entry clones purified with the QIAprep Spin Miniprep kit. **(C)** Screening of targeting-vector candidates after LR recombination. All the plasmids are seemingly correct, except that clone 10 has an extra band. **(D)** Targeting vectors pKU86-Hyg and pKU86-Puro purified with the QIAGEN Plasmid Maxi kit.

11. Vortex LR Clonase Plus enzyme, mix briefly. Add 4 μL to the components above and mix well by vortexing briefly twice.
12. Incubate at 25°C for 16 h.
13. Add 2 μL of 2 μg/μL proteinase K solution and incubate at 37°C for 10 min (*see* **Note 7**).
14. Transform 5 μL of the reaction into 50 μL of competent *E. coli*. Select for ampicillin-resistant clones on LB agar plates containing 50 μg/mL ampicillin (*see* **Note 8**).
15. Pick 10–20 colonies to isolate plasmids by the standard alkaline-SDS method, and electrophorese the circular plasmids on agarose gels (**Fig. 3C**). Check 2–3 candidate plasmids by digesting with appropriate restriction enzymes, followed by agarose gel electrophoresis (*see* **Note 9**).

16. Propagate, purify, and quantitate the targeting vector (**Fig. 3D**) (*see* **Note 11**).
17. Linearize the targeting vector by digesting with I-*Sce*I. Add the following components to a 1.5-mL microcentrifuge tube, and incubate at 37°C for 4 h to overnight (*see* **Note 12**):

Targeting vector	50 μg
10X I-*Sce*I buffer	40 μL
100X Bovine serum albumin (10 mg/mL)	4 μL
I-*Sce*I	15 units
Sterile water	400 μL

18. Add 0.4 mL of PCI and vortex well.
19. Centrifuge at 12,000*g* for 5 min at room temperature.
20. Transfer the upper aqueous phase to a new 1.5-mL microcentrifuge tube.
21. Add 0.4 mL of CI and vortex well.
22. Centrifuge at 12,000*g* for 5 min at room temperature.
23. Transfer the upper aqueous phase to a new 1.5-mL microcentrifuge tube.
24. Add 40 μL of 3 *M* sodium acetate and 0.9 mL of EtOH. Mix well.
25. Centrifuge at 12,000*g* for 5 min at room temperature.
26. Wash the DNA pellet three times with 0.5 mL of 70% EtOH.
27. After the third centrifugation, aspirate the supernatant using a clean sterile pipet tip and dry the pellet in a sterile hood.
28. Dissolve the pellet in TE buffer to a final DNA concentration of 2–4 μg/μL.

3.2. Electroporation of Targeting Vector Into Nalm-6 Cells

1. Maintain Nalm-6 cells in growth medium at 37°C in a humidified atmosphere of 5% CO_2 in air (*see* **Note 13**).
2. Harvest cells (~8 × 10^6 logarithmically growing cells) in a 50-mL Falcon tube, and wash twice with prewarmed saline G (*see* **Note 14**).
3. Resuspend the cells in fresh saline G (~40 μL), count the cells, and adjust the cell concentration to 1 × 10^8 cells/mL with fresh saline G.
4. Transfer 40 μL (4 × 10^6 cells) to a 1.5-mL microcentrifuge tube, add 4 μg (1–2 μL) of linearized targeting vector (pKU86-Hyg; **Fig. 2A**), and mix well by pipeting.
5. Transfer the sample into a 40-μL chamber (FTC-13; Shimadzu, Kyoto, Japan) (*see* **Note 15**).
6. Set the chamber to the GTE-1 electroporation apparatus (Shimadzu, Kyoto, Japan) and apply an exponential electric pulse to the sample under the following condition: height, 300–400 V; width, 50 μs; intervals, 1.0 s; number, twice.
7. Remove the chamber from the electroporation apparatus, and stand for 10–15 min at room temperature.
8. Transfer the sample to a 60-mm dish containing 6 mL of growth medium, and mix well by stirring gently.
9. Incubate for 20–22 h at 37°C in a humidified atmosphere of 5% CO_2 in air.

3.3. Colony Formation

1. Preparation of 2X ES medium. Add 1 vol of calf serum to 4 vol of 2.25X ES medium (*see* **Note 3**). Prewarm at 40°C before use (*see* **Note 16**).
2. Preparation of agarose medium. Add an equal volume of 0.33% (w/v) agarose solution to the prewarmed 2X ES medium, and mix well by stirring vigorously. Keep at 40°C until immediately before use.
3. Use an aliquot of the cell suspension (**Subheading 3.2., step 8**) to count the cells. Subsequently, use 100 µL to prepare a diluted cell suspension at the density of 200 cells/mL in growth medium (*see* **Note 17**).
4. To estimate the plating efficiency, add 0.5 mL (100 cells) of the diluted cell suspension and 4.5 mL of agarose medium to two to three 60-mm dishes. Mix thoroughly by shaking and swirling the dishes to distribute cells evenly (*see* **Note 17**).
5. To select for transfected clones, aliquot 1 mL (~10^6 cells) of the undiluted cell suspension into five 90-mm dishes. Add 40 µL of 100 mg/mL hygromycin B to each dish (*see* **Note 18**). The cell suspension and the drug solution should be put separately on the dish.
6. Add 9 mL of agarose medium to each 90-mm dish. Mix thoroughly by shaking and swirling the dishes to distribute cells evenly.
7. Stand at room temperature for 20–30 min to let the agarose harden.
8. Incubate at 37°C for 2–3 wk in a humidified atmosphere of 5% CO_2 in air, to allow colony formation (*see* **Note 19**).

3.4. Colony Isolation and Selection of Targeted Clones

1. Aliquot 0.5 mL of growth medium containing 0.4 mg/mL hygromycin B to each well of 48-well multiwell plates (*see* **Note 20**).
2. Pick 50–200 visible, isolated drug-resistant colonies with yellow or blue pipet tips, and transfer to hygromycin-containing medium. Suspend well by pipeting.
3. Culture at 37°C for 2–3 d in a humidified atmosphere of 5% CO_2 in air.
4. Transfer each cell culture to a 1.5-mL microcentrifuge tube (*see* **Note 21**) and centrifuge at 5600–6100*g* for 5–10 min at room temperature.
5. After discarding the supernatant, add 270 µL of lysis buffer and 1 µL of 10 mg/mL proteinase K. Incubate at 37°C overnight or at 55°C for 1 h.
6. Add 80 µL of saturated NaCl solution and mix well.
7. Add 0.9 mL of EtOH and mix well.
8. Centrifuge at 12,000*g* for 15 min at 4°C.
9. Wash the pellet twice with 0.5 mL of 70% EtOH.
10. Dissolve the pellet in 30–100 µL of TE buffer.
11. Perform PCR analysis to detect targeted clones. Use primers kku86-5 and universal primer B, or kku86-6 and universal primer A (**Fig. 2B**). The PCR condition is: denaturation at 94°C for 2 min, followed by 40 cycles of 94°C for 30 s, 68°C for 1 min, and 72°C for 2 min; and the final extension at 72°C for 5 min (*see* **Note 22**).

3.5. Cre-Mediated Excision of Selection Marker

1. Transfect Cre expression vector (4 μg) into Nalm-6 cells of interest, as described in **Subheading 3.2.**, to allow transient expression of Cre, a site-specific recombinase for loxP sequences.
2. Replate an aliquot (100 cells) of transfected cells into agarose medium without drug selection (*see* **Note 23**). This can be performed as described in **Subheading 3.3.** (Note that **Subheading 3.3., steps 5** and **6** are unnecessary).
3. Pick approx 50 visible, isolated colonies with yellow or blue pipet tips, and transfer to drug-free medium. Suspend well by pipeting.
4. Culture at 37°C for 2–3 d in a humidified atmosphere of 5% CO_2 in air.
5. Split each cell culture into two separate wells containing fresh growth medium with or without selection drug. Compare the growth of cells in drug-containing wells and control wells (no drugs) to select for clones that have lost the drug-resistance marker(s) (*see* **Note 24**).

4. Notes

1. All the PCR primers to amplify genomic fragments should contain four G residues at the 5′-end, followed by an appropriate *att*B sequence. Thus, the DNA sequences for *att*B4, *att*B1, *att*B2, and *att*B3 containing primers are: 5′-GGGGACAACTTTG-TATAGAAAAGTTG, 5′-GGGGACTGCTTTTTTGTACAAACTTG, 5′-GGGGAC AGCTTTCTTGTACAAAGTGG, and 5′-GGGGACAACTTTGTATAATAAAGTTG, respectively, followed by gene-specific sequences. In general, PCR amplification of highly GC-rich regions should be avoided. Also, care must be taken to ensure that the primers do not hybridize with repetitive DNA sequences such as Alu. It is strongly recommended to use the genome browser of the University of California at Santa Cruz (http://genome.ucsc.edu) to examine the organization of a human gene of interest and find the best strategy for knocking out the gene, particularly for PCR and Southern analysis. An example of such analysis for the *KU86* locus (also called *XRCC5*) is shown in **Fig. 4**.
2. Use *ccdB*-resistant *E. coli* strain (such as DB3.1 carrying the *gyrA462* gene) to propagate pDEST DTA-MLS as well as pDONR P4-P1R and pDONR P2R-P3, as these plasmids harbor the *ccdB* gene.
3. To avoid generation of white precipitates, it is advised not to prewarm 2.25X ES medium at more than 37°C before serum addition.
4. For PCR amplification of each arm, the use of high-fidelity *Taq* DNA polymerase is recommended.
5. For purification of PCR products, we routinely use Wizard® SV Gel and PCR Clean-Up System (cat. no. A9282, Promega, Madison, WI).
6. It is recommended to avoid repeated freeze–thaw cycles.
7. These reaction mixtures can be stored at –20°C for a week.
8. It is important to use *recA*$^-$ *endA*$^-$ *E. coli* strain with high competency (~10^9 c.f.u.). Also, F′ episome-containing *E. coli* should not be used, as the *ccdA* gene will prevent negative selection with the *ccdB* gene.
9. PCR screening using the original, *att*B-containing primers is also feasible; however, it should be kept in mind that some PCR-positive plasmids are not true recombinants.

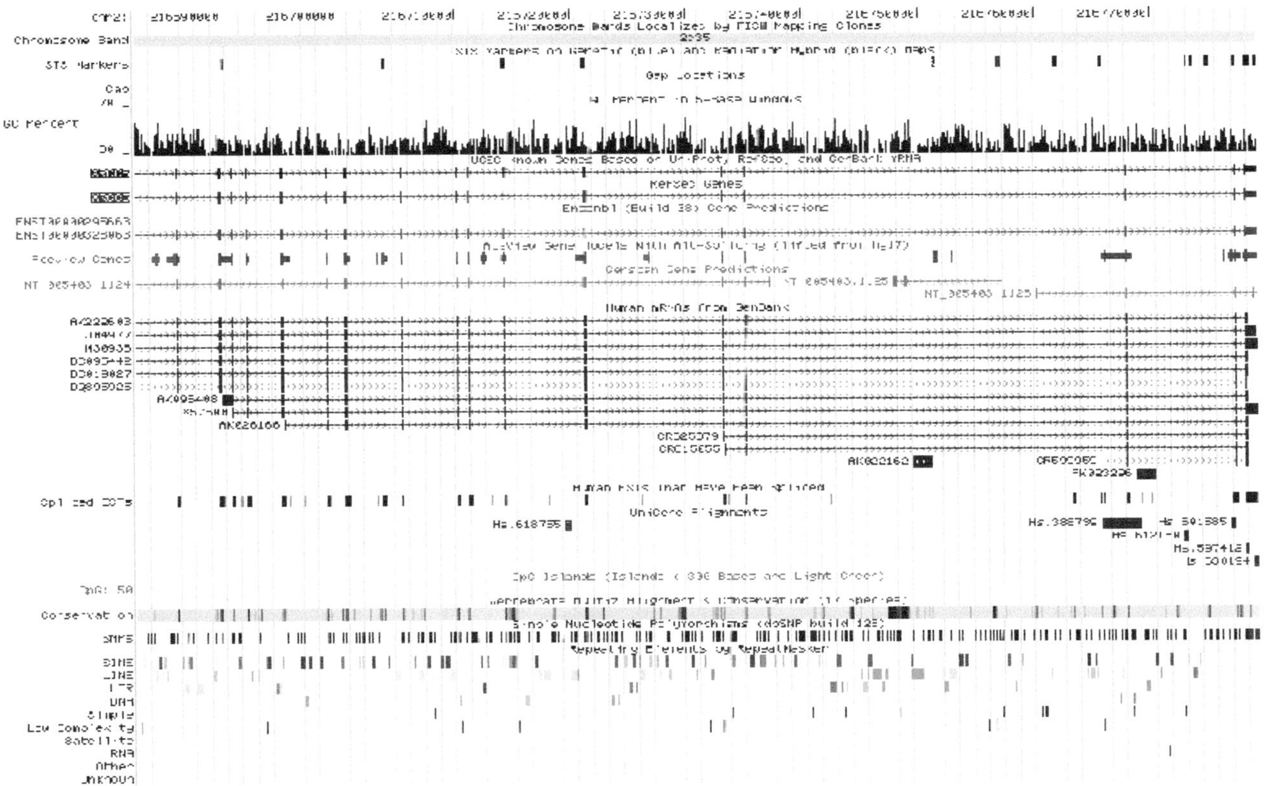

Fig. 4. Analysis of the human *KU86* locus using the genome browser of the University of California at Santa Cruz (http://genome.ucsc.edu/cgi-bin/hgTracks?position=chr2:216682377-216779248&hgsid=83084545&multiz17way=dense). Valuable information regarding the genomic organization and the locations of repetitive DNA sequences and GC-rich regions can be readily obtained from this website, which should help design PCR primers as well as the overall strategy for gene targeting.

10. We routinely use QIAprep® Spin Miniprep kit (cat. no. 27106, QIAGEN Inc., Valencia, CA). Final plasmid concentration should be 80–100 ng/μL.
11. For this purpose, we routinely use 300 mL overnight *E. coli* culture and QIAGEN® Plasmid Maxi kit (cat. no. 12163, QIAGEN Inc.).
12. Other restriction enzymes may be used for vector linearization; for example, *Pac*I, *Swa*I, *Asc*I, and *Pme*I are located at the multiple linearization sites and *Ahd*I in the *Amp*^{r} gene.
13. RPMI1640 medium (supplemented with 10% serum, 50 μ*M* 2-mercaptoethanol, and appropriate antibiotics) is also recommended for Nalm-6 cell culture.
14. Saline G may be replaced with PBS, especially when the Amaxa Nucleofector technology is used for transfection.
15. Care should be taken not to cause air bubbles.
16. 2X ES medium can be stored for several weeks at 4°C.
17. These steps (**Subheading 3.3., steps 3** and **4**) can be omitted if the calculation of plating efficiency is unnecessary. It is also important to note that colony formation can be performed simply by diluting the cell suspension and dividing cells into 96-well multiwell plates; so that each well contains approx 5000 cells per 200 μL of growth medium containing 0.4 mg/mL hygromycin B.
18. Add 25 μL of 200 μg/mL puromycin (filter-sterilized), instead of hygromycin B, if pKU86-Puro was used for gene targeting.
19. It is important not to move the agarose dishes during colony formation.
20. Aliquot 1 mL to each well when 24-well multiwell plates are used. For puromycin selection, use growth medium containing 0.5 μg/mL puromycin.
21. It is vitally important not to use all the cell culture for DNA isolation. Thus, each cell culture must be split into fresh growth medium before DNA isolation. In case PCR-based screening takes longer than expected, it is recommended to store each clone at −80°C after mixing with an equal volume of growth medium containing 20% dimethyl sulfoxide.
22. Although the genomic PCR-based method is effective in initial screening for targeted clones, it should be kept in mind that some PCR-positive clones do not have any disrupted allele, presumably because of abortive targeting events ***(16)***. Therefore, Southern blot analysis using external probes is indispensable for selection and verification of correctly targeted clones, where the intensity of a band corresponding to the disrupted allele should be the same as that of wild-type allele.
23. Overnight preincubation before replating (**Subheading 3.2., step 9**) can be omitted.
24. In addition to the drug sensitivity test, it is desirable to perform Southern blot analysis to verify the marker removal. Removal of drug-resistance marker(s) is useful for repeated use of the same targeting vector as well as for creating mutant cell lines lacking two or more genes.

Acknowledgments

The authors thank the lab members, especially Sairei So, Susumu Iiizumi, Yuji Nomura, and Koichi Uegaki for their advice and help. This work was supported in part by grants from Yokohama City University (Strategic Research Project, No. W18006; to N.A.), and by Grant-in-Aids from the Ministry of Education, Culture, Sports, Science, and Technology (MEXT) of Japan (to N.A. and H.K.).

References

1. Capecchi, M. R. (1989) Altering the genome by homologous recombination. *Science* **244,** 1288–1292.
2. Vasquez, K. M., Marburger, K., Intody, Z., and Wilson, J. H. (2001) Manipulating the mammalian genome by homologous recombination. *Proc. Natl. Acad. Sci. USA* **98,** 8403–8410.
3. Yanez, R. J. and Porter, A. C. (1998) Therapeutic gene targeting. *Gene Ther.* **5,** 149–159.
4. Iiizumi, S., Nomura, Y., So, S., et al. (2006) Simple one-week method to construct gene-targeting vectors: application to production of human knockout cell lines. *Biotechniques* **41,** 311–316.
5. Grawunder, U., Zimmer, D., Fugmann, S., Schwarz, K., and Lieber, M. R. (1998) DNA ligase IV is essential for V(D)J recombination and DNA double-strand break repair in human precursor lymphocytes. *Mol. Cell* **2,** 477–484.
6. So, S., Adachi, N., Lieber, M. R., and Koyama, H. (2004) Genetic interactions between BLM and DNA ligase IV in human cells. *J. Biol. Chem.* **279,** 55,433–55,442.
7. Uegaki, K., Adachi, N., So, S., Iiizumi, S., and Koyama, H. (2006) Heterozygous inactivation of human Ku70/Ku86 heterodimer does not affect cell growth, double-strand break repair, or genome integrity. *DNA Repair (Amst)* **5,** 303–311.
8. Adachi, N., So, S., Iiizumi, S., et al. (2006) The human pre-B cell line Nalm-6 is highly proficient in gene targeting by homologous recombination. *DNA Cell Biol.* **25,** 19–24.
9. Hurwitz, R., Hozier, J., LeBien, T., et al. (1979) Characterization of a leukemic cell line of the pre-B phenotype. *Int. J. Cancer* **23,** 174–180.
10. Wlodarska, I., Aventin, A., Ingles-Esteve, J., et al. (1997) A new subtype of pre-B acute lymphoblastic leukemia with t(5;12)(q31q33;p12), molecularly and cytogenetically distinct from t(5;12) in chronic myelomonocytic leukemia. *Blood* **89,** 1716–1722.
11. Filippini, G., Griffin, S., Uhr, M., et al. (1998) A novel flow cytometric method for the quantification of p53 gene expression. *Cytometry* **31,** 180–186.
12. Puck, T. T., Cieciura, S. J., and Robinson, A. (1958) Genetics of somatic mammalian cells. III. Long-term cultivation of euploid cells from human and animal subjects. *J. Exp. Med.* **108,** 945–956.
13. Lieber, M. R., Ma, Y., Pannicke, U., and Schwarz, K. (2003) Mechanism and regulation of human non-homologous DNA end-joining. *Nat. Rev. Mol. Cell Biol.* **4,** 712–720.
14. Li, G., Nelsen, C., and Hendrickson, E. A. (2002) Ku86 is essential in human somatic cells. *Proc. Natl. Acad. Sci. USA* **99,** 832–837.
15. Zhu, C., Bogue, M. A., Lim, D. S., Hasty, P., and Roth, D. B. (1996) Ku86-deficient mice exhibit severe combined immunodeficiency and defective processing of V(D)J recombination intermediates. *Cell* **86,** 379–389.
16. Aratani, Y., Okazaki, R., and Koyama, H. (1992) End extension repair of introduced targeting vectors mediated by homologous recombination in mammalian cells. *Nucleic Acids Res.* **20,** 4795–4801.

3

Robust Cell Line Development Using Meganucleases

Jean-Pierre Cabaniols and Frédéric Pâques

Summary

Cell line development for protein production or for the screening of drug targets requires the reproducible and stable expression of transgenes. Such cell lines can be engineered with meganucleases, sequence-specific endonucleases that recognize large DNA target sites. These proteins are powerful tools for genome engineering because they can increase homologous gene targeting by several orders of magnitude in the vicinity of their cleavage site. Here, we describe in details the use of meganucleases for gene targeting in Chinese hamster ovary-K1 cells, with a special emphasis on a gene insertion procedure using a promoter-less marker gene for selection. We have also monitored the expression of genes inserted by meganucleases-induced recombination, and show that expression is reproducible among different targeted clones, and stable over a 4 mo period. These experiments were conducted with the natural yeast I-*Sce*I meganuclease, but the general design and process can also be applied to engineered meganucleases.

Key Words: Cell line development; double-strand break; gene targeting; homologous recombination; I-*Sce*I; meganucleases; protein production.

1. Introduction

Homologous recombination is a powerful tool for genome engineering. Since the first gene targeting experiments in yeast more than 25 yr ago *(1,2)*, homologous recombination (HR) has been used to insert, replace, or delete genomic sequences in a variety of cells *(3–5)*. However, targeted events occur at a very low frequency in mammalian cells. The frequency of HR can be significantly increased by a specific DNA double-strand break (DSB) in the targeted locus *(6,7)*. Such DSBs can be delivered with meganucleases, sequence-specific endonucleases that recognize large DNA target sites (>12 bp). Because of their exquisite specificity, these proteins can cleave a unique chromosomal sequence without affecting global genome integrity. Natural meganucleases are essentially represented by homing endonucleases, a widespread class of proteins found in eukaryotes, bacteria, and archae *(8)*. Early studies of the I-*Sce*I and homothalic switching endonuclease

From: *Methods in Molecular Biology, vol. 435: Chromosomal Mutagenesis*
Edited by: G. Davis and K. J. Kayser © Humana Press Inc., Totowa, NJ

have illustrated how the cleavage activity of these proteins initiates HR events in living cells and demonstrated the recombinogenic properties of chromosomal DNA DSBs *(9,10)*. Since then, meganucleases-induced recombination has been successfully used for genome engineering purposes in bacteria *(11)*, mammalian cells *(6,7,12–14)*, mice *(15)*, and plants *(16,17)*.

However, the use of meganucleases-induced recombination has long been limited by the repertoire of natural meganucleases; although several hundreds of other meganucleases had been identified in between *(8)*, the diversity of cleavable sequences was too limited to address the complexity of genomes. In fact, only artificial loci, wherein an I-*Sce*I cleavage site had been introduced, could be engineered. Recently, the making of artificial meganucleases, based on homing endonucleases *(18,19)* or zinc-finger proteins *(20,21)*, has considerably enlarged the number of sequences that can be targeted. The production of such new proteins opens the door for the engineering of natural chromosomal sequences.

Here we present a general overview on how to use meganucleases to induce gene correction, gene insertion, or gene replacement. We provide precise protocols for the creation and characterization of gene targeting events, and give a brief survey of the impact of insertion or deletion size on the efficiency of the process, with a special emphasis on gene insertion. Gene insertion can be used, for example, to insert genes of interest in specific loci, for heterologous protein production. Recombinant therapeutic proteins are today mostly produced in mammalian cells such as Chinese hamster ovary (CHO), mouse SP2/0 and NS0 cells, or the human PerC.6 cell line, stably transfected with the gene of interest *(22)*. In the process of selecting highly expressing clones, the level and stability of protein expression are two major criteria. Furthermore, obtaining reproducible results from one clone to another would be an advantage in terms of screening efforts. These principles also apply to the generation of cells for screening of specific drug targets such as G protein-coupled receptors. In order, to illustrate the various advantages of meganucleases-induced gene targeting for this specific kind of application, we provide protocols to monitor the expression of inserted genes and data of expression stability following gene targeting.

In all the experiments described in **Subheading 3.**, we use the I-*Sce*I meganuclease, which remains today the "gold standard" for DSB-induced recombination, in terms of efficacy and specificity. However, the same protocols can be used to induce similar levels of gene correction or insertion with engineered meganucleases derived from the I-*Cre*I homing endonuclease, which were described in recent reports *(18,19)*.

2. Materials

2.1. Cell Culture and Transfection

1. Phosphate-buffered saline (PBS) (Invitrogen-Life Science, Carlsbad, CA).
2. Kaighn's modified F-12 medium (F12-K) (Invitrogen-Life Science) is supplemented with 2 m*M* L-glutamine, penicillin (100 UI/mL), streptomycin (100 μg/mL), amphotericin B

(Fongizone) (0.25 μg/mL) (Invitrogen-Life Science), and 10% fetal bovine serum (Sigma-Aldrich Chimie, St. Louis, MO).
3. Freezing medium: complete F12-K medium supplemented with 10% Dimethylsulfoxyde.
4. Puromycin dihydrochloride (Sigma-Aldrich Chimie). For CHO-K1, the concentration of 10 μg/mL is used.
5. Hygromycin B solution (Sigma-Aldrich Chimie). For CHO-K1, the concentration of 0.6 mg/mL is used.
6. Trypsin-EDTA solution (Invitrogen-Life Science).
7. Versene solution (Invitrogen-Life Science).

2.2. Molecular Characterization of Targeted Events by Southern Blot Using Nonradioactive Probes

2.2.1. Genomic DNA (gDNA) Preparation and Digestion

1. gDNA lysis buffer: 0.5% sodium dodecyl sulfate (SDS), 40 m*M* Tris-HCl, 40 m*M* EDTA, 200 m*M* NaCl, pH 7.5. Store at 4°C.
2. Proteinase K (20 mg/mL) (Eurobio, Les Ulis, France).
3. TE buffer: 10 m*M* Tris-HCl, 1 m*M* EDTA, pH 8.0.

2.2.2. DNA Electrophoresis and Transfer

1. Running buffer: 0.5X Tris-acetate EDTA: 20 m*M* Tris-acetate, 1 m*M* EDTA pH 8.0. Prepare a 50X stock solution.
2. Denaturation buffer: 0.5 *M* NaOH, 1.5 *M* NaCl. Store at room temperature.
3. Neutralization buffer: 0.5 *M* Tris-HCl (pH 7.4), 1.5 *M* NaCl. Store at room temperature.
4. Transfer buffer: 10X SSC (1.5 *M* NaCl, 0.15 *M* Na-citrate). Make 20X stock solution. Store at room temperature.
5. Hybond N+ membrane (Amersham, Little Chalfont, England).
6. 3 MM Chr Whatmann paper (Schleicher and Schuell, Maldstone, England).

2.2.3. Probe Labeling

1. Digoxygenin (DIG) DNA-labeling kit (Roche Diagnostics, Mannheim, Germany).
2. Nucleospin column (Macherey-Nagel, Düren, Germany).

2.2.4. Hybridization

1. Hybridization buffer: 0.5 *M* phosphate buffer pH 7.2, 7% SDS, 1 m*M* EDTA. Store at room temperature.

2.2.5. Wash and Anti-DIG Probing

1. Wash buffer: 40 m*M* phosphate buffer pH 7.2, 1% SDS. Store at room temperature.
2. Buffer I: 0.1 *M* maleic acid, 0.15 *M* NaCl, pH 7.5. Prepare a 10X stock solution. Adjust pH with NaOH. Store at room temperature. Add 0.3% Tween-20 before use.
3. Buffer II: buffer I supplemented with 10% (v/v) blocking reagent.
4. Buffer III: 100 m*M* Tris-HCl pH 9.5, 100 m*M* NaCl, 50 m*M* $MgCl_2$. Store at room temperature.

5. Alkaline phosphatase-conjugated anti-DIG Fab (Roche Diagnostics, Mannheim, Germany). Store at 4°C.
6. Blocking reagent 10X (Roche Diagnostics, Mannheim, Germany). Store at 4°C.
7. CDP star—chemiluminescence substrate for alkaline phosphatase—(Roche Diagnostics, Mannheim, Germany). Store at 4°C.

2.3. Protein Production

2.3.1. Measure of β-Galactosidase Activity

1. Lysis buffer: 10 m*M* Tris-HCl pH 7.5, 150 m*M* NaCl, 0.1% Triton X-100. Store at 4°C.
2. 100X Mg buffer: 100 m*M* $MgCl_2$, 35% β-mercaptoethanol. Store at room temperature.
3. Orthonitrophenyl-β-D-galactopyranoside (ONPG): prepare a 8 mg/mL solution in water. Store at –20°C.
4. 0.1 *M* Na_2HPO_4/NaH_2PO_4 pH 7.5.

2.3.2. Antibodies and FACS Reagents

1. Biotin-conjugated mouse antihuman CD4 monoclonal antibody (Becton Dickinson, San Jose, CA).
2. Biotin-conjugated mouse isotype control monoclonal antibody (Becton Dickinson).
3. Streptavidin-conjugated PhycoErythrin (Streptavidin-PE) (Becton Dickinson).
4. FACS buffer: PBS supplemented with 2% fetal bovine serum.

3. Methods

Meganucleases (and more generally gene targeting) can be used to trigger gene correction as well as gene insertion, deletion, or replacement. Obtaining the desired kind of events does not simply depend on a standard protocol, but also on the design of the targeting vector, also referred to as "repair matrix," because they actually provide a matrix or template for the DSB-repair process initiated by I-*Sce*I and other meganucleases. In this chapter, we will provide a few examples for the design of the repair matrix for gene correction, gene insertion, and gene replacement.

We take, as an example, experiments that were performed on a model cell line with a *puromycin-resistance* marker as target gene. This gene is under the transcriptional control of the human EF1α promoter, and about 1 kb of additional EF1α sequences, corresponding to the two first untranslated exons and first intron, are present between the promoter and the *puromycin-resistance* gene. Furthermore, a cleavage site for the natural endonuclease I-*Sce*I was inserted 132 bp downstream of the ATG of *puromycin-resistance* gene, thus inactivating this gene. The construct has been stably integrated in CHO-K1 cell genome in single copy. Then, different repair matrix were constructed, to modify this locus by gene targeting.

The first repair matrix was designed for the correction of the *puromycin-resistance* gene **(Fig. 1A)**. Gene correction should result in the removal of 22 bp including the I-*Sce*I cleavage site, and the consequent restoration of a functional

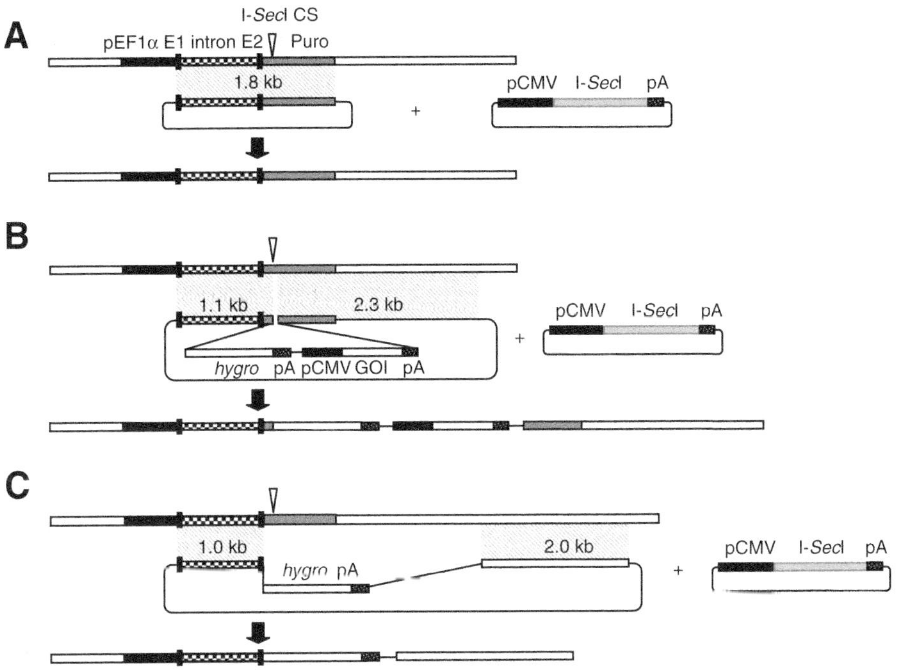

Fig. 1. Design of reporter system and repair matrix for gene correction (**A**), gene insertion (**B**), and gene replacement (**C**). The puromycin-resistance gene is interrupted by an I-*Sce*I cleavage site (CS) and under the control of the EFIα promoter. For gene correction, the repair matrix includes 1.8 kb of homologous sequence corresponding to the two untranslated exons (E1, E2) and first intron and a full length puromycin gene. For gene insertion, the repair matrix corresponds to a promoter-less hygromycin-resistance gene alone or with a complete expression cassette that is flanked by two stretches of homologous sequences (1.1 kb and 2.3 kb). The repair matrix for gene replacement is very similar to the insertion matrix except that (i) the 3′ homology sequence starts after the puromycin polyA signal (ii) all sequences from the puromycin-resistance gene have been removed from the 5′ homologous sequence. Homology lengths are depicted by hatched boxes. GOI stands for gene of interest (β-galactosidase or CD4 in this study).

puromycin-resistance cassette. The gene correction matrix includes a noninterrupted *puro*^{R} gene, with EF1α sequences in 5′ (these sequences include exon 1 and 2 and intron 1, but not the promoter). It shares a total of 1.8 kb of homology with the targeted locus, with 1.1 kb in 5′ of the I-*Sce*I site, and 0.7 kb in 3′.

Second, a series of three insertion vectors were designed by cloning inserts of various sizes in the *puromycin-resistance* gene (**Fig. 1B**). These repair matrix are intended to insert novel sequences within the targeted locus. The heterologous inserts are surrounded by 1.1 kb of homologous sequences in 5′ (the EF1α exons and intron), and 2.3 kb in 3′ (longer than in the gene correction vector). The simplest insertion vector includes a promoter-less *hygromycin-resistance* cassette of 1.3 kb, whereas the two others contain an additional expression cassette, coding for the *Escherichia coli* β-galactosidase enzyme or the human CD4 transmembrane protein, and resulting in total inserts length of 5.5 and 4.4 kb, respectively. Note that in these insertion vectors, the *hygromycin-resistance* gene is modified by addition in its 5′ end of the first 132 bp (44 first amino acids) of the *puromycin-resistance* gene (132 bp upstream of the I-*Sce*I site), resulting in the production of a fusion protein that confers resistance to hygromycin B.

Third, one can envision, in several applications, the total replacement of the *puromycin-resistance* gene by another selection marker. To meet this purpose, a gene replacement vector was produced by modifying the 3′ homologous sequence of the first gene insertion vector (**Fig. 1C**). In this construct, the 3′ homologous sequence (about 2 kb in length) started 2 kb after the end of the *puromycin-resistance* ORF, and all sequences from the *puromycin-resistance* gene (132 bp) have been removed from the 5′ homologous sequence (for a total deletion of 3 kb). The 5′ homologous sequence is also modified by removing all sequences from the *puromycin-resistance* gene.

In the experiments described in **Subheading 3.**, the repair matrix and the I-*Sce*I expression vector (**Fig. 1**) are cotransfected in our model cell line. For maximal efficiency, we use a vector wherein I-*Sce*I is placed under the control of the cytomeganlovirus promoter and described in a former report *(7)*. On targeted insertion, the functional *puromycin-* or *hygromycin-resistance* gene is placed downstream of the EF1α sequences, and transcribed from the EF1α promoter. In contrast, random insertion will not create an expressed *puromycin* or *hygromycin-resistance* cassette unless the cassette is fortuitously inserted just downstream of a functional promoter. As such "promoter trap" events are supposed to be very rare, our system allows for a considerable enrichment for targeted events among transformed puro^R or hygro^R cell lines. This kind of "refinement" is not compulsory, but can greatly simplify the task of the researcher.

3.1. Cell Transfection and Selection

1. All the transfection experiments were done with the Amaxa Electroporation system (Amaxa GmbH, Koeln, Germany), which ensures more than 50% transfection efficiency.
2. The adherent cells are washed once with PBS, then incubated with trypsin-EDTA solution for 5′ at 37°C and collected in a 15-mL conical tube (**Note 1**). After centrifugation at 300*g* for 5 min, the supernatant is discarded and the cell pellet is resuspended in 10 mL of complete F12K medium. Cells are numerated, centrifuged again, and adjusted at the concentration of 2×10^7cells/mL in Amaxa solution T.
3. 2–15 μg of plasmid DNA (**Note 2**) is added to 100 μL of cells (2×10^6 cells) in an Amaxa electroporation cuvet. The cells and DNA are gently mixed before insertion in the electroporation chamber of the Amaxa electroporation apparatus (**Note 3**). For CHO-K1 cells, the manufacturer recommends the use of program U-23.

4. After electroporation, 0.5 mL of prewarmed complete F12K medium is added to the cuvet and cells are gently transferred into a 10-cm culture dish containing 10 mL of prewarmed complete F12K medium. Dishes are incubated in a 37°C, 5% CO_2 humidified incubator for 24 h.
5. After a 24 h recovery period, transfected cells are washed once with PBS, then incubated with trypsin-EDTA solution for 5′ at 37°C and collected in a 15-mL conical tube. Cells are counted and cloned in 96-well plates at the density of 500 cells/well in complete F12K medium supplemented with the selecting reagent. Examples of selecting reagent: puromycin solution is added at the final concentration of 10 µg/mL. Hygromycin B solution is added at the final concentration of 0.6 mg/mL.
6. About 10 d later, puromycin ($puro^R$)- or hygromycin ($hygro^R$)-resistant clones can be counted. Positive clones are amplified by sequential passage onto 12-well plates, 6-well plates, and finally 10-cm dishes. The selecting agent is maintained through all these steps. At this point, clones are cryo-conserved in freezing medium.

The frequencies of $puro^R$ and $hygro^R$ cells are summarized in **Table 1**. These frequencies are established by dividing the number of wells containing $puro^R$ and $hygro^R$ cells by the total number of plated cells (For $puro^R$ cells, this number is an underestimation: if $puro^R$ cells occur at a frequency of 3×10^{-3}, and if 500 cells were plated per well, there might be two $puro^R$ clones in several). However, such phenotypes can result either from gene targeting, or from random integration of the repair matrix downstream of an active promoter. Indeed, $hygro^R$-resistant clones (but for unknown reasons, no $puro^R$ clone) were also obtained when the repair matrix were transfected without the I-*Sce*I expressing vector (**Table 1**). Therefore, the frequencies of targeted events can only be obtained after molecular characterization.

3.2. Molecular Characterization of Targeted Events by Southern Blot Using Nonradioactive Probes

To identify targeted events, the gDNA of $hygro^R$ and $puro^R$ clones can be analyzed by Southern blotting, using restriction enzymes that will discriminate targeted loci from nonrecombined ones. An example is shown on **Fig. 2**.

3.2.1. gDNA Preparation and Digestion

1. Confluent cells from 10-cm dish are collected in a 15-mL conical tube. Cells are washed once with PBS. After centrifugation (300*g*) supernatants are discarded and cell pellets are thoroughly vortexed.
2. 500 µL of gDNA lysis buffer is added to cell pellets and incubated at room temperature for 30 min. 15 µL of proteinase K solution (20 mg/mL) is added to the lysates. Incubate at 56°C overnight.
3. The next day, 10 mL of ethanol 85% is added. Mix gently then incubate for 10 min at room temperature. The gDNA pellet will sediment at the bottom of the tube.
4. DNA pellet is transferred in a 1.5-mL microfuge tube and centrifuged at 16,000*g* for 30 min at 4°C. Supernatants are discarded. DNA pellets are rinsed with 0.5 mL ethanol 70%, centrifuged again for 10 min at 16,000*g*. A maximum of ethanol is

Table 1
Summary of Gene Targeting Experiments

Nature of repair matrix	Gene correction (Fig. 1A)		Gene insertion (Fig. 1B)				Gene replacement (Fig. 1C)
Genes inserted	NA		*hygro*^{R}		*hygro*^{R} + CD4	*hygro*^{R} + *LacZ*	*hygro*^{R}
Size of insertion cassette (kb)	0		1.3		4.4	5.5	1.3
I-*Sce*I expressing vector	–	+	–	+	+	+	+
Frequency of hygro^R or puro^R clones	0	3×10^{-3}	10^{-4}	5×10^{-4}	4.7×10^{-4}	1.7×10^{-4}	5.5×10^{-4}
Total targeted events among hygro^R or puro^R clones	NA	30/30	0/27	29/37	4/12	14/20	9/36
Targeted events associated with random insertions	NA	0/30	NA	5/29	2/4	1/14	3/9
Frequency of targeted event among transfected cells	0	3×10^{-3}	0	3.9×10^{-4}	1.6×10^{-4}	0.7×10^{-4}	1.4×10^{-4}

NA: not applicable.

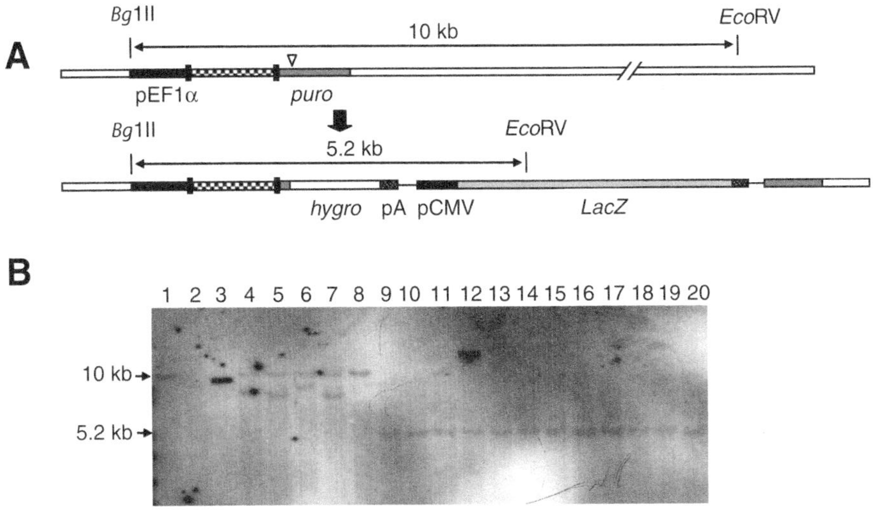

Fig. 2. Southern blot analysis of hygro^R clones using nonradioactive probes. Here, we present the analysis of clones obtained with the repair matrix containing the *hygro*^{R}-*LacZ* 5.5 kb insert. (**A**) Principle of analysis. gDNA is digested with restriction enzymes *Bgl*II and *Eco*RV. After electrophoresis and DNA transfer, the blot is hybridized with a probe corresponding to the first intron. In the parental clone, a 10 kb band corresponding to the nontargeted puro^R locus should be detected. On targeted insertion of the *hygro*^{R}::*LacZ* cassette, the 10 kb band disappears and should be replaced by a 5.2 kb band. In nontargeted clones, additional bands of various sizes correspond to random insertion of the repair matrix. (**B**) Southern blotting. **Lane 1**; control parental cell line, **lane 2**; CHO-K1 cell line before transformation with reporter cassette, **lane 3**; linearized repair matrix plasmid, **lanes 4–20**; hygro^R clones selected after transfection. Random insertion is observed in **lanes 4–7**. **Lane 8** likely corresponds to integration of the *hygro*^{R} ORF without the intron. Targeted insertion is observed in **lanes 9–11** and **13–20**. Targeted insertion can also be associated with random insertion, revealed by an additional band (**lane 12**).

removed. DNA pellets are dried for 5 min at 37°C (**Note 4**) before the addition of 100–200 μL of TE buffer. Resuspended DNA is incubated at 65°C until complete dissolution (**Note 5**). DNA concentration is measured at 260 nm with a spectrophotometer.

5. Usually, 10 μg of gDNA is digested overnight with a two- to fivefold excess of enzymes.

3.2.2. DNA Electrophoresis and Transfer

1. Digested gDNA are loaded on a 0.8% agarose gel in 0.5X Tris-acetate EDTA buffer and run at 5 V/cm for 1 h then 8.3 V/cm for 4–5 h for a total migration length of 12–14 cm (**Note 6**). Take a photograph.
2. The gel is soaked first in denaturation buffer for 30 min at room temperature under gentle agitation, then in neutralization buffer for 30 min and again in new neutralization buffer for 15°C. Finally, the gel is soaked in 10X SSC for 5°C.

3. The separated DNA is blotted onto Hybond N+ nylon membrane by capillarity in 10X SSC overnight. Briefly, the gel is positioned on a piece of Whatmann 3 MM paper that has been previously soaked in transfer buffer. The extremities of the paper are lying in a tank containing the transfer buffer. The piece of nylon membrane (previously soaked in ddH_2O) is placed on the top of the gel and air bubbles are removed by rolling a clean plastic pipet. Three sheets of Whatmann 3 MM paper are placed onto the membrane. The upward flow of buffer will be ensured by a stack of paper towels positioned on the top. Finally, a glass plate with a weight is applied to maintain a tight connection between the layers.
4. The next morning, the membrane is removed from the transfer system and incubated a few minutes in 40 m*M* Na_2HPO4. The membrane is then dried on a piece of Whatmann 3 MM paper and DNA is cross-linked to the membrane using a Ultraviolet Stratalinker (Stratagene) **(Note 7)**.

3.2.3. Probe Labeling

1. 1 µg of DNA probe (between 500 bp and 1 Kb long) is denatured for 10 min at 100°C in a water bath.
2. The denatured DNA is labeled with Klenow enzyme (12 IU) in the presence of random hexamer oligonucleotides, deoxynucleotides, and DIG-conjugated dUTP at 37°C overnight in a final reaction volume of 100 µL.
3. The next day, the probe is purified on a Nucleospin column (Macherey—Nagel, Düren, Germany) and aliquoted. The probe aliquots are stored at –20°C until use.

3.2.4. Hybridization

1. The membrane is soaked in ddH_2O for 5 min then placed into a hybridization bottle with 20 mL of hybridization buffer. The bottle is incubated for 1 h in a hybridization oven at 68°C.
2. The DIG-labeled probe (use about 15 ng/mL hybridization buffer) is denatured for 10°C in a water bath at 100°C, then is added to 20 mL of hybridization buffer. The hybridization is performed at 68°C overnight.

3.2.5. Wash and Anti-DIG Probing

1. The probe is recovered and stored at –20°C **(Note 8)**. The blot is washed two times in washing buffer (equilibrated at 68°C) at 68°C for 5 min then once in 100 mL of buffer I at room temperature under agitation on a retro orbital shaker (Certomat® RM, B. Braun Biotech International GmbH, Melsungen, Germany) (30 rpm).
2. The membrane is wrapped in a plastic bag and 20 mL of buffer II is added. This step corresponds to the blocking step. The membrane is shaken for 30 min at 250 rpm. Remove the buffer and add 20 mL of buffer II with 0,0375 U/mL of anti-DIG antibody. Incubate membrane for another 30 min (250 rpm).
3. The membrane is removed from the plastic bag and placed immediately in buffer I. The membrane is washed two times for 30 min under gentle agitation (30 rpm).
4. Finally, the membrane is incubated for 3 min in buffer III **(Note 9)**.
5. The membrane is placed on a plastic sheet. 3 mL of buffer III plus 10 µL of CDP star is carefully layered over the membrane and a second sheet of plastic is placed on top of it. After 5 min of incubation, excess of liquid is wiped out and the plastic sheets are sealed.

6. The membrane is exposed in a cassette to an X-ray film at room temperature. Time of exposure will depend on the probe efficiency. Usually 4 h exposure is enough but can be longer (overnight for example) (**Note 10**).

Results are summarized in **Table 1**. For gene correction, the frequency of puro^R clones reflects the frequency of targeted events, as only corrected puromycin gene can be selected. With the other cassettes, the frequency of gene targeting seems to decrease when the size of the insert increases. As shown on **Fig. 2** (Lane 12), targeted insertion can be associated with random insertion in the same clone, but these double events remain a minority, except with the 4.4 kb cassette (two out of four targeted clones). In addition, the replacement of the whole *puro*^{R} gene with additional 3′ sequences (3 kb) seems to be more demanding than the simple insertion of the *hygro*^{R} cassette. However, there is only a two-fold decrease in efficiency. Altogether, these results argue that large insertions (5.5 kb) and deletions (3 kb) can be obtained by meganucleases-induced gene targeting. Note that when the repair matrix were transfected without the I-*Sce*I expressing plasmid, we never could obtain any targeted clone (*see* **Table 1**).

3.3. Protein Production

Insertion of different heterologous sequences at the same locus can be interesting for protein production or drug screening purpose. The level of protein expression and its stability are both major criteria during cellular clone selection. Furthermore, reproducibility of expression would decrease the screening effort and allow for better planification. An insertion in the same locus would ensure reproducible levels of expression independently of the inserted sequences. As an example, we monitor the expression level of two different proteins (β-galactosidase and human CD4) over a 4 mo period after I-*Sce*I-induced gene insertion. In these experiments, only clones with targeted insertions and without additional random inserts were considered.

3.3.1. Measure of β-Galactosidase Activity

1. Cells are washed twice in PBS then incubated with 5 mL of trypsin-EDTA solution. After 5 min incubation at 37°C, cells are collected in a 15-mL conical tube and counted.
2. 2×10^6 cells are incubated with 200 μL of lysis buffer on ice for 30 min.
3. Cell lysates are centrifuged for 2 min at 10,000*g* at 4°C. Supernatants are transferred in a clean 1.5-mL microfuge tube.
4. 20 μL of 1/10th dilution of cell lysates are mixed with: 2 μL of Mg 100X buffer, 22 μL of ONPG (8 mg/mL) solution and 156 μL of 0.1 *M* Na_2HPO_4 pH 7.5 in a 96-well plate.
5. Plates are incubated at 37°C for 45 min. Optical density is read at 415 nm using a microplate reader (Model 550, BioRad, Hercules, CA).

The production level of β-galactosidase has been measured in this way for 4 different targeted clones during 16 wk. The graph shown on **Fig. 3A** displayed the

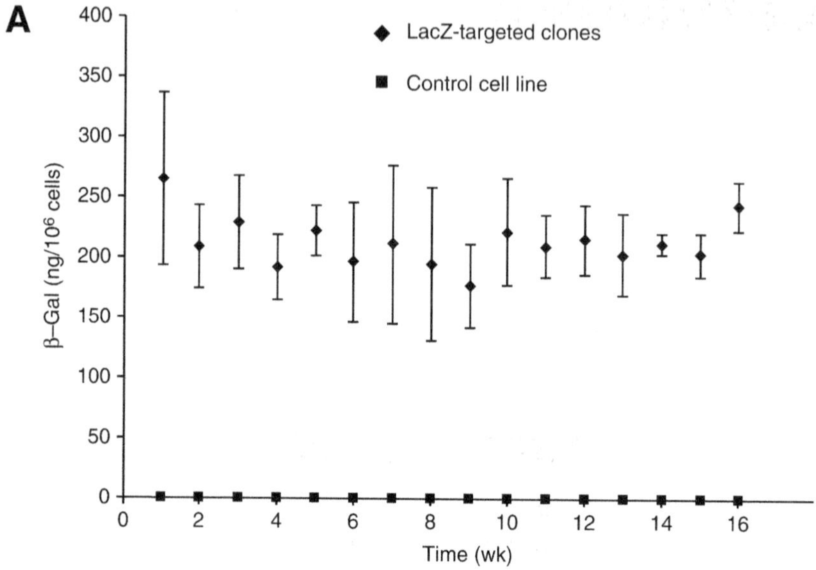

Fig. 3. *(Continued)*

mean level of expression for these 4 clones as compared with nonproducing cell line, showing a very small dispersion of the measurements.

3.3.2. FACS Detection of CD4-Expressing Cells

1. Cells are washed twice in PBS and incubated with 2 mL of Versene solution (**Note 11**). After 5 min incubation at 37°C, cells are collected in a 15-mL conical tube. The cells are counted.
2. 10^6 cells are transferred in 5-mL tube (Falcon, 2058) and centrifuged at 300*g* for 5 min at 4°C. Cells are washed once with FACS buffer. Cell pellets are resuspended in 20 μL of biotin-conjugated anti-CD4 or biotin-conjugated isotype control antibody. After 30 min of incubation on ice, cells are washed once in FACS buffer. Cell pellets are then incubated with 20 μL of streptavidin-conjugated PE for 30 min on ice and protected from light. The cells are washed once in FACS buffer and finally resuspended in 0.5 mL of FACS buffer.
3. The cells sample are analyzed on a FACS vantage II (BD Bioscience, San Jose, CA) using a 488 nm Ion-Argon laser. The emitted fluorescence (emission wavelength at ~580 nm) is collected in the fluorescence 2 channel.

The level of CD4 expression of a targeted clone has been analyzed by FACS for two different targeted clones during 16 wk. The pattern of CD4 expression for one of these clones is shown on **Fig. 3B**. For the other one the pattern is very similar (Data not shown). Thus, as for β-galactosidase, we demonstrated that the level of expression is very similar from clones to clones and that the expression is very stable over a long period of time.

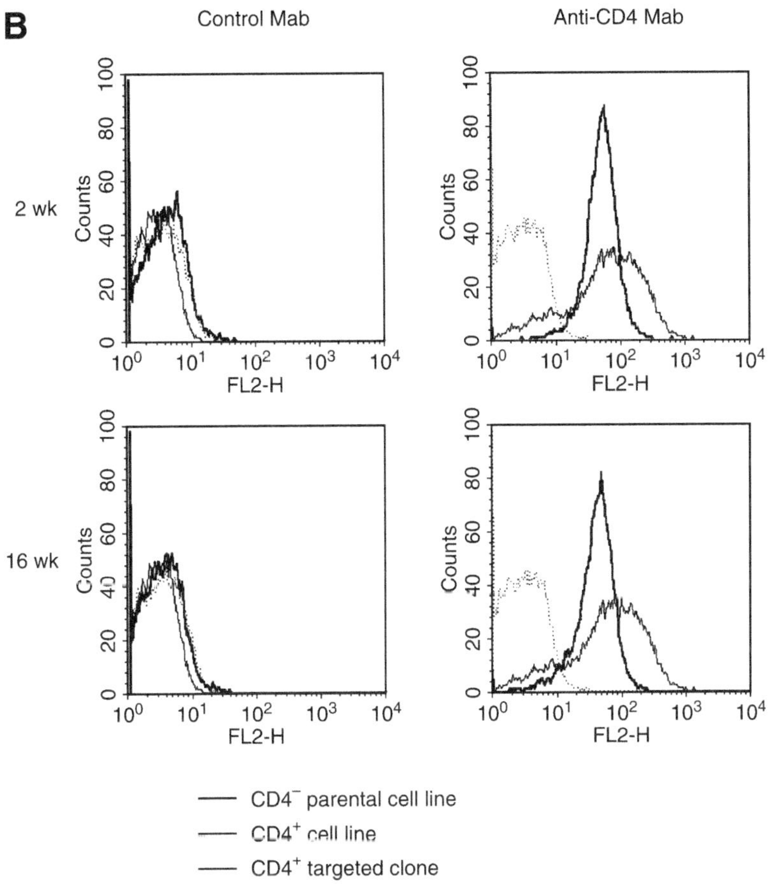

Fig. 3. Stability of protein expression in targeted clones. **(A)** β-galactosidase expression was assayed by the ONPG test for 4 independent targeted clones (♦) and for the parental cell line (■) over a period of 16 wk. The mean and standard deviation of expression levels for these 4 clones is plotted against time. **(B)** CD4-expression of a targeted clone was measured by FACS over a period of 16 wk. FACS displays for week 2 and 16 are shown. The CD4 expression of the targeted clone (bold line) is compared with the $CD4^-$, parental cells (light line) and to endogenous CD4 expressing cells (dotted line).

4. Notes

1. For optimal transfection efficiency, the cells should not be more than 80% confluent. For the same reason, we recommend to use cells with low passage number.
2. For optimal transfection efficiency, the plasmid DNA must be produced with an endotoxin-free preparation technique.
3. The cells should not stay more than 20 min in solution T.
4. gDNA pellet is quite difficult to resuspend in TE, unless if not dried completely.
5. The resuspension time may vary from sample to sample. It can take several hours. Do not hesitate to gently mix the DNA from time to time.

6. Trays, combs, and electrophoresis apparatus is carefully cleaned with soap and rinsed with distilled water. This step is crucial because dirty materials can impair the final result.
7. At this step, the membrane can be stored between two pieces of Whatmann paper for a long period of time.
8. The same probe can be used three to four times. When using a new aliquot, we recommend performing a first hybridization with a useless membrane. We noticed that the background is diminished when a probe has been used one or two times.
9. When performing the anti-DIG antibody revelation, it is important not to let the membrane dry up during the steps. Otherwise, a very high background will occur.
10. Hybridized membranes can be stored at –20°C in their plastic bag. Alternatively, they can be stripped by a short incubation in denaturation buffer (10 min) followed by two quick bath in distilled water first and then in 200 m*M* Na_2HPO_4, pH 7.2. The membrane is dried on Whatmann paper and stored at room temperature. This membrane can be reprobed but usually the reprobing does not give satisfactory results.
11. Because the CD4 protein is expressed at the cell surface, the Versene solution is used to collect the cells. This solution is preferred to the trypsin-EDTA solution because the trypsin can degrade surface molecules. As the Versene solution is less efficient than the trypsin-EDTA solution, it is recommended to gently scrap the cells after the 5 min incubation before the addition of medium. Furthermore, multiple samples are treated one at a time.

Acknowledgment

We thank Luc Mathis for critical reading of the manuscript.

References

1. Hinnen, A., Hicks, J. B., and Fink, G. R. (1978) Transformation of yeast. *Proc. Natl. Acad. Sci. USA* **75,** 1929–1933.
2. Rothstein, R. J. (1983) One-step gene disruption in yeast. *Methods Enzymol.* **101,** 202–211.
3. Thomas, K. R. and Capecchi, M. R. (1987) Site-directed mutagenesis by gene targeting in mouse embryo-derived stem cells. *Cell* **51,** 503–512.
4. Capecchi, M. R. (2001) Generating mice with targeted mutations. *Nat. Med.* **7,** 1086–1090.
5. Smithies, O. (2001) Forty years with homologous recombination. *Nat. Med.* **7,** 1083–1086.
6. Rouet, P., Smih, F., and Jasin, M. (1994) Introduction of double-strand breaks into the genome of mouse cells by expression of a rare-cutting endonuclease. *Mol. Cell Biol.* **14,** 8096–8106.
7. Choulika, A., Perrin, A., Dujon, B., and Nicolas, J. F. (1995) Induction of homologous recombination in mammalian chromosomes by using the I-SceI system of *Saccharomyces cerevisiae. Mol. Cell Biol.* **15,** 1968–1973.
8. Chevalier, B. S. and Stoddard, B. L. (2001) Homing endonucleases: structural and functional insight into the catalysts of intron/intein mobility. *Nucleic Acids Res.* **29,** 3757–3774.

9. Dujon, B., Colleaux, L., Jacquier, A., Michel, F., and Monteilhet, C. (1986) Mitochondrial introns as mobile genetic elements: the role of intron-encoded proteins. *Basic Life Sci.* **40,** 5–27.
10. Haber, J. E. (1995) In vivo biochemistry: physical monitoring of recombination induced by site-specific endonucleases. *Bioessays* **17,** 609–620.
11. Posfai, G., Kolisnychenko, V., Bereczki, Z., and Blattner, F. R. (1999) Markerless gene replacement in *Escherichia coli* stimulated by a double-strand break in the chromosome. *Nucleic Acids Res.* **27,** 4409–4415.
12. Sargent, R. G., Brenneman, M. A., and Wilson, J. H. (1997) Repair of site-specific double-strand breaks in a mammalian chromosome by homologous and illegitimate recombination. *Mol. Cell Biol.* **17,** 267–277.
13. Donoho, G., Jasin, M., and Berg, P. (1998) Analysis of gene targeting and intrachromosomal homologous recombination stimulated by genomic double-strand breaks in mouse embryonic stem cells. *Mol. Cell Biol.* **18,** 4070–4078.
14. Cohen-Tannoudji, M., Robine, S., Choulika, A., et al. (1998) I-SceI-induced gene replacement at a natural locus in embryonic stem cells. *Mol. Cell Biol.* **18,** 1444–1448.
15. Gouble, A., Smith, J., Bruneau, S., et al. (2006) Efficient in toto targeted recombination in mouse liver by meganuclease-induced double-strand break. *J. Gene Med.* **8,** 616–622.
16. Siebert, R. and Puchta, H. (2002) Efficient repair of genomic double-strand breaks by homologous recombination between directly repeated sequences in the plant genome. *Plant Cell* **14,** 1121–1131.
17. Puchta, H., Dujon, B., and Hohn, B. (1996) Two different but related mechanisms are used in plants for the repair of genomic double-strand breaks by homologous recombination. *Proc. Natl. Acad. Sci. USA* **93,** 5055–5060.
18. Arnould, S., Chames, P., Perez, C., et al. (2006) Engineering of large numbers of highly specific homing endonucleases that induce recombination on novel DNA targets. *J. Mol. Biol.* **355,** 443–458.
19. Smith, J., Grizot, S., Arnould, S., et al. (2006) A combinatorial approach to create artificial homing endonucleases cleaving chosen sequences. *Nucleic Acids Res.* **34,** E149.
20. Smith, J., Bibikova, M., Whitby, F. G., Reddy, A. R., Chandrasegaran, S., and Carroll, D. (2000) Requirements for double-strand cleavage by chimeric restriction enzymes with zinc finger DNA-recognition domains. *Nucleic Acids Res.* **28,** 3361–3369.
21. Urnov, F. D., Miller, J. C., Lee, Y. L., et al. (2005) Highly efficient endogenous human gene correction using designed zinc-finger nucleases. *Nature* **435,** 646–651.
22. Wurm, F. M. (2004) Production of recombinant protein therapeutics in cultivated mammalian cells. *Nat. Biotechnol.* **22,** 1393–1398.

4

Design and Testing of Zinc Finger Nucleases for Use in Mammalian Cells

Matthew Porteus

Summary

Homologous recombination is the most precise way to manipulate the genome. As a tool it has been used extensively in bacteria, yeast, murine embryonic stem cells, and a few other specialized cell lines but has not been available to researchers in other systems, such as for mammalian somatic cell genetics. Recently, work has shown that the creation of a gene-specific DNA double-strand break can stimulate homologous recombination by several thousand-fold in mammalian somatic cells. These double-strand breaks can now be created in mammalian genomes by zinc finger nucleases (ZFNs). ZFNs are artificial proteins in which a zinc finger DNA-binding domain is fused to a nonspecific nuclease domain. This chapter describes how to identify potential targets for ZFN cutting, to make ZFNs to cut this target site, and how to test whether the newly designed ZFNs are active in a mammalian cell culture-based system.

Key Words: Double-strand breaks; gene manipulation; gene targeting; gene therapy; homologous recombination; zinc finger nucleases.

1. Introduction

One of the most powerful ways to understand biological processes is to alter the genomic sequence of an organism and then observe the resulting phenotype. Genetic screens, for example, are based on using mutagens, such as chemicals, ultraviolet irradiation, or transposons to make random mutations in the genome and then identifying the progeny for interesting phenotypes. Another way to manipulate the genome is to introduce novel transgenes into the genome and observe the consequences. These transgenes can be integrated into genome using a variety of techniques, including viral- and transposon-based vectors, and usually integrate in an uncontrolled, if not random, fashion. However, the most precise way to manipulate the genome is by gene targeting using homologous recombination. Whereas investigators have used the term gene targeting for different things, for the purposes

From: *Methods in Molecular Biology, vol. 435: Chromosomal Mutagenesis*
Edited by: G. Davis and K. J. Kayser © Humana Press Inc., Totowa, NJ

of this chapter, gene targeting is used to describe the process of introducing new genetic material into the genome using homologous recombination.

In homologous recombination the genetic information from one piece of DNA is transferred from one DNA molecule to another homologous molecule of DNA; the mechanism of which has been reviewed elsewhere ***(1,2)***. Homologous recombination is an essential process and is used generally to repair DNA double-strand breaks (DSBs) in mitotic cells and to create genetic diversity in meiotic cells. In addition homologous recombination is used for specific processes such as mating-type switching in yeast and somatic hypermutation in chickens. Finally, homologous recombination has been co-opted by genetic parasites such as homing endonucleases to catalyze their spread ***(3)***. All of these natural occurrences of homologous recombination are initiated by the creation of a DNA DSB. In mammalian cells, for example, the creation of a DNA DSB stimulates homologous recombination by several thousand-fold ***(4,5)***. If one were able to create gene-specific DNA DSBs, researchers would also be able to co-opt the homologous recombination machinery to increase the efficiency of gene targeting. One method to create gene-specific DSBs is by the use of zinc finger nucleases (ZFNs) ***(6,7)***. ZFNs are artificial proteins in which a zinc finger (ZF) DNA-binding domain is fused to a nonspecific nuclease domain ***(8)***. If two ZFNs bind to DNA in the correct orientation, the nuclease domain dimerizes and then creates a DSB. ZFNs have been shown to create site-specific DSBs that stimulate homologous recombination in *Xenopus oocytes*, *Drosophila melanogaster*, plants, and in somatic mammalian cells ***(5,9–14)***. In this chapter, a method to identify, design, and test ZFNs for use in mammalian cells is described. In another chapter of this volume, Dana Carroll and his colleagues discuss the use of ZFNs in flies and worms. Here a method using overlap polymerase chain reaction (PCR) is described to assemble the zinc finger protein (ZFP) of the ZFN. Recently; however, a detailed protocol to assemble a ZF DNA-binding domain using restriction endonucleases without PCR has been described ***(15)***.

This chapter describes a method to assemble a ZFN using a "modular-assembly" approach. The modular-assembly approach to making ZFNs depends on two crucial features:

1. The modular nature of ZF binding to DNA.
2. The published data sets of individual ZFs and their cognate DNA-binding sites.

The modular nature of ZF binding was revealed by the crystal structure of the ZF domain ***(16,17)***. An individual ZF consists of 30 amino acids arranged in a ββα structure that is stabilized by chelating a single zinc ion. For each individual ZF, a portion of the α-helix, called the "recognition helix," lies in the major groove of DNA. Specific DNA binding is mediated by interactions of the recognition helix with the DNA bases. The amino acids that mediate DNA binding are numbered with respect to the beginning of the α-helix (–1 to 6). Each finger binds to a 3–4-bp target sequence. For purposes of modular-assembly; however, each finger is designed to bind a nonoverlapping nucleotide triplet. In a ZFP, the most amino-terminal finger is called "finger 1," the next "finger 2," and so on. In a three-finger protein, the most carboxy-terminal finger is "finger 3." In a three-finger ZFN the nuclease domain is linked to finger 3. A ZFP consists of a series of individual ZF domains. Thus, a three-finger protein is one that has three individual ZF domains and would be designed to bind a 9-bp site. The modular nature of ZF binding has two aspects to it:

1. Each individual finger seems to bind a nonoverlapping triplet independently of its neighboring finger.
2. Each finger makes contacts with each base of the triplet with a single amino acid. From these two modular aspects of ZF binding, two predictions were made. The first is that by altering the amino acid contact residues of an individual finger, you could create a new finger that binds to a different 3-bp sequence. The second is that by shuffling different individual ZFs, one could create a new ZFP with a new target site specificity. For a three-finger protein, for example, it would mean creating a protein with a novel 9-bp binding site. The testing, successes, and limitations of the modular model of ZF binding are well-described elsewhere ***(17,18)***.

The second crucial feature for the modular-assembly approach is that a variety of individual fingers have been published that bind to unique triplet-binding sites. For example, two groups have published data sets for individual fingers that bind to triplets of the form 5′-GNN-3′ ***(19–21)***. In addition, data sets for fingers that bind to 5′-ANN-3′ and 5′-CNN-3′ triplets have also been published ***(22,23)***.

Using the modular nature of binding and the published data sets, a number of different artificial transcription factors have been made ***(24)***. Almost all of these artificial transcription factors have been designed to recognize target sequences rich in 5′-GNN-3′ target site triplets. This bias may reflect that the published individual ZFs that recognize GNN triplets are of higher quality than those that bind to non-GNN triplets. Or it may reflect that fingers that bind GNN triplets are more modular in their binding and depend less on the context of the binding of the neighboring domains. Careful studies of ZF DNA binding have shown, for example, that binding is not completely modular and that binding of a given finger depends on its neighbors; i.e., there is "context" dependence ***(18)***. In the modular-assembly approach to designing new ZFPs, this context dependence is ignored and is an important caveat to the approach.

1.1. Overall Strategy

The use of ZFNs to stimulate gene targeting by homologous recombination involves the following steps.

1. A full ZFN-binding site must be identified within the gene of interest.
2. A pair of ZFNs must be designed and assembled.
3. The ZFN pair should be tested for activity using reporter assays.
4. A targeting construct that creates the desired genomic modification must be identified.
5. Targeting of the endogenous gene by cointroduction of the ZFNs to create a DSB in the target gene along with the targeting construct in order for gene targeting to occur.

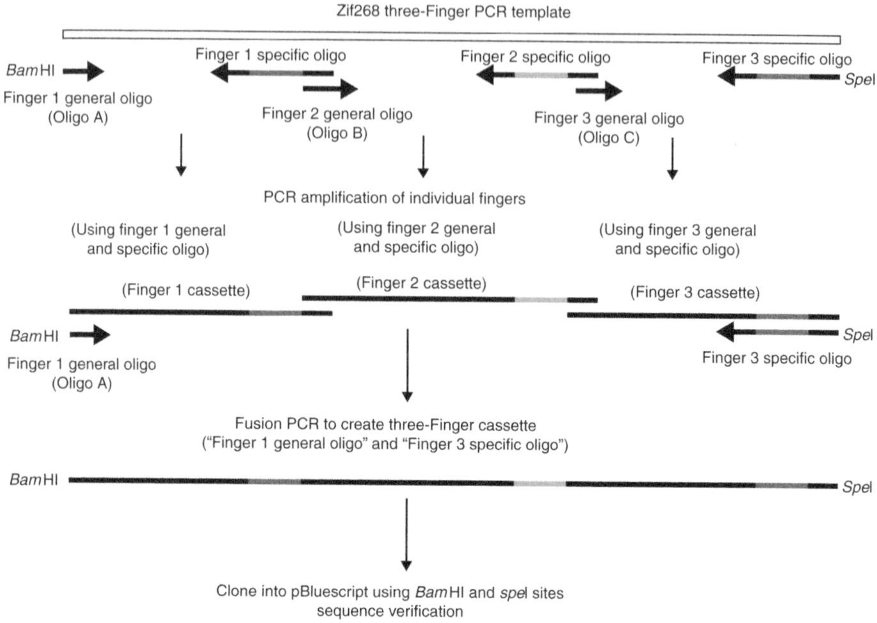

Fig. 1. Schematic overview of overlap PCR strategy to assemble new ZFP. The experimental details are in the text. The colored lines depict the unique recognition helix for each ZF.

In this chapter, protocols for the first three steps are described. Standardized protocols for the last two steps, perhaps the most interesting and important steps, are still being developed for ZFN-mediated gene targeting in mammalian cells.

2. Materials

1. Agarose gel (2.5%).
2. Calf intestinal alkaline phosphatase.
3. *Apa*I, *Bam*HI, and *Spe*I restriction endonucleases.
4. KOD polymerase kit (Novagen, cat. no. 71085–3, San Diego, CA).
5. Oligonucleotides for PCR amplification as described in **Fig. 1** and **Table 1**.
6. Plasmid: pBluescriptII (pBS) SK+ (Stratagene, San Diego, CA).
7. Plasmid: pcDNA6 (Invitrogen, Carlsbad, CA).
8. Gel purification kit (Qiagen, Valencia, CA).
9. PCR product purification kit (Qiagen).
10. *Escherichia coli* for transformation.
11. T7 primer for sequencing.
12. 2 *M* $CaCl_2$.
13. 2X HEPES-buffered saline (2X HBS): 50 m*M* HEPES pH 7.05, 10 m*M* Kcl, 12 m*M* Dextrose, 280 m*M* NaCl, and 1.5 m*M* Na_2HPO_4.
14. Midiprep plasmid DNA purification kit (Qiagen).
15. Dulbecco's modified eagle's media (DMEM).

Table 1
Oligonucleotides to Assemble New Three-Finger Protein

Finger 1 general oligo (oligo A)	5′-CAGTGGCGGCCGCTCTAGAAC-3′
Finger 2 general oligo (oligo B)	5′-CATATCCGCATCCATACC-3′
Finger 3 general oligo (oligo C)	5′-CACATCCGCACCCACACA-3′
Finger 1 specific oligo	5′-GTA TGG ATG CGG ATA TG **antisense for amino acid codons –1 to 6** AGA AAA GCG GCG ATC GC-3′
For example, GFP1-ZFN finger 1 (target triplet GGT)	5′-GTA TGG ATG CGG ATA TG **CCT CGT GAG GTG AGA AGA CTG** AGA AAA GCG GCG ATC GC-3′
Finger 2 specific oligo	5′-GTG TGG GTG CGG ATG TG **antisense for amino acid codons –1 to 6** ACT GAA GTT ACG CAT GC-3′
For example, GFP1-ZFN finger 2 (target triplet GAT)	5′-GTG TGG GTG CGG ATG TG **GCG GAC AAG GTT GCC ACC AGT** ACT GAA GTT ACG CAT GC-3′
Finger 3 specific oligo	5′-TTG ACT AGT TG GTC CTT CTG TCT TAA ATG GAT TTT GGT ATG **antisense for amino acid codons –1 to 6** GGC AAA CTT CCT CCC-3′
For example, GFP1-ZFN finger 3 (target triplet GAA)	5′-TTG ACT AGT TG GTC CTT CTG TCT TAA ATG GAT TTT GGT ATG **GCG GGC AAG GTT ACC CGA CTG** GGC AAA CTT CCT CCC-3′

Note: Need to pick codons so no *Bam*HI (GGATCC) or *Spe*I (ACTAGT) sites in oligonucleotides.

This table lists the oligonucleotides used in the overlap PCR approach described in the text and in **Fig. 1**. The finger-specific oligonucleotides are the antisense strand while the general oligonucleotides are the sense strand. To make the finger-specific oligonucleotides one must first determine the correct amino acids for the recognition helix for each finger, then deduce the optimal human codon for each amino acid of the recognition helix (7 amino acids leading to 21 nucleotides for each finger). The reverse-complement of this coding sequence is then determined and it is this reverse-complement 21-nucleotide sequence that is inserted into the location labeled "antisense fro amino acid codons –1 to 6." The sequence must then be checked to make sure that no *Bam*HI or *Spe*I sites are in it. If there are *Bam*HI or *Spe*I sites then an alternative codon must be chosen so that those sites are eliminated.

16. Bovine growth serum (Hyclone, Logan, UT).
17. 200 m*M* L-glutamine (sterile).
18. Penicillin–streptomycin (10,000 IU/mL penicillin and 10,000 μg/mL streptomycin) (sterile).

2.1. Equipment

1. Heating block or water bath at 37°C.
2. Thermal cycler.

3. Methods

3.1. Identifying a Target Site

The first step is to identify a potential ZFN target site within the gene of interest. A full ZFN target site consists of two half-sites that are inversely oriented with respect to each other and separated by a nucleotide spacer. A standard ZFN consists of three ZFs to bind a 9-bp target site. Theoretically, if one had individual ZFs that recognized all 64 different nucleotide triplets, then one could assemble a ZFP to recognize any target sequence. Individual ZFs have been published for all 16 different GNN and most of the CNN and ANN triplets and programs are available (www.zincfingers.org or www.zincfinger-tools.org) to help investigators create ZFPs to bind specific sequences ***(19–23)***. In our hands we have only been successful at using modular-assembly to make three-finger ZFNs active against sites with the following structure 5′-GNNGNNGNN-3′ where N can be any base ***(12)***. Others have been successful in making ZFNs to sites that are different from this consensus but the efficiency at using modular-assembly to make active ZFNs to target such sites remains to be determined ***(14)***.

In general, we have had the most experience at targeting sequences in which the individual ZFN binding sites (half-sites) are separated by a 6 nucleotide spacer, although highly efficient targeting was achieved at sites that were separated by only 5 nucleotides ***(13)***.

In summary, one should search the gene of interest for a ZFN full consensus target site of the following form: 5′-NNCNNCNNCnnnnnnGNNGNNGNN-3′. This ZFN full consensus site consists of two half-sites of the form 5′-GNNGNNGNN-3′ oriented inversely from each other separated by a 6 nucleotide spacer. The site is oriented in this fashion because the nuclease domain is attached to the C-terminal finger (finger 3) that binds the most 5′ triplet. It is in this orientation that the nuclease domain from each ZFN can dimerize and then cut DNA. Carroll and his colleagues show a schematic of this organization in their chapter in this volume. In the humanized green fluorescent protein (GFP) gene, for example, the target site 5′-ACCATCTTC ttcaag GACGACGGC-3′ was identified ***(12)***. To target this site, one ZFN would be designed to bind to the target sequence 5′-GAAGATGGT-3′ and the other ZFN to bind to the target sequence 5′-GAAGATGGT-3′. Most modern sequence analysis software is capable of searching a given sequence for the consensus site shown in the starting of this paragraph.

3.2. Designing a ZFP

Once a full ZFN target site is identified in the gene of interest the next step is to design the pair of ZFNs to target that specific site. The first step in that process is to break the full consensus-binding site into the two ZFN half-sites. In the example given in the aforementioned paragraph, the GFP full consensus binding site 5′-ACCATCTTC ttcaag GACGACGGC-3′ is broken into two half-sites: 5′-GAAGATGGT-3′ (GFP-left) and 5′-GAAGATGGT-3′ (GFP-right). These two half-sites are then broken down further into triplets with each triplet being the binding site for an individual ZF. For example, GFP-left would be broken down into 5′-GAA-3′, 5′-GAT-3′, and 5′-GGT-3′. As ZFPs bind DNA in an "antiparallel" fashion, finger 1 would bind the 3′ triplet (5′-GGT-3′), finger 2 the middle triplet (5′-GAT-3′), and finger 3 would bind the 5′ triplet (5′-GAA-3′). Once the cognate triplet-binding sites are identified, the next step is to determine the amino acid content of the DNA recognition helix for each finger. For GNN target sites, there are two possible sources for recognition helices ***(19–21)***. We have primarily used the data from Liu et al. but others have used the data sets from Segal et al. and Dreier et al. to successfully design ZFNs ***(12,14,19–21)***. Continuing to use GFP-left as an example and using the data set from Liu et al. ***(19)***, one would identify amino acids (using the standard 1-letter code) for the recognition helix for finger 1 to be QSSHLTR, for finger 2 to be TSGNLVR, and for finger 3 to be QSGNLAR (*see* **Note 1**).

3.3. Creating a ZFN

An overlapping PCR strategy is used to assemble the new ZFPs (schematized in **Fig. 1**). Each finger is amplified independently using a general primer at the 5′ end and a finger-specific primer at the 3′ end. Details of the design of the oligonucleotides are in **Table 1**. The Zif268 backbone is used as a template for each PCR reaction as there is enough heterogeneity in the backbone to allow assembly of the three-fingers in the correct order in the final step. In some artificial ZFNs constructs, such as QQR-ZFN, the nucleotide sequence surrounding the recognition helix of each finger is identical, which makes it nearly impossible to assemble the fingers in the correct orientation by overlap PCR. Each finger fragment is amplified so that it has a 15-bp overlap with its neighboring finger. The individual fingers are then assembled using an overlap PCR strategy (schematized in **Fig. 1**) (*see* **Note 2**). The PCR product is digested with *Bam*HI and *Spe*I and cloned into pBS that has been digested with *Bam*HI/*Spe*I. The three-finger cassette is then sequenced to verify that no errors were created in the PCR process. We generally use the KOD polymerase (Novagen) as it is both a high-fidelity polymerase thus reducing the probability of unintended mutations during the amplificaton, and because it performs well in the overlap PCR reaction. The sequences of oligonucleotides used in this process are shown in **Table 1**.

3.3.1. Protocol to Assemble New ZFN Using Overlap PCR

1. PCR reaction 1 (amplification of individual ZFs).

Zif-ZFN DNA template	100 ng
10X polymerase buffer	5 µL
$MgCl_2$ (25 m*M*)	2 µL
dNTP (2 m*M* each)	7.5 µL
Finger 1 (or 2 or 3)-specific oligo	30 pmoles
Finger 1 (or 2 or 3) general oligo	30 pmoles
KOD polymerase	2 U

dH_2O to final volume of 50 mL.

Amplify using the following program.

1	94°C for 5 min
2	94°C for 30 s
3	50°C for 45 s
4	72°C for 60 s
Go to **step 2** and perform 14 times	
5	72°C for 5 min
6	4°C forever

2. Purify individual fingers on 2.5% agarose gel into 30 µL buffer using Qiagen gel purification kit.
   a. Expected size of finger 1 = 161 bp.
   b. Expected size of finger 2 = 100 bp.
   c. Expected size of finger 3 = 124 bp.
3. PCR reaction 2 (assembly of three-finger protein).

Finger 1 fragment	10 µL
Finger 2 fragment	10 µL
Finger 3 fragment	10 µL
10X polymerase buffer	5 µL
$MgCl_2$ (25 m*M*)	2 µL
dNTP (2.5 m*M* each)	7.5 µL
Finger 1 general oligo	30 pmoles
Finger 3-specific oligo	30 pmoles
KOD polymerase	2 U

dH_2O to final volume of 50 mL.

Cycling parameters: same as for PCR reaction 1.

4. Purify PCR product using PCR purification kit (Qiagen).
5. Digest the PCR product with *Bam*HI and *Spe*I.
6. Purify the digested PCR product on 2.5% agarose gel into 30 µL using Qiagen gel purification kit (expected fragment size is about 312 bp).

7. Clone the digested/purified PCR fragment into *Bam*HI/*Spe*I digested pBS to create pBS-ZF.
8. Sequence pBS-ZF to confirm the new three-finger protein.

Once the assembled three-finger ZF is sequence verified, it is then ready to be cloned into a mammalian expression vector while simultaneously adding the *Fok*I nuclease domain. The procedure for that is described in **Subheading 3.3.2.**

3.3.2. Protocol to Create ZFN Expression Vector from New ZF

1. Digest pcDNA6 (Invitrogen) with *Bam*HI/*Apa*I (2 h at room temperature followed by 2 h at 37°C). Treat with calf intestinal alkaline phosphatase and isolate the 5.1-kb fragment by gel purification. This is the vector fragment.
2. Digest pBS-ZF with *Bam*HI/*Spe*I and isolate the approx 310-bp fragment by gel purification. This is the ZF fragment.
3. Digest GFP-ZF1 with *Spe*I/*Apa*I (2 h at room temperature followed by 2 h at 37°C). Isolate the approx 590-bp fragment by gel purification. This is the nuclease fragment.
4. Ligate the three fragments together using a molar ratio of vector:nuclease fragment: ZF fragment of 1:4:4 and standard procedures.
5. Transform *E. coli* using standard procedures.
6. Identify positive clones by *Bam*HI/*Apa*I digest (2 h at room temperature followed by 2 h at 37°C) and analyze by 0.8% agarose gel electrophoresis. Correct clones should have a 5.1-kb band and a 900-bp band.
7. Sequence correct clones with T7 primer to confirm that the ZF is intact and the nuclease domain was cloned in-frame.
   a. This cloning strategy will create a ZFN that has a 5 amino acid linker between the terminal histidine of the third ZF and the first residue of the nuclease domain. These ZFNs have been shown to be most active on sites that are separated by 6 bp.

3.4. Testing of a ZFN

The activity of the ZFN seems to be directly related to the quality of the ZFP. Before using the ZFN to target an endogenous gene, it should be tested using a reporter assay. To test the ZFP more directly, one can use transcriptional reporter assays either in mammalian cells ***(14)*** or in bacteria ***(15,25)***. In our lab, we are currently adapting such transcriptional reporters as a first screen of our newly designed ZFPs but have routinely tested ZFNs using a mammalian cell-based GFP gene targeting reporter assay ***(6,12)***. This assay is described in **Subheadings 3.4.1.–3.4.3.**

The GFP gene targeting assay is based on the conversion of a mutated GFP gene that has been integrated as a single copy into the genome of a mammalian cell into the wild-type version after the introduction of a targeting construct that would correct the genomic mutation. This conversion only occurs efficiently if a DNA DSB is created in the mutated genomic target gene. To test a new ZFN, the 9-bp ZFN target site is inserted into the GFP gene and oriented inversely and separated by 6 bp from a Zif268 target site. To test, a ZFN that targets the sequence

5′-GCCGCCGCC-3′, for example, one would insert into the GFP gene the following sequence: 5′-GGCGGCGGCtctagaGCGTGGGCG-3′. The first half of the site is the reverse-complement of the target site of the new ZFN, the lower case letters are the 6-bp spacer, and the second half of the site (5′-GCGTGGGCG-3′) is the binding site for Zif268. Adjacent to the insertion of the full ZFN site is the recognition site for the I-*Sce*I endonuclease (*Sce*) (5′- GGGATAACAGGGTAAT-3′). *Sce* serves as a positive control and allows us to compare the activity of the ZFN pair with *Sce*. After this plasmid is made, it is electroporated into 293-S cells (not G418 resistant) and clones are identified that have integrated the reporter into the genome and actively transcribe the reporter (*see* **Notes 3** and **4**). We identify these clones by cell surface expression of CD8α, a gene that is not normally expressed in 293-S cells but is driven from the same promoter as the mutated GFP gene through an internal ribosomal entry site as part of the reporter construct. Once a cell line is established, it is then transfected with a:

1. GFP expression plasmid (to measure transfection efficiency) plus the ZFN expression plasmid plus the Zif268-ZFN expression plasmid.
2. The GFP-targeting construct (repair substrate) plus a *Sce* expression plasmid.
3. The GFP repair substrate plus the ZFN expression plasmid alone.
4. The GFP repair substrate plus the Zif268-ZFN expression plasmid alone.
5. The GFP repair substrate plus the new ZFN expression plasmid and the Zif268-ZFN expression plasmid.

Two days after transfection well 1 is analyzed by flow cytometry to determine the transfection efficiency. Three days after transfection the remaining wells are analyzed by flow cytometry to determine the rate of gene targeting. We usually analyze 100–200,000 cells per condition. Protocol for each of these steps is as follows.

3.4.1. Construction of GFP Reporter Plasmid

1. Digest parental reporter plasmid (pPC17) with *Xho*I and *Hin*dIII and isolate the approx 9 kb fragment by gel purification. Do not treat with calf intestinal alkaline phosphatase.
2. Order target-site oligonucleotides in which one half-site is the binding site for Zif-ZFN (a ZFN with known activity and heterodimerizes with ZFNs made by modular-assembly) and the other half site is the new target site. Design the pair of oligonucleotides so that, when annealed, they create *Xho*I and *Hin*dIII compatible overhangs. Place an *Eco*RI site 5′ to the full ZFN site in order to provide a new restriction site for later analysis. This site will be used to determine, which clones have inserted the annealed oligonucleotide correctly. For example, to test HGBZF1-ZFN (target site of 5′-GAGGTTGCT-3′), we ordered oligonucleotides with the following sequence:
   A: 5′-AGCT GAATTC CGCCCACGC ggatcc GAGGTTGCT-3′
   B: 5′-TCGA AGCAACCTC ggatcc GCGTGGGCG GAATTC-3′
3. Anneal oligonucleotides A and B using standard procedures and ligate into the purified reporter plasmid vector (pPC17 *Xho*I/*Hin*dIII) using a molar ration of vector: annealed oligo of 1:100.

4. Transform the ligation into *E. coli* using standard procedures.
5. Analyze colonies by *Eco*RI digest and agarose gel electrophoresis. Clones that contain the new oligonucleotide will have fragments of 4400, 2100, 1700, and doublet at 400 bps.
6. Sequence clones with appropriate fragment sizes with primer A220A (5′-ACCGGCAAGCTGCCCGTGCCCTGG-3′) to confirm single-copy insertion of the oligonucleotide.
7. Using standard procedures, midi prep sequence confirmed clones to prepare DNA of quality and quantity to create reporter cell line.

3.4.2. Making GFP Reporter Cell Line

We make our reporter lines in HEK-293S cells for the following reasons. These cells are easily grown and easily transfectable. In contrast to 293T cells, 293-S cells are also sensitive to G418 so can be used to make cell lines using the neomycin gene as a selectable marker (*see* **Note 4**). We make single copy integrants of the reporter transgene using electroporation.

1. Grow HEK-293S to mid-log phase in 10-cm plate using full media (DMEM/10% bovine growth serum (HyClone, Logan UT)/2 m*M* L-glutamine/Pen-strep).
2. Trypsinize the cells and wash once in 10 mL of serum-free DMEM. Resuspend at 10^7 cells/mL in serum-free DMEM.
3. Remove 400 μL of cells and mix with 10 μg of super-coiled reporter plasmid.
4. Incubate cells with plasmid for 5 min on ice and then electroporate using a BTX ECM399 electroporator at 150 V.
5. Allow cells to recover for 5 min on ice and then gently add to 10 mL of full media in a 10 cm plate.
6. 24 h after electroporation, add G418 to a final concentration of 500 μg/mL (active).
7. Change media with G418 every 3–4 d gently to avoid disrupting any colonies.
   a. G418 usually takes 48–72 h to begin killing cells. If cells are too confluent, G418 can take longer to kill cells.
8. After 2 wk of G418 selection, individual colonies are clearly evident. Pick individual colonies and expand in a 24-well plate.
9. Identify clones that transcribe the mutant GFP gene by either Northern or reverse transcriptase-PCR to look for GFP transcript or by staining with anti-CD8 antibody.
   a. To stain for cell surface expression with anti-CD antibody:
      i. Trypsinize cells briefly.
      ii. Replate 50% of the cells in a 24-well plate.
      iii. To remaining, wash once in 1 mL of phosphate buffered saline (PBS).
      iv. Resuspend in 100 mL PBS and 10 mL of phycoerytherin-conjugated anti-CD8 antibody (Becton-Dickinson).
      v. Incubate on ice for 10 min in dark.
      vi. Add 1 mL PBS and pellet.
      vii. Resuspend in 400 mL of PBS and analyze by flow cytometry. Clones that have high, consistent levels of CD8 expression should be used as the reporter.

3.4.3. Testing New ZFN in ZFN Reporter Cell Line by Calcium Phosphate Transfection

1. The day before transfection, seed 16 wells of a 24-well plate with 125,000 cells of the reporter cell line.
2. Just before transfection replace each well with 500 μL of fresh full DMEM.
3. Transfect each of the wells with the following plasmids using a standard calcium phosphate procedure. We have found that other transfection procedures (including commercial lipid reagents) work just as well, although the amounts of plasmid may vary depending on the reagent used. We use the calcium phosphate technique because it is cheaper than using commercial lipid reagents.
   a. Well 1: 200 ng of GFP expression plasmid + 200 ng of I-*Sce*I expression plasmid.
   b. Wells 2–4: 200 ng of I-*Sce*I expression plasmid + 200 ng of GFP repair donor (pRS2700).
   c. Well 5: 200 ng of GFP expression plasmid + 200 ng of Zif-ZFN expression plasmid.
   d. Wells 6–8: 200 ng of Zif-ZFN expression plasmid + 200 ng of GFP repair donor (pRS2700).
   e. Well 9: 200 ng of GFP expression plasmid + 200 ng of new ZFN expression plasmid.
   f. Wells 10–12: 200 ng of new ZFN expression plasmid + 200 ng of GFP repair donor (pRS2700).
   g. Well 13: 200 ng of GFP expression plasmid + 100 ng of Zif-ZFN expression plasmid + 100 ng of new ZFN expression plasmid.
   h. Wells 14–16: 100 ng of Zif-ZFN expression plasmid + 100 ng of new ZFN expression plasmid + 200 ng of GFP repair donor (pRS2700).
4. Add calcium phosphate precipitate to cells and incubate for 8–16 h.
5. Remove media and calcium phosphate precipitate and add 1 mL of full DMEM.
   a. We do not wash with PBS as we find that washing PBS often washes off the loosely adherent 293 cells.
6. 48 h after transfection, analyze wells 1, 5, 9, and 13 by flow cytometry to determine transfection efficiency (*see* **Note 5**).
7. 72 h after transfection, analyze remaining cells for percentage of cells that are GFP-positive by flow cytometry.
   a. We analyze by flow cytometry rather than fluorescent microscopy because it is faster, more sensitive, and more quantitative.
8. Determine the rate of gene targeting by normalizing the percentage of GFP-positive cells in the gene targeting wells to the transfection efficiency. If the new ZFN is active, the normalized rate of gene targeting should be greater than 10^{-4} (*see* **Notes 6** and **7**).

4. Notes

1. When assembling a ZFP (**Subheading 3.2.**), the first attempt should use individual fingers from the same data set. If the assembled ZF is not active, then one should consider trying another data set or mixing fingers from different data sets. The investigators should anticipate that a pair of ZFNs to a single target site may not be active and ideally should plan to develop pairs to several potential sites.

2. The efficiency of assembling the ZFP is increased by gel purification of fragments for each PCR amplification (**Subheading 3.3.1.**). After amplification of the individual fingers, the three fingers can usually be assembled in a single PCR reaction. If this is not successful, then one can try assembling the fingers in a step-wise fashion by first joining finger 1 with finger 2 to create a finger 1–2 product and then adding finger 3 to it.
3. Cell lines should be split 1:20 every 4 d for maintenance (**Subheading 3.4.**).
4. In creating the GFP reporter cell line, G418 takes about 2–4 d to kill untransfected cells (**Subheading 3.4.2.**). If the cells are too confluent, the G418 does not effectively kill untransfected cells. Thus, G418 selections should be done when the cells are not confluent.
5. In gene targeting experiments, if the transfection efficiency is less than 5%, it is usually difficult to get reproducible results (**Subheading 3.4.3.**). If transfection efficiencies are not above 10% using the calcium phosphate precipitation technique, then one should try transfecting using lipid-based commercial reagents such as Lipofectamine 2000 (Invitrogen, Carlsbad, CA). We have found that, method of transfection has little impact on the rate of gene targeting using this system.
6. The GFP gene targeting reporter line should give >2000 targeting events per million transfected cells when I-*Sce*I is used as the nuclease (**Subheading 3.4.3.**). If the line does not give such rates, then monoclonal lines should be examined until one is found that gives such a rate.
7. If the new ZFN are not active in the GFP gene targeting reporter system, then one should confirm that the ZFN is being expressed (**Section 3.4.3.**). The ZFNs described here have a FLAG tag at the amino-terminus and using ZFN expression can be determined by Western analysis using an anti-FLAG antibody (M2 from Sigma, St. Louis, MO).

Acknowledgments

The work in the Porteus lab is supported by a Basil O'Connor Starter Award from the March of Dimes, a career development award from the Burroughs-Wellcome Fund, and grants K08 HL070268 and R01 HL079295 from the National Institute of Health.

References

1. Sung, P. and Klein, H. (2006) Mechanism of homologous recombination: mediators and helicases take on regulatory functions *Nat. Rev. Mol. Cell Biol.* **7,** 739–750.
2. West, S. C. (2003) Molecular views of recombination proteins and their control. *Nat. Rev. Mol. Cell Biol.* **4,** 435–445.
3. Chevalier, B. S. and Stoddard, B. L. (2001) Homing endonucleases: structural and functional insight into the catalysts of intron/intein mobility. *Nucleic Acids Res.* **29,** 3757–3774.
4. Jasin, M. (1996) Genetic manipulation of genomes with rare-cutting endonucleases. *Trends Genet.* **12,** 224–228.
5. Porteus, M. H. and Baltimore, D. (2003) Chimeric nucleases stimulate gene targeting in human cells. *Science* **300,** 763.

6. Durai, S., Mani, M., Kandavelou, K., Wu, J., Porteus, M. H., and Chandrasegaran, S. (2005) Zinc finger nucleases: custom-designed molecular scissors for genome engineering of plant and mammalian cells. *Nucleic Acids Res.* **33,** 5978–5990.
7. Porteus, M. H. and Carroll, D. (2005) Gene targeting using zinc finger nucleases. *Nat. Biotechnol.* **23,** 967–973.
8. Kim, Y. G., Cha, J., and Chandrasegaran, S. (1996) Hybrid restriction enzymes: zinc finger fusions to Fok I cleavage domain. *Proc. Natl. Acad. Sci. USA* **93,** 1156–1160.
9. Bibikova, M., Carroll, D., Segal, D. J., et al. (2001) Stimulation of homologous recombination through targeted cleavage by chimeric nucleases. *Mol. Cell Biol.* **21,** 289–297.
10. Bibikova, M., Beumer, K., Trautman, J. K., and Carroll, D. (2003) Enhancing gene targeting with designed zinc finger nucleases. *Science* **300,** 764.
11. Wright, D. A., Townsend, J. A., Winfrey, R. J., Jr., et al. (2005) High-frequency homologous recombination in plants mediated by zinc-finger nucleases. *Plant J.* **44,** 693–705.
12. Porteus, M. H. (2006) Mammalian gene targeting with designed zinc finger nucleases. *Mol. Ther.* **13,** 438–446.
13. Urnov, F. D., Miller, J. C., Lee, Y. L., et al. (2005) Highly efficient endogenous human gene correction using designed zinc-finger nucleases. *Nature* **435,** 646–651.
14. Alwin, S., Gere, M. B., Guhl, E., et al. (2005) Custom zinc-finger nucleases for use in human cells. *Mol. Ther.* **12,** 610–617.
15. Wright, D. A., Thibodeau-Beganny, S. A., Sander, J. D., et al. (2006) Standardized reagents and protocols for engineering zinc finger nucleases by modular assembly. *Nature Protocols* **1,** 1637–1652.
16. Pavletich, N. P. and Pabo, C. O. (1991) Zinc finger-DNA recognition: crystal structure of a Zif268-DNA complex at 2.1 A. *Science* **252,** 809–817.
17. Pabo, C. O., Peisach, E., and Grant, R. A. (2001) Design and selection of novel Cys2His2 zinc finger proteins. *Annu. Rev. Biochem.* **70,** 313–340.
18. Wolfe, S. A., Nekludova, L., and Pabo, C. O. (2000) DNA recognition by Cys2His2 zinc finger proteins. *Annu. Rev. Biophys. Biomol. Struct.* **29,** 183–212.
19. Liu, Q., Xia, Z., Zhong, X., and Case, C. C. (2002) Validated zinc finger protein designs for all 16 GNN DNA triplet targets. *J. Biol. Chem.* **277,** 3850–3856.
20. Segal, D. J., Dreier, B., Beerli, R. R., and Barbas, C. F., 3rd. (1999) Toward controlling gene expression at will: selection and design of zinc finger domains recognizing each of the 5′-GNN-3′ DNA target sequences. *Proc. Natl. Acad. Sci. USA* **96,** 2758–2763.
21. Dreier, B., Segal, D. J., and Barbas, C. F., 3rd. (2000) Insights into the molecular recognition of the 5′-GNN-3′ family of DNA sequences by zinc finger domains. *J. Mol. Biol.* **303,** 489–502.
22. Dreier, B., Beerli, R. R., Segal, D. J., Flippin, J. D., and Barbas, C. F., 3rd. (2001) Development of zinc finger domains for recognition of the 5′-ANN-3′ family of DNA sequences and their use in the construction of artificial transcription factors. *J. Biol. Chem.* **276,** 29,466–29,478.

23. Dreier, B., Fuller, R. P., Segal, D. J., et al. (2005) Development of zinc finger domains for recognition of the 5′-CNN-3′ family DNA sequences and their use in the construction of artificial transcription factors. *J. Biol. Chem.* **280,** 35,588–35,597.
24. Segal, D. J., Beerli, R. R., Blancafort, P., et al. (2003) Evaluation of a modular strategy for the construction of novel polydactyl zinc finger DNA-binding proteins. *Biochem.* **42,** 2137–2148.
25. Hurt, J. A., Thibodeau, S. A., Hirsh, A. S., Pabo, C. O., and Joung, J. K. (2003) Highly specific zinc finger proteins obtained by directed domain shuffling and cell-based selection. *Proc. Natl. Acad. Sci. USA* **100,** 12,271–12,276.

5

Gene Targeting in *Drosophila* and *Caenorhabditis elegans* With Zinc-Finger Nucleases

Dana Carroll, Kelly J. Beumer, J. Jason Morton, Ana Bozas, and Jonathan K. Trautman

Summary

Zinc-finger nucleases (ZFNs) are promising new tools for enhancing the efficiency of gene targeting in many organisms. Because of the flexibility of zinc finger DNA recognition, ZFNs can be designed to bind many different genomic sequences. The double-strand breaks they create are repaired by cellular processes that generate new mutations at the cleavage site. In addition, the breaks can be repaired by homologous recombination with an exogenous donor DNA, allowing the experimenter to introduce designed sequence alterations. We describe the construction of ZFNs for novel targets and their application to targeted mutagenesis and targeted gene replacement in *Drosophila melanogaster* and *Caenorhabditis elegans*.

Key Words: *Caenorhabditis elegans*; DNA repair; *Drosophila*; fruit flies, gene targeting; homologous recombination; mutagenesis; nematodes; nonhomologous end joining; zinc finger; zinc-finger nuclease.

1. Introduction

At some level all geneticists working with higher eukaryotes envy yeast researchers for the ease with which gene replacements can be made in that organism *(1)*. Gene targeting procedures have been developed for some other systems, including mouse *(2)* and *Drosophila (3)*. Although these are certainly effective, they yield the desired products at low frequency and only after considerable experimental effort.

We have been developing zinc-finger nucleases (ZFNs) as tools to enhance the efficiency of targeted mutagenesis and gene replacement and to extend targeting procedures to a wider range of organisms *(4)*. The underlying principles are first, the observation that double-strand breaks (DSBs) in DNA stimulate localized mutagenesis and homologous recombination (HR) in many cell types *(5)*, and

From: *Methods in Molecular Biology, vol. 435: Chromosomal Mutagenesis*
Edited by: G. Davis and K. J. Kayser © Humana Press Inc., Totowa, NJ

second, that modulation of a DNA-binding domain consisting of zinc fingers (ZFs) can direct proteins to desired chromosomal targets ***(6–8)***.

ZFNs are hybrid proteins ***(9)***: they have a DNA-binding domain made up of Cys_2His_2 ZFs at the N-terminus and a nonspecific cleavage domain derived from the restriction endonuclease *Fok*I at the C-terminus (**Fig. 1**). ZFs bind to consecutive triplets in DNA in a nearly independent manner ***(10)***, so new recognition domains can be assembled from existing fingers for specific triplets; and fingers have been identified for many of the 64 possible 3-bp sequences ***(11–14)***. The cleavage domain has to dimerize to be active ***(15,16)***; because the dimer interface is very weak, this amounts to a requirement for two ZFNs bound to nearby sites. This is very advantageous, as the active cleavage reagent is assembled at the target and is, at least in principle, not active at other sites. For flies and worms, we typically produce ZFNs with three ZFs, but for organisms with larger genomes, adding more fingers can give additional specificity ***(17)***.

Success of various types has been achieved with ZFNs in a number of different systems ***(17–27)***. Both endogenous and synthetic substrates have been attacked, and sequence alterations have been recovered based on two different modes of DSB repair (*see* **Fig. 1**). One is of the type we generally mean when discussing gene targeting—i.e., replacement of pre-existing sequences with ones provided in designed donor DNA through HR. The other results from inaccurate nonhomologous end joining (NHEJ). The former allows introduction of carefully designed genetic changes whereas the latter leads largely to small insertions and deletions that are not under the control of the experimenter, but frequently create null alleles. Both modes of repair are essentially universal, so the utility of ZFN-induced gene targeting should be quite broad. Here, we focus on the two organisms used in our own lab: the fruit fly, *Drosophila melanogaster* ***(19,20,22)***, and the nematode, *Caenorhabditis elegans* ***(27)***. We assume that researchers proposing to work with these organisms either have themselves or have access to the basic expertise in manipulating them. Excellent compilations on the biology of and experimental methods for flies ***(28,29)*** and worms ***(30,31)*** are available.

1.1. Experimental Design

The first step in targeting a new gene is to select that gene and choose the type(s) and location of alterations desired. The gene or gene region is then searched for sequences that look like good candidates for ZF binding, and the corresponding ZFNs are produced.

We make new ZF combinations by assortment of existing sequences ***(32)***. Fingers for many DNA triplets have been described. Some of these show excellent discrimination against related sequences in assays on naked DNA, and some do not. When searching a new gene for targetable sites, we try to include only triplets for which there are "good" fingers ***(32)***. In our experience, the optimum target consists of two 9-bp sequences that will be bound by ZFNs separated by 6 bp

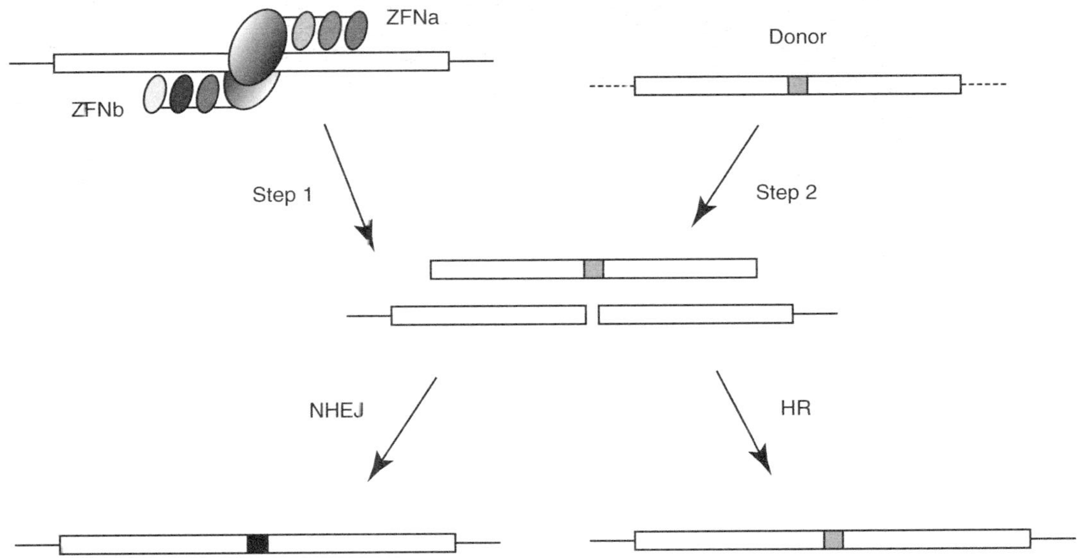

Fig. 1. Schematic diagram of ZFN-induced gene targeting. The open bar represents the target gene of interest. A pair of three-finger ZFNs is designed to bind a sequence within the target. In the top left diagram, individual ZFs are shown as small ovals (shaded uniquely to indicate that they typically bind different DNA triplets), and the cleavage domains are shown as larger, ovals. Step 1 shows cleavage of the target by the ZFN pair. For targeted gene replacement, a donor DNA is provided (step 2), which carries the desired alteration in target sequence (shaded box). The ZFN-induced break can be repaired by HR with the donor, yielding the desired replacement, or by NHEJ, producing novel sequence changes (black box) at the cleavage site.

(21). Cleavage occurs within this 6-bp spacer, leaving 4-nt 5′ overhangs. It is important to realize two things about ZF binding sites: (1) the sequences of the component 9-bp sites run in opposite directions on opposite strands, as usually written and (2) the N-terminal finger binds the 3′ triplet, so the protein and DNA orientations are "backward" compared with how we usually consider them. The coding sequences for new ZF sets are then linked to coding sequences for the nuclease domain.

Once a pair of ZFNs has been constructed, the proteins must be delivered to the appropriate cells in the organism of choice. This can be the most challenging aspect of any protocol, because the requirements for effective expression will likely differ among organisms and cell types. We describe below three different approaches for *Drosophila* (**Subheading 4.**) and one for *C. elegans* (**Subheading 5.**); and we mention additional methods that are under development.

If a donor DNA is included, it will be designed to carry the sequence alterations that are desired ultimately in the target. Several aspects of donor design must be considered, and it is not certain that characteristics of one system will translate simply to others. First, a method for delivering the donor must be chosen. In *Drosophila*, the Golic lab (Department of Biology, University of Utah, Salt Lake City) developed a very effective procedure for generating linear donor DNAs *in situ* based on excision with FLP recombinase (derived from the *S.cerevisiae* 2μ plasmid) and cleavage with I-*Sce*I meganuclease *(3,33)*. Whereas linear molecules are excellent substrates for HR, some injection procedures have achieved good (perhaps better) success with circular donors *(34)*. Second, the size of the donor must be selected. How much homology between donor and target is required to achieve efficient gene replacement? Although this issue has not been thoroughly studied following ZFN cleavage, some information is available for repair of DSBs in *Drosophila* following P element excision *(35)*. The answer may depend considerably on the cells involved and the method of donor DNA introduction. Third and related to the second, what constraints are there on the relationship between the location of the alteration in the donor and the ZFN-induced cut in the target? Conversion tracts vary from tens to perhaps a few hundred base pairs in mammalian cells *(36)* to several kilobase in *Drosophila (35)*, but again this may vary depending on experimental parameters. Can the efficiency of incorporation of an alteration distant from the cut site be improved by extending the length of the donor distal to the alteration? This point has not been explored in any detail either.

2. Materials

2.1. General Materials

1. User-defined synthetic oligonucleotides.
2. *Escherichia coli* cells: DH5α.
3. pENTR-ZFN (e.g., pENTRben1AFN) or pENTR-NLS-ZFN. The latter carries a nuclear localization sequence (NLS), which is crucial for the *Drosophila* RNA injection method. These plasmids and their full nucleotide sequences can be obtained from

the Carroll Lab (Department of Biochemistry, University of Utah School of Medicine, Salt Lake City; www.biochem.utah.edu/carroll).

4. Standard bacterial plates and media and suitable antibiotics.
5. T4 DNA ligase (New England BioLabs, M0202S; Ipswich, MA).
6. *Taq* DNA polymerase (New England BioLabs, M0267L); Platinum *Taq* polymerase (high fidelity; Invitrogen, 11304-011, Carlsbad, CA). We use the standard *Taq* polymerase for most purposes and the high-fidelity version only for long polymerase chain reaction (PCR) products, like donor DNAs that are several kilobase in length.
7. PCR reagents and thermal cycler.
8. Qiagen MinElute columns and Qiagen Plasmid Maxiprep kits (Valencia, CA).

2.2. *Materials for* Drosophila

1. pCaSpeR-hs-DEST. This is a P element vector that has been modified so that it also serves as a Gateway Destination vector. Expression of the insert is driven by a *Drosophila* hsp70 heat-shock promoter. It is the recipient for ZFN-coding sequences in the heat-shock method. It is marked with a w^+ gene.
2. pP{WhiteOut2}. This is another P element vector, made in Jeff Sekelsky's lab (Department of Biology, University of North Carolina, Chapel Hill, http://sekelsky.bio.unc.edu) that serves as carrier for the donor DNA. The multiple cloning site is flanked by recognition sites for I-*Sce*I and FLP. It is marked with a w^+ gene that lies outside the FLP and I-*Sce*I sites.
3. Donor DNA. The nature of this component will depend entirely on the gene being targeted and the alteration desired. It must have homology with the target on both sides of the ZFN cut site, and the ZFN site in the donor should be modified to prevent its cleavage. (This adjustment may not be necessary *[17]*, but seems prudent.) It must be endowed with restriction sites at its ends compatible with the pP{WhiteOut2} vector.
4. *Drosophila* embryo injection equipment. This includes an upright or inverted microscope, micromanipulator (e.g., Narishige model MN-153; East Meadow, NY) needle puller, and glass capillary tubes for making needles.
5. Dissecting microscope for screening flies. A source of CO_2 and simple equipment for anesthetizing flies while screening, is also very useful (Genesee Scientific, San Diego, CA, http://www.flystuff.com).
6. Aquagenomic DNA extraction reagent (MultiTarget Pharmaceuticals, Salt Lake City, UT; http://www.aquaplasmid.com) for preparing genomic DNA from single adult flies.
7. *E. coli* strains: DB3.1 (Invitrogen) for propagating pCaSpeR-hs-DEST and other DEST vectors (required to provide resistance to the product of the *ccdB* gene); DH5α.
8. LR Clonase II Enzyme Mix (Invitrogen).
9. *Drosophila* strains (*Drosophila* Stock Center, http://flystocks.bio.indiana.edu/).
   a. w^{1118} recipient for transformations.
   b. Balancer stocks for mapping:
      i. $w^{1118}/Dp(1{:}Y)\ y^+$; $CyO/nub^1\ b^1\ noc^{Sco}\ lt1\ stw^3$; $MKRS/TM6B,\ Tb^1$.
      ii. CyO/noc^{Sco}; $TM6/MRS$.
   c. Stocks carrying heat-inducible FLP and I-*Sce*I genes.
      i. $y^1\ w^*$; $P\{ry[+t7.2]=70FLP\}11\ P\{v[+t1.8]=70I\text{-}SceI\}2B\ noc^{Sco}/CyO,\ S^2$ (on Chr. 2).
      ii. $y^1\ w^*$; $P\{ry[+t7.2] = 70FLP\}11\ P\{v[+t1.8] = 70I\text{-}SceI\}4A/TM6$ (on Chr. 3).
   d. Stock expressing a Gal4-VP16 transcriptional activator in the germline.
      i. *P{GAL4::VP16-nos.UTR}MVD.*

10. pP{UASp-DEST} (pPW) (http://www.ciwemb.edu/labs/murphy/Gateway%20vectors.html). This destination plasmid places an insert received from an Entry vector under the control of a Gal-UASp, which has been optimized for expression in the female germline *(37)*. It is available from the *Drosophila* Genomics Resource Center, University of Indiana, http://dgrc.cgb.indiana.edu/.
11. pCS2-DEST. This is a version of the in vitro transcription vector pCS2 that has been modified to serve as a destination vector. It is the recipient for ZFN-coding sequences in the injection method.
12. In vitro transcription kit with SP6 RNA polymerase. We have used both AmpliCap SP6 High-yield Message Maker (Epicentre, Madison, WI) and mMessage mMachine SP6 (Ambion, Austin, TX) with good results.
13. Standard molecular biology supplies and equipment, including agarose gel electrophoresis.

2.3. Materials for C. elegans

1. Basic supplies and equipment for nematode husbandry and injection.
2. Fluorescence microscope set up to detect green fluorescent protein (GFP).
3. pJM1-ZFN DNA *(27)*. This plasmid contains a ZFN-coding sequence from which the ZF segment can be excised by digestion with *Nde*I + *Spe*I and replaced with new ZF sequences from **Subheading 3.1., steps 5** or **9** above. For nematodes we currently do not use the Gateway system, but may switch to it in the future. Expression of the ZFN is driven by the nematode 16–48 heat-shock promoter.
4. Carrier DNA (1-kb ladder, Invitrogen) and GFP marker plasmid (pPD118.33, *Pmyo*-2::GFP) for injections.

3. Methods

3.1. Methods for ZFN Construction

Three very detailed protocols for constructing coding sequences for new ZF combinations have been published recently. One involves assembly of pre-existing one-finger modules in sequential cloning steps *(38)*. Chapter 4 describes production and assembly of new one-finger modules by PCR using an existing three-finger template. We prefer to make new three-finger sets from synthetic oligonucleotides, as this allows complete control over framework sequences and codon preference *(32)*. We describe this approach briefly here. Once the ZF coding sequences have been obtained, they are cloned in-frame with the nuclease domain. We usually choose to make the initial ZFN inserts in a Gateway Entry vector (Invitrogen), which facilitates transfer to any of a number of Destination vectors designed to promote expression in various cells and organisms.

1. Choose a gene to be targeted and search it for sequences that look like good ZF-binding sites in proper relative position. The search can be performed at either of the two websites designed for the purpose: one maintained by Carlos Barbas' lab (Scripps Research Institute, La Jolla, CA) *(39)*, http://zincfingertools.org, and one by the Zinc Finger Consortium, http://www. zincfingers.org. The features of a promising site for

three-finger ZFNs are inverted 9-bp sequences made up of triplets for which good fingers exist separated by 6 bp. In some instances the 6-bp spacer will include or overlap a restriction enzyme recognition sequence, which allows molecular screening for NHEJ products *(27)*, but this will not always be possible. Additional details are given in **ref.** *32*.

2. Design amino acid sequences of the ZFs that will bind each of the component 9-bp sites (ZFNa and ZFNb in **Fig. 1**).
3. Design coding sequences for these ZF combinations.
4. Synthesize long oligonucleotides corresponding to the designed coding sequences, and combine them by PCR. The product for each three-finger subsite should be 283-bp long, if our exact protocol is followed.
5. Gel purify, and if necessary, reamplify the 283-bp PCR product.
6. Ligate into pENTR-ZFN (or pENTR-NLS-ZFN) in place of the existing ZF coding sequence. This requires cleaving vector and PCR product with *Nde*I and *Spe*I, recovering the desired fragments, and ligation. Detailed DNA sequences and reaction conditions are given in **ref.** *32*.
7. Transform into competent *E. coli* cells—for example, DH5α—and select on plates containing kanamycin (50 μg/mL).
8. Pick individual colonies and screen for the correct inserts—for example, by colony PCR *(40)* using one primer specific for the desired insert and another complementary to vector sequences. Ultimately, the inserts should be verified by sequencing.
9. Purify the verified plasmid DNAs—for example, by a maxi-prep procedure using a Qiagen kit.

3.2. Methods for Drosophila

3.2.1. General Considerations

Our original method for gene targeting with ZFNs in *Drosophila* involves integrating the ZFN-coding sequences and the donor DNA into the fly genome using P element vectors and driving ZFN expression with a heat-inducible promoter *(19,22)*. We have also placed the ZFNs under Gal4-upstream activating sequence (UAS) control and shown that we can generate new mutations at high frequency by driving them with a Gal4-VP16 activator that is expressed in the germline. Although very effective, these are rather laborious procedures that require construction of the various cloned DNAs, introducing them individually into flies, mapping, then bringing together all of the necessary components. More recently, we have begun to have success with a simplified protocol that involves direct injection of synthetic mRNAs for the ZFNs into *Drosophila* embryos. This procedure has not yet been optimized, but we present all three approaches in the hope that other researchers will be motivated to help explore critical parameters.

In these protocols, the ZFNs can be expressed in the presence or absence of donor DNA. Without donor, targeted cleavage leads to production of localized mutations through NHEJ *(19,22)*. In some instances, this may be sufficient, if the goal is simply to make a null mutation in the target gene. With the injection procedure, it is easy to prepare an injection mix that contains mRNAs for the two ZFNs, with or without a designed donor DNA. In the heat-shock method, there is a substantial difference in the requirements for mutagenesis by NHEJ and gene replacement by HR.

For simple mutagenesis, all that is needed are individual flies carrying an inducible transgene for each of the two ZFNs. Crossing these parents generates offspring with both ZFNs, and when these are induced, cleavage and mutagenesis should result. For gene replacement, the goal is to produce flies that carry both ZFN-coding sequences on a single chromosome, plus the donor DNA on a different chromosome, and genes for FLP and I-*Sce*I, also under heat-shock promoter control ***(19,20)***. Fortunately, pairs of the latter two genes are available on chromosomes 2 and 3 ***(41)***, and these can be crossed into the desired stocks. We always carry the donor construct in a separate stock from FLP and I-*Sce*I, as we have found that leaky expression of one or both of these enzymes leads to destruction of the donor with time.

3.2.2. Heat-Shock Method

1. Run Clonase reactions, following the supplier's instructions, using the pENTR-ZFNa and -ZFNb constructs from **Subheading 3.1., step 9** and the pCaSpeR-hs-DEST vector. Reaction conditions we use are described in **ref.** ***32***.
2. Transform the products into *E. coli* DH5α and select for resistance to ampicillin (100 μg/mL).
3. Screen individual colonies—for example, by colony PCR—and verify by DNA sequencing. In our experience, the Clonase reaction is very efficient, so nearly all the colonies have the desired inserts.
4. Purify DNA from verified clones for each of the two new ZFNs. Use the Qiagen Maxiprep Kit (or similar) to obtain sufficiently pure DNA.
5. Clone the designed donor DNA into pP{WhiteOut2}; screen and verify colonies as above. Purify DNA from one verified clone.
6. Transform each of the above plasmid DNAs individually into *Drosophila* embryos using standard methods (e.g., http://www.molbio.wisc.edu/carroll/methods/miscellaneous). This involves making injection mixes with your DNAs (600–800 μg/mL) and the P transposase expression plasmid pπ25.1wc (100 μg/mL) in 0.1X phosphate-buffered saline buffer. Inject approx 200 w^{1118} embryos with each DNA individually: pCaSpeR-hs-ZFNa, pCaSpeR-hs-ZFNb, and pP{WhiteOut2}-donor.
7. Collect eclosing adults and cross to w^{1118} partners. Offspring from these crosses with pigmented (not white) eyes are successful transformants.
8. Map the transgenes to one of the *Drosophila* chromosomes using mapping stocks, such as $w^{1118}/Dp(1{:}Y)\ y^{+}$; $CyO/nub^{1}\ b^{1}\ noc^{Sco}\ It1\ stw^{3}$; $MKRS/TM6B,\ Tb^{1}$.
9. If you want to generate targeted mutations by NHEJ after ZFN cleavage, you can simply cross flies carrying the ZFNa and ZFNb transgenes. For example, w^{1118}/Y; ZFNa/CyO × w^{1118}; ZFNb/CyO. Place two to four parents of each type in a single vial. Three or four days later, remove the parents and place the vials in a 37°C water bath for 1 h, then return it to room temperature or a 25°C incubator. During the heat shock, press the cotton plug part way into the vial, so the larvae cannot crawl above the water level and escape the heat shock.

10. When flies eclose from these vials, their offspring can be screened for new mutations in the target gene. If the mutant phenotype is known, the flies can be crossed to known mutants. Otherwise, a molecular screen is possible using PCR. We have obtained frequencies of mutagenesis in up to approx 10% of all progeny of such crosses ***(19)***, so such a screen is feasible. We use the aquagenomic DNA extraction kit to prepare DNA from individual flies for PCR.
11. For gene replacements, we first place the two ZFN transgenes on the same chromatid by recombination in the female germline. Cross flies carrying the individual transgenes that have been mapped to the same chromosome—for example, chromosome 2. Screen progeny of females by PCR to identify those carrying both transgenes using primers specific to unique ZFNa and ZFNb sequences. The frequency of recombination depends on the distance between the transgenes on the genetic map, but is rarely too low to be useful.
12. Create the following stocks (this assumes the ZFN, FLP, and I-*Sce*I transgenes are on chromosome 2 and the donor on chromosome 3, but analogous situations apply to other locations): (FLP) (I-*Sce*I)/(ZFNa) (ZFNb) and Donor/*MKRS*. Cross individuals from these stocks in individual vials and heat shock as in **step 9** above.
13. Screen for new mutations in the target gene as in **step 10** above. There will inevitably be NHEJ mutations. To find the desired gene replacements, you will need to design a PCR or other assay that identifies these specifically. We have incorporated diagnostic restriction sites into our donors, the presence of which can easily be detected in PCR products ***(19,20)***.

3.2.3. UAS Method

1. Run Clonase reactions, as in **Subheading 3.2.2., step 1** above, using the pENTR-ZFNa and -ZFNb constructs from **Subheading 3.1., step 9** and the pP{UASp-DEST} vector.
2. Transform DH5α, verify the resulting clones, and make DNA as in **Subheading 3.2.2., steps 2–4** above.
3. Transform each of the above plasmid DNAs individually into *Drosophila* embryos as in **Subheading 3.2.2., step 6** above. Follow **Subheading 3.2.2., steps 7–8** to recover flies with the UASp-ZFNa and UASp-ZFNb transgenes and map them to specific chromosomes.
4. Recombine UASp-ZFNa and UASp-ZFNb transgenes onto the same chromatid, as described in **Subheading 3.2.2., step 11** above.
5. For targeted mutagenesis through NHEJ, cross flies carrying the two ZFN genes with flies with Gal4-VP16*nos*. Induction occurs in cells in which the *nos* gene is normally expressed, so no heat shock is necessary.
6. Screen for new mutants as in **Subheading 3.2.2., step 10** above.

3.2.4. Injection Method

1. Run Clonase reactions, as in **Subheading 3.2.2., step 1** above, using the pENTR-NLS-ZFNa and -ZFNb constructs from **Subheading 3.1., step 9** and the pCS2-DEST vector. (It is also possible to insert annealed oligonucleotides encoding an NLS into pENTR-ZFNa and -ZFNb, if the NLS vector was not used initially.)

2. Transform DH5α, verify the resulting clones, and make highly purified DNA as in **Subheading 3.2.2., steps 2–4** above.
3. Linearize the plasmids by digestion with *Not*I, which cuts distal to the 3′-UTR.
4. Transcribe the pCS2-NLS-ZFNa and pCS2-NLS-ZFNb DNAs using the AmpliCap or mMessage mMaker kit. Examine the resulting RNAs by agarose gel electrophoresis to ensure that they are full length.
5. Prepare an injection mix that has 350–800 μg/mL of each RNA. If gene replacement is desired, include the donor DNA in a circular plasmid at 1500 μg/mL.
6. Inject ~300 *Drosophila* embryos with this mix. When adults eclose, screen them for new targeted mutations as described in **Subheading 3.2.2., step 10** above.

3.3. Methods for C. elegans

At the time of this writing, we have achieved high frequencies of ZFN-induced somatic cleavage and mutagenesis of both a synthetic and a pre-existing genomic target in nematodes *(27)*; but we have yet to accomplish germline ZFN expression or targeted gene replacement. Nonetheless, we present procedures for somatic cleavage, as they are useful for studying aspects of DSB repair, and they represent a significant step on the route to germline gene targeting in the worm. As suggested in **ref.** *27*, several approaches to overcome germline silencing of ZFN transgenes suggest themselves and are under exploration in our lab currently.

1. Digest the gel-purified PCR product from **Subheading 3.1., step 5** and pJM1-ZFN separately with *Nde*I + *Spe*I.
2. Gel purify the large fragment from both digests.
3. Mix and ligate the purified fragments.
4. Transform the products into *E. coli* DH5α and select for resistance to ampicillin (100 μg/mL).
5. Screen individual colonies—for example, by colony PCR—and verify by DNA sequencing.
6. Purify DNA from verified clones for each of the two new ZFNs.
7. Digest both of the ZFN plasmids and pPD118.33 with *Sca*I to linearize them.
8. Prepare an injection mix containing each of the *Sca*I-digested ZFN plasmids at 5 μg/mL, pPD118.33 at 1 μg/mL, and 1-kb ladder at 100 μg/mL.
9. Inject this mix into the syncytial gonad of approx 10 hermaphrodite worms.
10. Screen offspring of the injected worms for progeny showing GFP expression in their pharynx, indicating successful formation of an extrachromosomal array. Monitor GFP in subsequent generations to ensure that the array is stable.
11. Transfer 8–10 nematodes at larval stages L2 and L3 to an agar plate. Wrap the plate in Parafilm (Pechiney Plastic Packaging Co., Chicago, IL), place in a 35°C water bath for 1 h, then return to room temperature and remove the Parafilm.
12. On the following day, repeat the 1-h 35°C heat shock. Return to room temperature for several hours.
13. Isolate total DNA by freezing the worms individually at –80°C in 3 μL of single worm lysis buffer (50 m*M* of KCl, 10 m*M* of Tris-HCl [pH 8.3], 2.5 m*M* of $MgCl_2$, 0.45% of NP-40, 0.45% of Tween-20, 0.01% gelatin, 200 μg/mL of proteinase K). Lyse the worms at 65°C for 1 h, followed by 95°C for 15 min.

14. To assess sequence changes in the target, it can be amplified by PCR with specific primers and examined by cloning and sequencing or other methods appropriate to the specific target *(27)*.
15. Alternatively, if a phenotypic change is anticipated following ZFN cleavage and repair, this can be evaluated after the heat shocks.

4. Notes

1. One issue with the ZFNs has been toxicity owing to off-target cleavage ***(4,17,19,22,25)***. In *Drosophila* we perform a simple test for lethality by heat-shocking flies carrying each ZFN transgene singly. Flies heterozygous for the ZFN sequence and its associated w^+ marker are crossed to w^{1118} partners, and their progeny are heat shocked as in **step 9**, **Subheading 3.2.2**. Just before heat shock, the parents are transferred to a fresh vial where they continue to produce offspring that serve as controls. The numbers of w^+ and w^- flies are counted in the induced and control vials. A deficiency of w^+ offspring after heat shock indicates a problem with ZFN-induced lethality. In such a case, it is often possible to find a lower heat-shock temperature that allows survival and retains some efficacy ***(19,22)***.
2. In the *Drosophila* heat-shock method, we typically perform the heat induction 3 or 4 d after initiating a cross (**Subheading 3.2.2., step 9**). One experiment on the timing of the heat shock showed that equivalent frequencies of targeting were obtained after induction on the second, third or fourth day, but a substantially a lower frequency when induced after only 1 d (A.B., unpublished results). This was true in both males and females, and we presume it reflects the capabilities of germline cells at different stages of development.
3. In standard P element transformations, it is common to lose 25–40% of the injected animals at hatching and owing to sterility of the adults. When injecting ZFN mRNAs, we generally see a higher level of loss (70–90%), although we have not demonstrated that this is because of cleavage activity.
4. It is important to be able to distinguish the modified target from residual donor when assaying a gene replacement experiment by PCR (**Subheading 3.2.2., step 13**). One solution is to design PCR primers so that one is in target sequences outside those present in the donor. If this is impractical, you can propagate the altered target to the next generation and choose offspring in which the donor chromosome has segregated away. This can be done by monitoring loss of the associated w^+ marker or by judicious use of balancers.
5. RNA is much more sensitive to degradation than DNA, so we make fresh mRNA preparations quite frequently and check them by electrophoresis within 1 d of injection. We store RNAs at –80°C under ethanol to minimize degradation.
6. In the *Drosophila* RNA injection experiments, it is absolutely necessary to endow the ZFNs with a nuclear localization sequence. This has not been required in the heat shock or UAS methods, perhaps because of the timing and/or duration of ZFN induction. If germline precursor cells are dividing while ZFN proteins are present, they could gain access to the genomic target when the nuclear membrane breaks down at mitosis.
7. In the injection procedure, we are still exploring optimum concentrations for the mRNAs and the donor DNA. If the ZFNs induce significant lethality, it may be necessary to reduce one or both RNA concentrations. It is possible that donor concentrations higher than we have used to date would increase the yield of gene replacements.

8. In all the *Drosophila* methods ZFN induction occurs during the premeiotic stage of germ cell development. As a consequence, we often recover multiple offspring carrying the same new mutation from a single parent. For HR products, it is not possible to tell whether or not these result from independent events. In the case of NHEJ products, the spectrum of sequence changes is sufficiently broad that clusters of the same mutation almost certainly result from mitotic divisions and ultimate gamete formation following a single mutagenic event in a precursor cell.
9. Expected results for *Drosophila*. From the heat-shock method, we have obtained frequencies of targeted germline mutagenesis in the *ry* gene owing to NHEJ as high as 14% of all offspring of heat-induced parents ***(19)***. In the presence of an excised linear donor, the frequency of targeted gene replacement can be as high as 15% of all offspring. These frequencies were obtained in the female germline; in males the frequencies were somewhat lower. Using the UAS method, we have achieved very high levels of NHEJ mutagenesis in the *ry* gene (K.J.B. and J.K.T., unpublished results). All of the flies, both males and females, which carried the UASp-ZFN pair and the Gal4-VP16*nos* driver gave mutant progeny; and more than half of all offspring were new mutants. To date we have not included a donor DNA in this protocol. We have also had good success with the injection method, using ZFN RNAs for the *ry* gene. As noted earlier, the survival is considerably lower than with the standard DNA injections for transformation, usually in the range of 20–30%. Among the flies surviving to be assayed for germline mutations, about 40% give mutant progeny, with similar results for males and females. The number of mutants per productive parent is typically quite large, with some individuals yielding only *ry* offspring. In two experiments with a coinjected circular donor DNA, 7% of the mutants were the result of gene replacement, and the rest were NHEJ products.
10. If there is any question about whether both ZFN expression plasmids are incorporated into an array in nematodes (**Subheading 3.3., step 8**), PCR assays can be designed with one primer corresponding to a unique sequence within the ZF coding region and another within vector sequence. If primers specific for both ZF sets give good amplification, both ZFN plasmids are present. If you are still nervous, you can even sequence some such products to be certain.
11. Expected results for *C. elegans*. As described in **ref. *27***, we have recovered high levels of ZFN-induced NHEJ mutations in both a synthetic target on an extrachromosomal array and in a pre-existing genomic target. In both cases, the frequency of mutations was approx 20% of PCR-amplified targets.
12. If you want to test the ability of new ZFNs to cut their designed target, this can be done by creating a synthetic version of that target and assaying cleavage and mutagenesis in worms ***(27)***. Make synthetic oligonucleotides that represent the ZFN target and clone them into an arbitrary vector. Coinject this target along with the two ZFN expression plasmids to form an array. After heat induction and DNA isolation as in **Subheading 3.3., steps 9** and **10**, the region containing the synthetic target can be amplified. A simple way to design the assay is to include a unique restriction site in the 6 bp between the two ZFN-binding sites and to test if a fraction of the PCR products has become resistant to that enzyme ***(27)***. If so, this indicates that the ZFNs made the expected cut, and repair by NHEJ led to sequence alterations in the target.

Acknowledgments

We are grateful to Srinivasan Chandrasegaran, who constructed the first ZFNs; to Kent and Mary Golic, who helped us introduce the ZFN technology to *Drosophila*; to Erik Jorgensen, and Wayne Davis, who collaborated on the *C. elegans* applications. We also thank others who have worked on the ZFNs in our lab in the past, particularly Marina Bibikova and David Segal. Our work has been supported by the US National Institutes of Health through Research Grants GM58504, GM65173, GM078571 to D.C., and Training Grant in Genetics T32 GM07464 to J.J.M.; also by a seed grant from the University of Utah to D.C. and through core facilities funded in part by the University of Utah Cancer Center Support Grant.

References

1. Rothstein, R. J. (1983) One-step gene disruption in yeast. *Methods Enzymol.* **101,** 202–211.
2. Capecchi, M. R. (1989) Altering the genome by homologous recombination. *Science* **244,** 1288–1292.
3. Rong, Y. S. and Golic, K. G. (2000) Gene targeting by homologous recombination in *Drosophila. Science* **288,** 2013–2018.
4. Porteus, M. H. and Carroll, D. (2005) Gene targeting using zinc finger nucleases. *Nat. Biotechnol.* **23,** 967–973.
5. Johnson, R. D. and Jasin, M. (2001) Double-strand-break-induced homologous recombination in mammalian cells. *Biochem. Soc. Trans.* **29,** 196–201.
6. Blancafort, P., Segal, D. J., and Barbas, III, C. F. (2004) Designing transcription factor architectures for drug discovery. *Mol. Pharmacol.* **66,** 1361–1371.
7. Durai, S., Mani, M., Kandavelou, K., Wu, J., Porteus, M. H., and Chandrasegaran, S. (2005) Zinc finger nucleases: custom-designed molecular scissors for genome engineering of plant and mammalian cells. *Nucleic Acids Res.* **33,** 5978–5990.
8. Papworth, M., Kolasinska, P., and Minczuk, M. (2006) Designer zinc-finger proteins and their applications. *Gene* **366,** 27–38.
9. Kim, Y. -G., Cha, J., and Chandrasegaran, S. (1996) Hybrid restriction enzymes: zinc finger fusions to *Fok*I cleavage domain. *Proc. Natl. Acad. Sci. USA* **93,** 1156–1160.
10. Pabo, C. O., Peisach, E., and Grant, R. A. (2001) Design and selection of novel Cys_2His_2 zinc finger proteins. *Annu. Rev. Biochem.* **70,** 313–340.
11. Dreier, B., Beerli, R. R., Segal, D. J., Flippin, J. D., and Barbas III, C. F. (2001) Development of zinc finger domains for recognition of the 5′-ANN-3′ family of DNA sequences and their use in the construction of artificial transcription factors. *J. Biol. Chem.* **276,** 29,466–29,478.
12. Dreier, B., Fuller, R. P., Segal, D. J., et al. (2005) Development of zinc finger domains for recognition of the 5′-CNN-3′ family DNA sequences and their use in construction of artificial transcription factors. *J. Biol. Chem.* **280,** 35,588–35,597.
13. Liu, Q., Xia, Z. Q., Zhong, X., and Case, C. C. (2002) Validated zinc finger protein designs for all 16 GNN DNA triplet targets. *J. Biol. Chem.* **277,** 3850–3856.

14. Segal, D. J., Dreier, B., Beerli, R. R., and Barbas, III, C. F. (1999) Toward controlling gene expression at will: selection and design of zinc finger domains recognizing each of the 5′-GNN-3′ DNA target sequences. *Proc. Natl. Acad. Sci. USA* **96,** 2758–2763.
15. Bitinaite, J., Wah, D. A., Aggarwal, A. K., and Schildkraut, I. (1998) *Fok*I dimerization is required for DNA cleavage. *Proc. Natl. Acad. Sci. USA* **95,** 10,570–10,575.
16. Smith, J., Bibikova, M., Whitby, F. G., Reddy, A. R., Chandrasegaran, S., and Carroll, D. (2000) Requirements for double-strand cleavage by chimeric restriction enzymes with zinc finger DNA-recognition domains. *Nucleic Acids Res.* **28,** 3361–3369.
17. Urnov, F. D., Miller, J. C., Lee, Y. -L., et al. (2005) Highly efficient endogenous gene correction using designed zinc-finger nucleases. *Nature* **435,** 646–651.
18. Alwin, S., Gere, M. B., Gulh, E., et al. (2005) Custom zinc-finger nucleases for use in human cells. *Mol. Ther.* **12,** 610–617.
19. Beumer, K., Bhattacharyya, G., Bibikova, M., Trautman, J. K., and Carroll, D. (2006) Efficient gene targeting in *Drosophila* with zinc finger nucleases. *Genetics* **172,** 2391–2403.
20. Bibikova, M., Beumer, K., Trautman, J. K., and Carroll, D. (2003) Enhancing gene targeting with designed zinc finger nucleases. *Science* **300,** 764.
21. Bibikova, M., Carroll, D., Segal, D. J., Trautman, J. K., Smith, J., Kim, Y. -G., and Chandrasegaran, S. (2001) Stimulation of homologous recombination through targeted cleavage by chimeric nucleases. *Mol. Cell. Biol.* **21,** 289–297.
22. Bibikova, M., Golic, M., Golic, K. G., and Carroll, D. (2002) Targeted chromosomal cleavage and mutagenesis in *Drosophila* using zinc-finger nucleases. *Genetics* **161,** 1169–1175.
23. Lloyd, A., Plaisier, C. L., Carroll, D., and Drews, G. N. (2005) Targeted mutagenesis using zinc-finger nucleases in *Arabidopsis*. *Proc. Natl. Acad. Sci. USA* **102,** 2232–2237.
24. Porteus, M. H. (2006) Mammalian gene targeting with designed zinc finger nucleases. *Mol. Ther.* **13,** 438–446.
25. Porteus, M. H. and Baltimore, D. (2003) Chimeric nucleases stimulate gene targeting in human cells. *Science* **300,** 763.
26. Wright, D. A., Townsend, J. A., Winfrey, R. J., Jr., et al. (2005) High-frequency homologous recombination in plants mediated by zinc-finger nucleases. *Plant J.* **44,** 693–705.
27. Morton, J., Davis, M. W., Jorgensen, E. M., and Carroll, D. (2006) Induction and repair of zinc-finger nuclease-targeted double-strand breaks in *Caenorhabditis elegans* somatic cells. *Proc. Natl. Acad. Sci. USA* **103,** 16,370–16,375.
28. Ashburner, M., Golic, K. G., and Hawley, R. S. (2005) *Drosophila*. A Laboratory Handbook, Cold Spring Harbor Laboratory Press, Cold Spring Harbor, NY.
29. Greenspan, R. J. (2004) Fly pushing: the theory and practice of *Drosophila* genetics, Cold Spring Harbor Laboratory Press, Cold Spring Harbor, NY.
30. Riddle, D. L., Blumenthal, T., Meyer, B. J., and Priess, J. R. (eds.) (1997) *C. elegans* II, Cold Spring Harbor Laboratory Press, Cold Spring Harbor Laboratory Press, Cold Spring Harbor, NY.
31. Wood, W. B. (ed.) (1988) The Nematode *Caenorhabditis elegans*, Cold Spring Harbor Laboratory Press, Cold Spring Harbor Laboratory Press, Cold Spring Harbor, NY.

32. Carroll, D., Morton, J. J., Beumer, K. J., and Segal, D. J. (2006) Design, construction and *in vitro* testing of zinc finger nucleases. *Nat. Protoc.* **1,** 1329–1341.
33. Gong, W. J. and Golic, K. G. (2003) Ends-out, or replacement, gene targeting in *Drosophila. Proc. Natl. Acad. Sci. USA* **100,** 2556–2561.
34. Keeler, K. J., Dray, T., Penney, J. E., and Gloor, G. B. (1996) Gene targeting of a plasmid-borne sequence to a double-strand DNA break in *Drosophila melanogaster. Mol. Cell. Biol.* **16,** 522–528.
35. Gloor, G. B. (2002) The role of sequence homology in the repair of DNA double-strand breaks in *Drosophila. Adv. Genet.* **46,** 91–117.
36. Elliott, B., Richardson, C., Winderbaum, J., Nickoloff, J. A., and Jasin, M. (1998) Gene conversion tracts from double-strand break repair in mammalian cells. *Mol. Cell. Biol.* **18,** 93–101.
37. Rorth, P. (1998) Gal4 in the Drosophila female germline. *Mech. Dev.* **78,** 113–118.
38. Wright, D. A., Thibodeau-Beganny, S., Sander, J. D., et al. (2006) Standardized reagents and protocols for engineering zinc finger nucleases by modular assembly. *Nat. Protoc.* **1,** 1637–1652.
39. Mandell, J. G. and Barbas, III, C. F. (2006) Zinc finger tools: custom DNA-binding domains for transcription factors and nucleases. *Nucleic Acids Res.* **34,** W516–W523.
40. Sambrook, J. and Russell, D. W. (2001) Molecular Cloning: A Laboratory Manual, Cold Spring Harbor Press, Cold Spring Harbor, NY.
41. Rong, Y. S. and Golic, K. G. (2001) A targeted gene knockout in *Drosophila. Genetics* **157,** 1307–1312.

6

Orpheus Recombination

A Comprehensive Bacteriophage System for Murine Targeting Vector Construction by Transplacement

Knut Woltjen, Kenichi Ito, Teruhisa Tsuzuki, and Derrick E. Rancourt

Summary

In recent years, methods to address the simplification of targeting vector (TV) construction have been developed and validated. Based on in vivo recombination in *Escherichia coli*, these protocols have reduced dependence on restriction endonucleases, allowing the fabrication of complex TV constructs with relative ease. Using a methodology based on phage-plasmid recombination, we have developed a comprehensive TV construction protocol dubbed Orpheus recombination (ORE). The ORE system addresses all necessary requirements for TV construction; from the isolation of gene-specific regions of homology to the deposition of selection/disruption cassettes. ORE makes use of a small recombination plasmid, which bears positive and negative selection markers and a cloned homologous "probe" region. This probe plasmid may be introduced into and excised from phage-borne murine genomic clones by two rounds of single crossover recombination. In this way, desired clones can be specifically isolated from a heterogeneous library of phage. Furthermore, if the probe region contains a designed mutation, it may be deposited seamlessly into the genomic clone. The complete removal of operational sequences allows unlimited repetition of the procedure to customize and finalize TVs within a few weeks. Successful gene-specific clone isolation, point mutations, large deletions, cassette insertions, and finally coincident clone isolation and mutagenesis have all been demonstrated with this method.

Key Words: Bacteriophage-λ; embryonic stem cell; γ; gene targeting vector; library screening; phage–plasmid recombination; supF; suppressor tRNA; transplacement mutagenesis.

1. Introduction

Gene targeting in murine embryonic stem (ES) cells is an extremely potent method of introducing rationally designed modifications into the mouse genome, and has completely revolutionized the field of functional genomics ***(1)***. The complexities of standard targeting vector (TV) construction, compounded by the

From: *Methods in Molecular Biology, vol. 435: Chromosomal Mutagenesis*
Edited by: G. Davis and K. J. Kayser © Humana Press Inc., Totowa, NJ

demands for more complex mutation types, has been a major bane for mouse geneticists. To address these TV construction issues, we have developed techniques which have been compiled into a single recombination-based TV construction system using bacteriophage-λ as a contiguous scaffold—Orpheus recombination (ORE)*.

Unlike many static plasmid or bacterial artificial chromosome systems in recombination-proficient bacteria, we have found that transient bacteriophage infections in *rec*$^+$ bacterial hosts allow the cloning and maintenance of large regions of homology, which normally succumb to rearrangements. Recombination of λ genomes with homologous elements in circular plasmids appears to be dependent on the *Escherichia coli* RecABCD pathway ***(2,3)***, an event which is enhanced nearly 1000-fold by the λ-phage recombination adept with plasmids (*rap*) gene ***(4)***. We have taken advantage of this in vivo recombination phenomenon to produce simple TVs through a double crossover selection scheme ***(5)***. From our original trials with double crossover phage–plasmid recombination, we have moved to more efficient single crossover mechanisms using the small probe plasmid πANγ (*see* **Fig. 1A**) ***(6)***, a derivative of πAN13 ***(7)***. The positive–negative selection potential of πANγ has allowed us to develop simple methods of seamless mutagenesis ***(6,8,9)*** and improved recombination-based library screening techniques ***(10,11)***. To further streamline TV construction, we have prepared a partially completed TV library of genomic clones from the R1 ES cell line in λTK (*see* **Fig. 1B**) ***(11)***. The basic recombination and selection principles of the ORE system are outlined in **Fig. 2**.

In our hands, ORE has been applied to the specific isolation of genomic clones from the λTK library for approx 200 genes. We have also been successful with various downstream genomic clone modifications such as the deposition of unique restriction sites ***(6,8,9)*** or *lox*P elements, removal of introns without coding region disruptions, and finally the insertion of selection/disruption cassettes to finalize replacement TVs ***(9)***. It is also important to highlight the potential to combine both retro recombination screening (RRS) and transplacement mutagenesis (TM) protocols to coincidentally isolate and mutagenize clones from the λTK library ***(9)***, theoretically allowing TVs to be completed in one step. The simplicity of the procedure allows multiple TVs to be processed in parallel.

---

*Why Orpheus? Syrinx (the nymph pursued by the Pan, who changed her shape to that of reeds to escape the God) and Charon (the spectral figure who ferries the dead across the river Styx), two well-known bacteriophage-λ cloning vectors, are the parents of the λTK phage vectors used in our recombination system. In keeping with this naming tradition, we have dubbed the system "Orpheus." Although Orpheus was the only mortal to enter and depart from the underworld alive, he left a part of himself behind, a suitable parallel to the integrative and excisive recombination used to deposit designed modifications in phage-born clones.

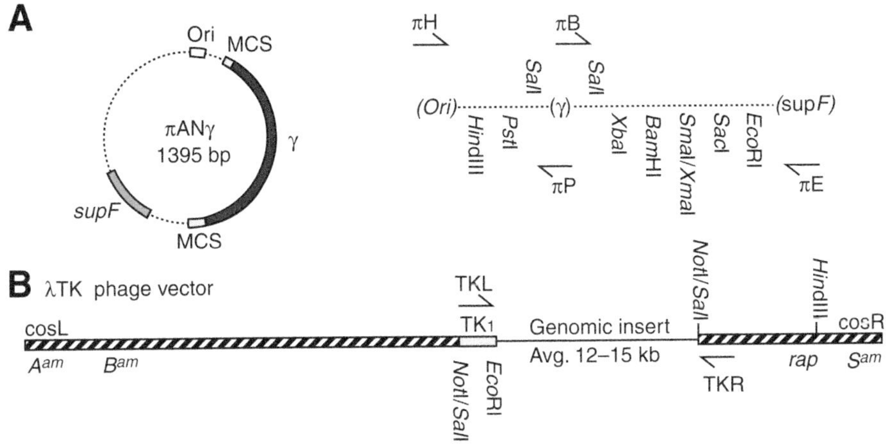

Fig. 1. Phage and plasmid components for TV construction using the ORE system. **(A)** Diagram of the πANγ probe plasmid and associated multiple cloning sites. The plasmid contains positive (*supF*) and negative (γ) genetic markers used to monitor the integrative and excisive recombination events, respectively. The positions of universal screening primers (πH, πP, πB, and πE) are indicated. **(B)** The λTK phage vector. Essential genetic markers and landmark restriction endonuclease sites are shown (*see* text for details). Inclusion of the herpes simplex virus thymidine kinase gene (TK1) allows phage clones to be used as replacement TVs following the addition of a eukaryotic selection/disruption cassette. The positions of universal screening primers (TKL and TKR) are indicated. Sequence information for πANγ and the λTK vector are available on request.

The ORE protocols described herein relate specifically to the isolation and modification of murine genomic clones for the purpose of TV construction. However, the core ORE system is not limited to these ends alone, and may be used to generate mutant cDNA clones, modify promoter elements, construct transgenes or minigenes, and beyond. The principles of ORE may be adapted to TV construction in a variety bacteriophage vectors. Protocols included for standard manipulation and analysis of bacteriophage-λ vectors and their DNA clones are relevant to all of these applications.

2. Materials

2.1. Probe Plasmid Preparation

1. πANγ plasmid ***(6)***.
2. Library screening or mutagenic probe DNA (polymerase chain reaction [PCR] product, restriction fragment, cDNA fragment, or synthetic oligonucleotide).
3. Restriction endonucleases (Invitrogen, Carlsbad, CA) for standard plasmid cloning and recombinant DNA analytical procedures.
4. T4 DNA ligase (Invitrogen) for standard plasmid cloning procedures.

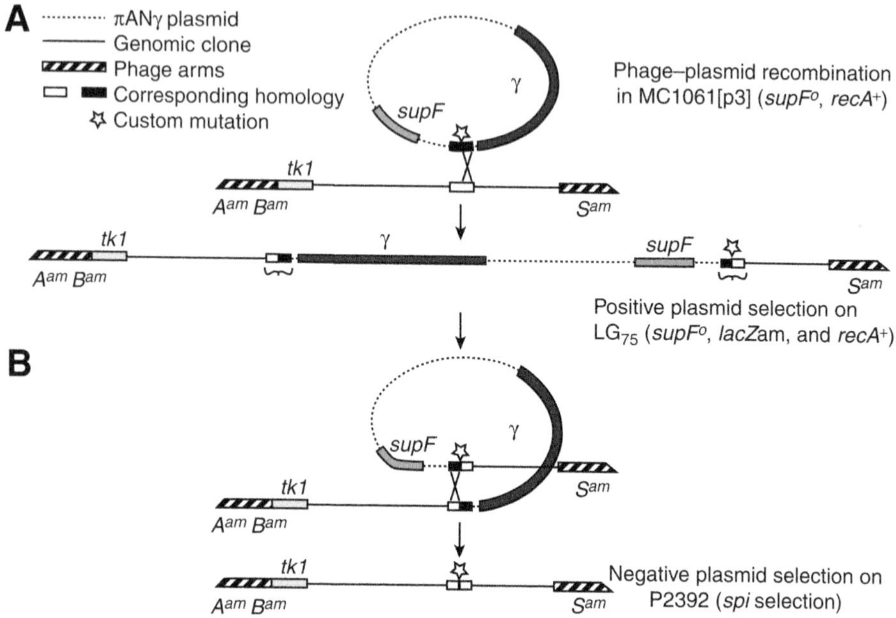

Fig. 2. Principles of the recombination and selection steps for RRS and TM. **(A)** Pure (TM) or heterogeneous (RRS) phage clones are passaged over the MC1061(p3) host containing a $\pi AN\gamma$ probe-plasmid. In the case of TM, this probe region bears a mutation for transplacement. The plasmid is integrated into the phage by single crossover homologous recombination, and the event is selected by *supF* suppression of phage and host amber mutations. Integration of the plasmid into the phage results in a duplication of the region of homology, flanking plasmid operational sequences. **(B)** Condensation and plasmid excision occurs spontaneously through a second single crossover recombination event between the duplicated homology regions. Phage having undergone condensation and plasmid loss are capable of growth on a P2 lysogen. During TM applications, if the second recombination event occurs contra-lateral to the integration event (~50% of the time for balanced-homology probes) the subtle mutation is seamlessly deposited into the target phage. During RRS, the genomic clone is returned to its native state, purified from the library pool. Note that the phage and plasmid are not drawn to scale.

5. 20 mg/mL of glycogen.
6. 3 *M* NaOAc (pH 5.2), autoclaved.
7. 100% Ethanol and 70% ethanol.
8. Sterile water, micron filtered and autoclaved.
9. 10% Glycerol solution, autoclaved.
10. MC1061(p3) ($recA^+$; $supF^o$; p3[kan^R; amp^Ram; tet^Ram]) electrocompetent cells (*see* **Notes 1** and **2**).
11. Bacterial electroporator for plasmid transformation (Gene Pulser, Bio-Rad, CA).
12. 0.2-cm gap bacterial electroporation cuvets (Gene Pulser, Bio-Rad).
13. Luria Bertani (LB) liquid medium.

14. LB solid medium (15 g/L bacto-agar), autoclave, cool to 50°C before addition of antibiotics.
15. Tetracycline solution: 12.5 mg/mL in 70% ethanol, store at –20°C, light-sensitive.
16. Ampicillin solution: 25 mg/mL in water, filter-sterilize, store frozen at –20°C.
17. Kanamycin solution: 25 mg/mL in water, filter-sterilize, store frozen at –20°C.
18. Synthetic oligonucleotide primers for plasmid PCR screening and sequencing (5′→3′): πH-TTTGTGATGCTCGTCAGG; πP-GCTTGATCTCAGTTTCAG; πB-GCTGGCTGAACGTGTCGGCATG; πE-GGCGCATCATATCAAATG.
19. Standard recombinant *Taq* polymerase PCR system.
20. Plasmid mini-prep kit (Qiagen, Valencia, CA).
21. Dimethylsulfoxide (DMSO).
22. 1.5-mL Microfuge tubes.
23. 100-mm Petri dishes.
24. Glass or plastic pipets (10 mL).
25. Glass culture tubes, autoclaved.
26. 37°C cabinet and shaking incubators.

2.2. Standard Bacteriophage Manipulation Techniques

1. 10 m*M* of $MgSO_4$ solution.
2. SM buffer (per liter): 5.8 g of NaCl, 2 g of $MgSO_4 \cdot 7H_2O$, 50 mL of 1 *M* Tris-HCl (pH 7.5), and 5 mL of 2% gelatin solution, autoclave and aliquot.
3. LBM (LB-Mg^{2+} low salt) agar (per liter): 5 g of NaCl, 5 g of yeast extract, 10 g of bacto-tryptone, 1 g of $MgCl_2 \cdot 6H_2O$, and 15 g of bacto-agar, autoclave.
4. LBM (LB-Mg^{2+} low salt) top agar (per liter): 5 g of NaCl, 5 g of yeast extract, 10 g of bacto-tryptone, 1 g of $MgCl_2 \cdot 6H_2O$, and 7.5 g of bacto-agar, autoclave, liquefy in microwave and equilibrate to 48°C before use.
5. Chloroform.
6. 48°C water bath or heat-block to equilibrate molten top-agar.
7. 5¾ in. glass Pasteur pipets and small pipet bulbs.
8. 15- and 50-mL polypropylene tubes.

2.3. Phage–Plasmid Recombination

1. For RRS applications: ES cell genomic library in bacteriophage (e.g., λTK R1 library *[11]*; *see* **Notes 3** and **4**).
2. For TM applications: purified phage genomic or cDNA clone (*see* **Note 4**).
3. LG75 (*recA*$^+$; *supF*^{o}; *lacZ*^{am}) plating cells.
4. LE392 (*recA*$^+$; *supE*; *supF*) plating cells.
5. P2392 (P2 lysogen of LE392) plating cells.
6. *Optional:* DK21 (*recA*$^+$; *supF*^{o}; *lac*Zam; *dnaB*^{am}; *P1ban*) plating cells.
7. LBM (LB-Mg^{2+} low salt) medium (per liter): 5 g of NaCl, 5 g of yeast extract, 10 g of bacto-tryptone, and 1 g of $MgCl_2 \cdot 6H_2O$, autoclave.
8. Isopropyl-β-D-1-thiogalactopyranoside (IPTG) solution: 20 mg/mL in water, filter-sterilize and store frozen at –20°C.
9. 4-Chloro-5-bromo-3-indolyl-β-D-galactopyranoside (X-Gal) solution: 20 mg/mL in dimethylformamide, store at –20°C, light-sensitive.

2.4. Bacteriophage DNA Isolation

1. Polyethylene glycol (PEG8000).
2. Sodium chloride (NaCl).
3. 40% PEG8000; 2.5 *M* NaCl solution.
4. 10 mg/mL of RNaseA.
5. 10 mg/mL of DNaseI.
6. 10 mg/mL of proteinase K.
7. Cesium chloride (CsCl).
8. TM dialysis buffer (per liter): 50 mL 1 *M* Tris-HCl (pH 7.8), 5 mL of 1 *M* $MgSO_4$, autoclaved.
9. 10% of Sodium dodecyl sulfate (SDS) solution, autoclaved.
10. 0.5 *M* Ethylene diamine tetra acetic acid (EDTA) (pH 8.0), autoclaved.
11. Buffer-saturated phenol, chloroform and phenol:chloroform mixtures (50:50).
12. 10 m*M* Tris-HCl (pH 8.0).
13. Ultracentrifuge and 5-mL ultracentrifuge tubes (Beckman, CA) for CsCl gradient separation of λ-phage particles.
14. 18-Gauge needle and 2–5 mL syringes.
15. Dialysis tubing and clamps.

2.5. Bacteriophage DNA Analysis

1. Synthetic oligonucleotide primers for λTK library clone end sequencing (5′→3′): TKL-GGGGTTTGCTCGACATTGGG; TKR-AACACTCGTCCGAGAATAAC.
2. Restriction endonucleases (Invitrogen). Suggested: *Eco*RI, *Hin*dIII, *Not*I, and *Sal*I.

3. Methods

3.1. Probe Plasmid Preparation

The nature of the probe region is dependant on the stage of ORE (RRS or TM), as well as the design of the mutation to be deposited. In RRS applications, we often use genomic PCR for the preparation of gene-specific probe regions. For clones of average representation in the λTK library, 250–500 bp of homology is sufficient for high-frequency homologous recombination. Using genomic alignment tools such as Basic Local Alignment and Search Tool (*see* http://130.14.29.110/BLAST/) to score probe regions for identity and repetitive DNA element avoidance is recommended to reduce the chance of false-positives. Design the genomic PCR primers to contain restriction endonuclease sites for simplified πANγ cloning. Note that the fragments may also be cloned blunt into the *Sma*I site of the πANγ multiple cloning site (*see* **Fig. 1A**). Additional sources of probe DNA for RRS include restriction fragments, or contiguous regions of gene-specific cDNA clones (*see* **Note 5**) ***(11)***.

For TM applications using a pure phage population, an overall homology length of 50 bp (25-bp flanking each side of the mutation) is sufficient for recombination ***(6,8)***. It is important to note that this value is unaffected by the size of the intervening insertion or deletion (*see* **Note 6**). Probe regions for small changes such as base pair mutations, amino acid substitutions, splice site modifications, and restriction enzyme recognition site insertions are easily prepared using complementary synthetic oligos. Design oligos to include overhangs, which will simplify the πANγ cloning procedure. For larger insertions (such as selection/disruption cassettes for gene targeting) or deletions, probes may be prepared using long-oligo PCR, such that the homology region is included at the ends of the 5′ and 3′ primers. Alternatively, more complicated mutagenic probe elements may be prepared by ligation or fusion of two PCR amplicons *(9)*. For coincident RRS and TM applications, probes must bear sufficient homology lengths flanking the mutation site to ensure gene-specific clone isolation.

1. Set up a ligation reaction containing 25–50 ng of the appropriately digested πANγ plasmid (*see* **Note 7**), and a threefold molar amount of the probe fragment. Add 2 μL of 5X reaction buffer, and 1 μL of T4 DNA ligase (Invitrogen). Increase the reaction volume to 10 μL with sterile water. Incubate at room temperature for 1–4 h for sticky end or 4–16°C for at least 4 h for blunt end ligation.
2. Add 1 μL of glycogen (20 mg/mL) and increase the total volume to 100 μL. Precipitate the reaction with 10% vol 3 *M* NaOAc and 2 vol 100% ethanol. Wash in 70% ethanol, air-dry briefly, and resuspend the sample in 10 μL of sterile water. Increase the volume to 40 μL with 10% glycerol. Chill on ice.
3. Add 10–20 μL of MC1061(p3) electrocompetent cells (*see* **Note 1**) and transfer to a precooled 0.2-cm-gap cuvet. Electroporate using the following conditions: 2.5 mV, 400 Ω, 25 μFD.
4. Allow the cells to recover through growth for 30–45 min at 37°C (shaking, 260 rpm) in 500 μL of LB medium.
5. Plate the cell suspension on LB-AKT (ampicillin-25 μg/mL; kanamycin-50 μg/mL; tetracycline-12.5 μg/mL) (*see* **Notes 8** and **9**). Grow overnight at 37°C.
6. Circle, number, and pick clones from the plate (*see* **Note 10**). Touch the clones slightly with a 2-μL micropipet tip, and rinse directly into a PCR tube containing 10 μL of sterile water (*see* **Note 11**).
7. Set up the PCR colony screen as follows: master mix (per reaction): 1 μL of dNTPs (10 μ*M* each), 1 μL of each primer (πH + πP or πB + πE; 10 pmol/μL, **Subheading 2.1., step 18**), 2.5 μL of PCR buffer (10X), 0.75 μL of $MgCl_2$ (50 m*M*), 0.25 μL of *Taq* polymerase (Invitrogen), and 8.5 μL of sterile water. Add 15 μL to each colony suspension (final volume 25 μL).
8. Standard PCR screening conditions are: 95°C for 5 min (95°C for 30 s, 56°C for 30 s, and 72°C for 90–120 s) × 30 total cycles, 72°C for 5 min, 4°C hold.
9. Resolve the products on a gel, and highlight positive clones.
10. Repick choice-positives into liquid LB-AKT (3 mL) for overnight culture (37°C, 260 rpm).
11. *Optional*: mini-prep 2 mL of the culture using the Qiagen plasmid mini-prep kit. Confirm the insert by restriction analysis.
12. *Optional*: send 150–200 ng πANγ plasmid DNA for sequencing with the appropriate primers (*see* **Note 12**).
13. Keep the remaining 1 mL of culture for storage of positive clones as frozen permanents (in 7% DMSO).

3.2. Standard Bacteriophage Manipulation Techniques

3.2.1. Preparation of Host Plating Cells

1. Inoculate a 10 mL culture of LB medium in a 50-mL polypropylene tube with an isolated colony of the appropriate host strain, and grow the bacterial culture at 37°C in a shaking incubator (~260 rpm) to an $OD_{600} = 0.8–1$ (~16 h).
2. Pellet the cells by centrifugation at 3000*g* for 10–15 min.
3. Remove the supernatant and resuspend the pellet in 4 mL of 10 m*M* $MgSO_4$.
4. Store the plating cells at 4°C. Plating cells prepared and stored in this fashion may be used for up to 4 wk.

3.2.2. Bacteriophage Plating

1. Incubate 100 µL of appropriate plating cells with 100 µL of the phage suspension in a glass culture tube at room temperature for 15 min.
2. Add 4 mL LBM molten (48°C) top agar, and mix gently.
3. Pour the mixture onto the surface of a prepoured 100-mm LBM bacterial plate, allow the top agar to solidify at room temperature for approx 5 min, and grow inverted at 37°C overnight.

3.2.3. Bacteriophage Plaque Isolation

1. Pick *isolated* plaques with a short glass pipet and pipet bulb. Core the plaques and expel the entire agar plug into 500 µL of SM buffer in a 1.5-mL microfuge tube.
2. Add a few drops (~50 µL) of chloroform.
3. Elute the phage with gentle rocking at room temperature for at least 1 h.

3.2.4. Preparation of a High-Titer Phage Stock

1. Prepare a phage culture in a 50-mL polypropylene tube by incubating 100 µL of plating cells with 5–10% (25–100 µL) of a plaque eluate for 15 min at room temperature.
2. Add 10 mL of LBM medium, and incubate shaking (260 rpm) at 37°C for 6–9 h, or overnight. Clearing of the culture or formation of stringy/flocculent bacterial debris is a good indication of lysis.
3. Add 200 µL chloroform and shake (260 rpm) for 10 min at 37°C to complete lysis and maximize phage yield.
4. Pellet cellular debris at 3000*g* for 20 min. Transfer the supernatant into new 15-mL polypropylene tube and store at 4°C (*see* **Note 13**).

3.2.5. Bacteriophage Titer Determination

1. Prepare a 10-fold dilution series in SM buffer spanning expected range relevant to the stock, which is being quantified (*see* **Note 14**).
2. Plate 100 µL of each phage dilution on the appropriate host strain (usually the unrestrictive host LE392). Note that plating 100 µL acts as an additional 10^{-1} dilution.
3. Count the resulting plaques and multiply by the inverse of the dilution factor to obtain an estimate (*see* **Note 15**). Bacteriophage titers are expressed as plaque-forming units per milliliter (PFU/mL).

3.3. Phage–Plasmid Recombination

The basic principles of phage-plasmid recombination required for both RRS and TM are shared (*see* **Fig. 2**). The differences in the protocols lie primarily in the design of the probe regions (*see* **Subheading 3.1.**), and the phage pool, which is used for primary infection. Owing to these similarities, one common ORE protocol has been provided for both RRS and TM applications. Please take note of the subtle differences as they have been indicated.

1. Subculture the appropriate MC1061(p3; πANγ) strain directly from an isolated colony or the frozen permanent in 2 mL LB-AKT (ampicillin-25 μg/mL; kanamycin-50 μg/mL; tetracycline-12.5 μg/mL) (*see* **Note 8**). Grow shaking (260 rpm) overnight at 37°C.
2. Pellet 1 mL of culture at 8000*g* for 2 min. Remove the supernatant and resuspend the cells in 100 μL of 10 m*M* $MgSO_4$. *Optional:* process the remaining 1 mL of bacterial culture for reanalysis of the plasmid.
3. Mix 100 μL of MC1061(p3; πANγ) cells with 100 μL of the λTK library (for RRS, 3×10^8 PFU/mL) or purified phage clone (for TM, 1×10^5 PFU/mL) in a glass culture tube. Let stand at room temperature for 15 min.
4. Add 4 mL LBM-A (ampicillin-50 μg/mL) liquid medium and grow shaking (260 rpm) at 37°C for 2–6 h (*see* **Note 16**).
5. Add 100 μL of chloroform and shake (260 rpm) for 10 min at 37°C.
6. Remove cellular debris by centrifugation in a 15-mL polypropylene tube at 3000*g* for 20 min. Transfer the supernatant to a new 15-mL tube. Store at 4°C for later screening trials, if required.
7. To screen for plasmid integration, mix 10 μL of lysate (*see* **Note 17**) with 100 μL of LG75 plating cells (*see* **Note 18**) and plate in 4-mL LBM-top agar containing 40 μL each of X-Gal and IPTG (20 mg/mL stock concentration) (*see* **Note 19**).
8. Pick isolated blue phage plaques (*see* **Notes 20–22**).
9. Blue plaques may be screened directly for condensation and plasmid excision by plating on P2392 (*see* **Note 23**). Generally, 10 μL of eluate will be sufficient to provide isolated plaques on P2392.
10. Pick isolated phage clones (*see* **Note 20**), and generate a high-titer stock.

3.4. Bacteriophage DNA Isolation

Two separate protocols are outlined for the preparation of bacteriophage-λ DNA, based on the requirements of downstream applications. DNA prepared on small scale (*see* **Subheading 3.4.1.**) typically yields approx 50 μg of DNA and is of sufficient quality and quantity for multiple restriction digests and sequencing reactions. Phage DNA prepared on large scale (*see* **Subheading 3.4.2.**) is of significantly higher purity and quantity (up to 1 mg), and is typically used for preparation of the final TV for direct applications in murine ES cell electroporations.

3.4.1. Small-Scale DNA Preparation

1. Prepare a 10 mL of high titer stock of bacteriophage (*see* **Subheading 3.2.4.**).
2. Add 15 μL each of RNase and DNase (10 mg/mL stocks) and incubate at 37°C for 30 min.

3. Add 2.5 mL of 40% PEG; 2.5 *M* NaCl. Precipitate at 4°C for at least 2 h. Phage precipitation may be performed overnight.
4. Pellet the phage at 3000*g* for 10 min. Remove the supernatant, invert the tube, and drain well. Wipe the lip of the tube to eliminate any remaining media, and gently resuspend the phage pellet in 500 μL of SM.
5. Extract the suspension once with an equal volume of chloroform to remove excess PEG.
6. Add 5 μL of 10% SDS and 5 μL of 0.5 *M* EDTA. Incubate at 68°C for 15 min.
7. Extract with an equal volume of phenol/chloroform. Repeat until no protein is visible at the interphase (two to three times).
8. Add 50 μL of 3 *M* NaOAc and 1 mL of 100% ethanol. Inversion of the sample here should provide a visible DNA precipitate.
9. Pellet the DNA at 12,000*g* for 5 min. Wash with 70% ethanol, and respin at 12,000*g* for 2 min.
10. Remove all the ethanol, and air-dry. Resuspend in 50 μL 10 m*M* Tris-HCl, pH 8.0 containing 2.5 μg RNaseA.

3.4.2. Large-Scale DNA Preparation

1. Preadsorb phage (8×10^7–1.6×10^8 PFU) to bacteria (1.6×10^{10} cells, or approx 8 mL of LE392 plating cells, *see* **Subheading 3.2.1.**) in a 50-mL tube at room temperature for 15 min. This infection is optimized to a multiplicity of infection of 1:100–1:200 (phage:cell ratio).
2. Add the phage:cell mixture to 400 mL of prewarmed LBM medium and incubate shaking at 37°C, 260 rpm for 6–9 h, or until obvious lysis occurs.
3. Add 1.5 mL of chloroform and 23.4 g of NaCl (final concentration of ~1 *M*). Continue shaking incubation for 10–30 min.
4. Pellet the bacterial debris at 8000*g* for 5–10 min. Transfer the supernatant to a new vessel.
5. Add 40 g of PEG8000 and mix to dissolve. Precipitate the phage particles at 4°C for at least 2 h.
6. Collect the phage particles by centrifugation at 5000*g* for 10 min.
7. Resuspend the phage pellet in 3–4 mL of SM buffer. Wash the vessel with an additional 2 mL of SM buffer and pool in a 15-mL tube.
8. Add an equal volume of chloroform and invert gently to mix. Separate the phases by centrifugation at 3000*g* for 10–15 min. A thick white band of PEG8000 should form at the interphase.
9. Remove the (upper) aqueous phase and 0.75 g of CsCl to each milliliter of phage suspension.
10. Transfer the CsCl mixture to two 5-mL ultracentrifuge tubes, and top up with a 0.75 g/mL of CsCl solution. Ensure the tubes are closely balanced. Spin the tubes at 55,000*g* for 4 h (VTI 65.2 rotor). The phage will appear as a pearl-blue band.
11. Gently remove the phage from the tubes in the smallest volume possible (<1 mL total) through side-puncture with an 18-gauge needle.
12. Dialyze the phage thrice against a 1000-fold volume of TM buffer to remove CsCl.
13. Transfer the dialyzed phage to 1.5-mL microfuge tube, and add 10% of SDS to a final concentration of 0.2% and 0.5 *M* EDTA to a final concentration of 5 m*M*. Incubate at 68°C for 15 min.

14. Extract with an equal volume of phenol/chloroform. Repeat until no protein is visible at the interphase (two to three times).
15. Add 5 *M* of NaCl to a final concentration of 0.2–0.25 *M* and a twofold volume of 100% ethanol. Inversion of the sample here should provide a visible DNA precipitate.
16. Pellet the DNA at 12,000*g* for 5 min. Wash with 70% ethanol, and respin at 12,000*g* for 2 min.
17. Remove all the ethanol, and air-dry. Do not over dry the pellet. Resuspend gently in 200 μL of 10 m*M* Tris-HCl, pH 8.0. Increase the volume to obtain a standard concentration of 1 μg/μL.

3.5. Bacteriophage DNA Analysis

In this section, we suggest a variety of analytical methods, which will be useful for the characterization of genomic library clones or confirmation of mutation deposition during transplacement mutagenesis. RRS isolation of phage from a genomic library will typically provide a series of overlapping clones. To determine the end-points of a particular genomic clone, sequencing of phage DNA with universal primers (*see* **Fig. 1B**) is standard. Characterization of the TM clones may be achieved using phage particle PCR or standard restriction digestion of phage DNA. Finally, we have included our protocol for phage TV preparation for gene targeting experiments.

3.5.1. PCR From Phage Particles

1. Set up a phage PCR screen as follows: master mix (per reaction): 1 μL of dNTPs (10 μ*M* each), 1 μL of each primer (custom design for mutation detection; 10 pmol/μL), 2.5 μL of PCR buffer (10X), 0.75 μL of $MgCl_2$ (50 m*M*), 0.25 μL of *Taq* polymerase (Invitrogen), and 17.5 μL of sterile water.
2. Add 24–1 μL of phage eluate in SM (final volume 25 μL).
3. Standard PCR screening conditions are: 95°C for 5 min (95°C for 30 s, 56°C for 30 s, and 72°C for 90–120 s) × 30 total cycles, 72°C for 5 min, 4°C hold. Adjust according to the specific requirements of the screen.

3.5.2. λTK Phage DNA Sequencing

1. Typically, 1 μg of phage DNA (500 ng–2 μg) is sufficient for strong sequencing reads (*see* **Note 24**).
2. Use the TKL (left arm) and TKR (right arm) primers at standard concentrations for sequencing (3.2 pmol/μL).
3. The specifics of the cycle sequencing reaction will depend on the method used.

3.5.3. Restriction Analysis of Bacteriophage DNA

1. Digest 250–500 ng samples of phage DNA using standard restriction endonuclease conditions (*see* **Note 25**).
2. Before loading the samples, heat to 72°C for 5 min and quick cool on ice to denature the λ-*cos* sites.
3. Separate by gel electrophoresis.

4. To simplify mapping and clone confirmation, compare the restriction patterns with a predicted map clone generated *in silico*, using the phage vector landmarks as reference points (*see* **Fig. 1B**).

3.5.4. Preparation of Final TVs for ES Cell Electroporation

1. Prepare phage DNA on a large scale (*see* **Subheading 3.4.2.**).
2. For each ES cell electroporation, digest 150 μg of TV DNA with either *Not*I or *Sal*I (threefold excess enzyme is recommended) to remove the phage arms (*see* **Fig. 1B**). Ensure that the total volume of the reaction is more than 150 μL. Mix well, as pure phage DNA is concatamerized and quite viscous.
3. Allow the digestion to proceed for at least 3 h at 37°C. The reaction may be performed overnight. The solution should loose viscosity as the phage DNA is cleaved. Extract with an equal volume of phenol:chloroform, and precipitate the digested DNA with 1/10th vol of 3 *M* NaOAc and a twofold volume of 100% ethanol. Wash the pellet well with 70% ethanol. Do not over-dry the DNA pellet.
4. Resuspend the digested DNA in water, phosphate-buffered saline, or an electroporation buffer compatible with your ES cell electroporation conditions.

4. Notes

1. An electrocompetent cell stock is prepared from 500 mL of log-phase bacterial culture (OD_{600} = 0.5–0.8). Chill the culture on ice, pellet the cells, and wash in 500 mL of ice-cold sterile water. Repeat wash. Wash the cells once in 250 mL of ice-cold 10% glycerol. Pellet and resuspend in 2 mL of ice-cold 10% glycerol. Store at –80°C in small (40 μL) aliquots. Although we have exclusively used electroporation for plasmid transformation, it is possible that MC1061(p3) cells may also be made chemically competent and used as such.
2. The amber mutation in the *amp*^{R} gene of the p3 episome is subject to relatively high rates of reversion. Before the generation of electrocompetent MC1061(p3) stocks, it is suggested that isolated colonies selected on kanamycin are screened for antibiotic sensitivity on ampicillin. Additional screening for tetracycline sensitivity is not normally required.
3. The λTK library was constructed using standard methods from mouse genomic DNA isolated from the R1 ES cell line *(11)*. The λTK bacteriophage cloning vector is a derivative of λSyrinx2A *(7)*.
4. The ORE system is applicable to the modification DNA cloned into the most common bacteriophage-λ vectors (e.g., λDASH, λFIX, and λGEM), with following genetic properties required for compatibility: *rap*$^+$; *gam*$^-$; *supF*^{o}; and *lacZ*$^-$. Notable exceptions are λgt10, which is *gam*$^+$ and λgt11, which is *rap*$^-$; *gam*$^+$. In the case of bacteriophage cloning vectors, which do not harbor amber mutations in essential genes, positive selection for πANγ plasmid integration is carried out on the host DK21 (*see* **Note 17**).
5. The use of intron-spanning cDNA fragments for RRS clone isolation is not recommended, as it has the propensity to result in an intronic deletion in the condensed genomic clones.
6. Although the frequency of recombination is not affected by the size of insertions or deletions, the size of these types of mutations is restricted by the packaging limits of

bacteriophage-λ (minimum = 78% or 37,830 bp, maximum = 105% or 50,925 bp, representing a variation of 13,095 bp). Thus, for λTK, we can expect that inserts ranging from 4850–17,945 bp in length may be packaged efficiently. Given an average genomic clone size of 12.3 kb, cassettes up to approx 5.6 kb in length may be easily inserted into these vectors, as well as being able to suffer deletions of up to approx 7.5 kb.

7. Dephosphorylation of the vector with phosphatase is recommended if the cloning scheme is symmetrical.
8. When using tetracycline for selection, ensure that LB and not LBM medium is used. Tetracycline is inactivated by Mg^{2+} ions.
9. Plating 20% and 80% of the cell suspension on two separate plates typically provides sufficient isolated colonies for screening.
10. When screening MC1061(p3;πANγ) clones from LB-AKT plates, we have generally found that abnormally large or small colonies do not harbor either parental or recombinant πANγ plasmid. Such clones are assumed to represent amber reversion events (*see* **Note 2**).
11. During the PCR and gel electrophoresis the selection plate may be kept at 37°C to re-establish colonies, which were overpicked.
12. Sequencing of the insert region is recommended if the probe was generated by genomic PCR.
13. High-titer bacteriophage stocks are stable at 4°C for 1–2 mo without a significant drop in titer. Permanent frozen stocks at –80°C may be prepared by the addition of 7% DMSO to an aliquot of high titer stock or plaque eluate. Note that freeze/thaw of such a stock may reduce the titer by as much as one order of magnitude. For revival, use a sterile toothpick to scrape a small piece of ice into a microfuge containing SM, titer, and reamplify if necessary. Return the remainder of the stock immediately to –80°C.
14. The average titer of a phage culture is dependant on the individual bacteriophage clone and the host used for amplification. As a rule of thumb, average-sized plaques picked into SM typically yield 10^5–10^7 PFU/mL; however, small plaques may produce stocks of only 10^3–10^4 PFU/mL. High titer bacteriophage stocks often yield greater than 10^9 PFU/mL, occasionally as much as 10^{12} PFU/mL.
15. A rapid method to estimate the order of magnitude for a given phage stock is the spot titration. Pour 4 mL of molten LBM top agar containing 100 µL of plating cells onto an LBM plate. Once the top agar has solidified, spot 10 µL of each dilution directly onto the surface. A standard 100-mm plate can hold 6–10 spots. Allow the liquid to adsorb into the solid media, invert, and incubate the plate at 37°C. This method consumes less time and materials than full plate titers; however, it is also less accurate. Note that plating 10 µL acts as an additional 10^{-2} dilution.
16. It is recommended that the length of bacteriophage lytic culture in MC1061 (p3;πANγ) is kept to a minimum. For highly represented phage clones we have managed successful RRS after only 3 h culture time, although 4–6 h is usually optimal. During TM protocols, a culture time of 2 h is usually sufficient to obtain positive clones on LG75. Although we have not personally observed genomic clone rearrangements during phage passage through any ORE hosts, such events are theoretically possible during prolonged exposure to RecA.

17. This value (10 μL) is a standard volume, which acts as a reference for making plating adjustments. The frequency of blue plaque appearance is directly related to the quality and length of the probe, as well as the representation of target phage in the library. Thus, lysate volume adjustments or replica plates may be necessary to obtain the desired clone density. For TM, recombination frequencies are usually 10^{-2}, whereas RRS is more variable, with frequencies ranging from 10^{-5} to 10^{-8}. If desired, approximate recombination frequencies may be determined at this stage by division of the LG75 phage titer by the titer of the total phage population on LE392.
18. *Optional use of DK21*: the DK21 host is required for screening of probe plasmid recombination with nonamber phage vectors (*see* **Note 4**). Because of the absolute requirement of *supF* for phage growth, the use of DK21 may also be applied to reduce the background of nonamber *supF°* phage seen in the λTK library (*see* **Note 20**). This characteristic may also be used to enhance screening for low-level recombination events ***(12)***. Note that bacteriophage-λ plaques on the host DK21 often appear small and weakly stained for β-galactosidase activity. The use of λ-tryptone medium ***(13)*** can somewhat alleviate this phenotype by slowing the growth of the bacterial host; however, we have found that it is not always beneficial or necessary for routine applications.
19. Stock solutions of IPTG and X-Gal may be premixed in the appropriate ratios and stored in a light-tight container at –20°C for at least 2–3 mo with no apparent loss in activity.
20. The number of phage clones to be picked from LG75 or P2392 selection stages differs significantly for RRS and TM applications. As TM begins with a pure phage population, the blue plaques seen on LG75 are homogenous. A single isolate is sufficient at this stage. On the other hand, blue plaques derived from RRS may represent a variety of overlapping phage clones. Choose at least 5–10 plaques to maximize the range of genomic coverage. For P2392 selection, condensed RRS clones from a single blue plaque are homogenous, requiring the selection of only one clone for each P2392 screen. However, TM deposits mutations at a frequency of approx 50% (assuming the homology flanking the mutation is balanced). Pick multiple plaques at this step to screen for mutation transplacement. For coincident clone isolation and mutagenesis using RRS/TM probes, multiple plaques must be picked and screened at each selection stage.
21. Because of the use of a nonirradiated packaging extract during the production of the R1λTK library, a background of *supF°*, nonamber phage appear as white plaques on a LG75 lawn ***(11)***. This issue is λTK library-specific, and may not be seen with libraries constructed using more recent packaging extract preparations. The level of this background may be controlled somewhat by the use of DK21 plating cells (*see* **Note 17**). We recommend that isolated blue plaques from λTK RRS screens are confirmed to be background-free by secondary plating on LG75 (*see* **Note 21**).
22. It is imperative to choose bacteriophage-λ plaques, which are well isolated to reduce clone cross-contamination. It is also good practice, especially if one is not convinced of purity, to perform a dilution and secondary isolation of key phage clones.
23. In the original published ORE applications ***(6,8,9,11,12)***, it was standard procedure to "relax" phage clones *through* passage through the nonrestrictive bacterial host

LE392. We have since found this step to be completely dispensable (and potentially deleterious for RRS protocols). LE392 relaxation has been observed to allow the unchecked passage of low-homology recombinant phage during RRS, stemming from poor RRS probe design. As a rule of thumb, if the direct plating of blue plaques on P2392 does not readily provide viable phage (as occasionally seen with low-frequency isolates on LG75), these phage are most likely not true homologous recombinants. Relaxation through LE392 allows these rogue phage to excise the plasmid and grow to high titer, providing false-positives on P2392.

24. Owing to a need for high-throughput sequencing of clone ends from RRS isolates, we have made efforts to eliminate the requirement for small-scale phage DNA preparation (*see* **Subheading 3.4.1.**). Briefly, circularized phage DNA is prepared as follows: preadsorb 10 μL of LE392 plating cells with phage at 1:1 or 3:1 (phage:cell ratio). Add 1 mL of LBM medium and incubate without shaking at 37°C for 35 min. Collect the cells by centrifugation and perform standard alkaline lysis (100 μL of Qiagen plasmid mini-prep buffers P1, P2, and P3 may be used here). Extract the supernatant once with phenol:chloroform and precipitate DNA by the addition of 2 vol of ethanol. Pellet the DNA precipitate, and resuspend in 1–2 μL of 10 m*M* Tris-HCl (pH 8.0). This phage DNA preparation (0.5 μL) may be used as a template for amplification using commercially available ϕ29 DNA polymerase (TempliPhi Amplification Kit, GE Healthcare, NJ). Note that DNA prepared using standard methods is concatamerized and is therefore an inefficient template for ϕ29 polymerase.
25. For rare cutting enzymes (one to two cuts), 125–250 ng is sufficient for visualization. For intermediate cutters (three to six cuts) use 500 ng, and for more frequent cuts, use 750 ng–1 μg. The panel of restriction endonucleases used for mapping is clone specific; however, for genomic clones we typically suggest *Eco*RI, *Hin*dIII, *Bam*HI, and combinations thereof. If the phage DNA will be used as a gene TV, confirmation of predicted *Sal*I and *Not*I sites at this stage is recommended.

Acknowledgments

We would like to acknowledge the individuals who have assisted at various levels with development of the Orpheus system: Chieko Aoyama, Fiona Mansergh, Kim Melton, Yoshimichi Nakatsu, Sandi Nishikawa, and Brad Thomas. We would also like to thank all of the scientists who have applied the Orpheus system in their own research, providing valuable feedback on its utility and required areas of improvement. This work has been supported through funding from AHFMR, CIHR, NSERC, and Genome Canada.

References

1. Capecchi, M. R. (2005) Gene targeting in mice: functional analysis of the mammalian genome for the twenty-first century. *Nat. Rev. Genet.* **6,** 507–512.
2. King, S. R. and Richardson, J. P. (1986) Role of homology and pathway specificity for recombination between plasmids and bacteriophage lambda. *Mol. Gen. Genet.* **204,** 141–147.
3. Shen, P. and Huang, H. V. (1986) Homologous recombination in *Escherichia coli*: dependence on substrate length and homology. *Genetics* **112,** 441–457.

4. Poteete, A. R., Fenton, A. C., and Wang, H. R. (2002) Recombination-promoting activity of the bacteriophage lambda Rap protein in *Escherichia coli* K-12. *J. Bacteriol.* **184,** 4626–4629.
5. Tsuzuki, T. and Rancourt, D. E. (1998) Embryonic stem cell gene targeting using bacteriophage lambda vectors generated by phage-plasmid recombination. *Nucleic Acids Res.* **26,** 988–993.
6. Unger, M. W., Liu, S. Y., and Rancourt, D. E. (1999) Transplacement mutagenesis. A recombination-based in situ mutagenesis protocol. *Nucleic Acids Res.* **27,** 1480–1484.
7. Lutz, C. T., Hollifield, W. C., Seed, B., Davie, J. M., and Huang, H. V. (1987) Syrinx 2A: an improved lambda phage vector designed for screening DNA libraries by recombination in vivo. *Proc. Natl. Acad. Sci. USA* **84,** 4379–4383.
8. Woltjen, K., Unger, M. W., and Rancourt, D. E. (2002) Transplacement mutagenesis. A recombination-based in situ mutagenesis protocol. *Methods Mol. Biol.* **182,** 189–207.
9. Aoyama, C., Woltjen, K., Mansergh, F. C., Ishidate, K., and Rancourt, D. E. (2002) Bacteriophage gene targeting vectors generated by transplacement. *Biotechniques* **33,** 806–810.
10. Seed, B. (1983) Purification of genomic sequences from bacteriophage libraries by recombination and selection in vivo. *Nucleic Acids Res.* **11,** 2427–2445.
11. Woltjen, K., Bain, G., and Rancourt, D. E. (2000) Transplacement mutagenesis. A recombination-based in situ mutagenesis protocol. *Nucleic Acids Res.* **28,** E41.
12. Thomas, B., Woltjen, K., and Rancourt, D. E. (2003) Deep screening of recombination proficient bacteriophage libraries. *Biotechniques* **34,** 36–38.
13. Kurnit, D. M. and Seed, B. (1990) Improved genetic selection for screening bacteriophage libraries by homologous recombination in vivo. *Proc. Natl. Acad. Sci. USA* **87,** 3166–3169.

7

Transposon-Mediated Mutagenesis in Somatic Cells

Identification of Transposon-Genomic DNA Junctions

David A. Largaespada and Lara S. Collier

Summary

Understanding the genetic basis for tumor formation is crucial for treating cancer. Forward genetic screens using insertional mutagenesis technologies have identified many important tumor suppressor genes and oncogenes in mouse models of human cancer. Traditionally, retroviruses have been used for this purpose, allowing the identification of genes that can cause various forms of leukemia or lymphoma with murine leukemia viruses or mammary cancer with mouse mammary tumor viruses. Recently, the *Sleeping Beauty* transposon system has emerged as a tool for cancer gene discovery in mouse models of human cancer. Transposons mobilized in the mouse soma can insertionally mutate cancer genes, and the transposon itself serves as a molecular "tag," which facilitates candidate cancer gene identification. We provide an overview of some general issues related to use of *Sleeping Beauty* for cancer genetic studies and present here the polymerase chain reaction-based method for cloning transposon-tagged sequences from tumors.

Key Words: Cancer genetics; linker-mediated PCR; mouse transgenesis; *Sleeping Beauty*; somatic mutagenesis; transposon.

1. Introduction

The *Sleeping Beauty* (SB) transposon system is a useful tool for gene delivery or insertional mutagenesis in model genetic organisms (reviewed in **refs. *1–3***). The SB system consists of two parts—the transposon vector DNA and the enzyme that mobilizes this DNA, the transposase. Both the transposon and transposase must be in the same cell for transposition to occur. The original SB transposase enzyme is designated SB10, and was created by reconstruction of an active enzyme gene from defunct, mutated copies of Tc1/*mariner* family transposase genes cloned from various species of Salmonid fish ***(4)***. The minimal transposon vector DNA consists of a left and a right outer inverted repeat/direct repeat element (IRDR) that are the binding sites for the transposase enzyme. Transposons containing the

From: *Methods in Molecular Biology, vol. 435: Chromosomal Mutagenesis*
Edited by: G. Davis and K. J. Kayser © Humana Press Inc., Totowa, NJ

original IRDRs built from sequences from Salmonid fish are designated "pT" vectors ***(4)***. Changes have been introduced into the IRDRs to increase transposition rates, and transposons based on this second generation of IRDRs are designated "pT2" vectors ***(5)***. A variety of insertional mutagenesis applications have been reported for SB including germline mutagenesis in the mouse and zebrafish ***(6–15)***, and most recently, somatic mutagenesis for cancer gene discovery in the mouse ***(16,17)***.

SB-based somatic mutagenesis for cancer gene discovery is accomplished by breeding transposon transgenic mice to transposase transgenic mice to generate transposon/transposase in doubly transgenic mice. Transposons are mobilizing in every somatic cell in which transposase is expressed, mutagenizing cancer genes and promoting tumor formation in these doubly transgenic mice. Once SB-induced tumors have been obtained, high-throughput methods for cloning transposon insertions are necessary to begin identifying candidate cancer genes. This chapter will detail the molecular biology protocols necessary for cloning transposon-DNA genomic junctions for SB transposons based on the "pT" or "pT2" versions of the IRDRs. Although beyond the scope of this chapter, methods must also be in place for automated "trimming" and genome mapping of these junctions. In addition, statistical analysis of insertions must be performed to identify common sites of insertion (CIS), which are regions of the genome that harbor multiple independent transposon insertions from multiple independent tumors. Although in a somatic cell, transposons are theoretically mobilizing to new locations throughout the genome, only those that land in or near a tumor suppressor gene or oncogene should promote tumorigenesis. Nevertheless, in any given tumor cell there will also be "passenger" insertions that have occurred stochastically and are not relevant to tumorigenesis. Statistical analysis must be performed to identify regions of the genome that are mutated by transposon integrations in multiple independent tumors more frequently than expected by random chance. These CISs indicate that a cancer gene is likely to be nearby because of the repetitive selection for transposon integrations in the chromosomal region in multiple different tumors. Statistical analysis of insertions and CIS identification is beyond the scope of this chapter, but we refer readers to several papers that address the topic ***(16–19)***. We describe here a protocol for cloning transposon-genomic DNA junctions that is based on our modifications of methods described by Karl Clark, Adam Dupuy, and the laboratory of Shawn M. Burgess ***(11,16,20)***. A brief description of the methods used to create cancer models using SB precedes this protocol.

1.1. Creating Cancer Models Using SB

1.1.1. Transposon Design

Mobilization of highly mutagenic transposons from chromosomally resident concatomers in the mouse soma has been used to initiate or accelerate tumor

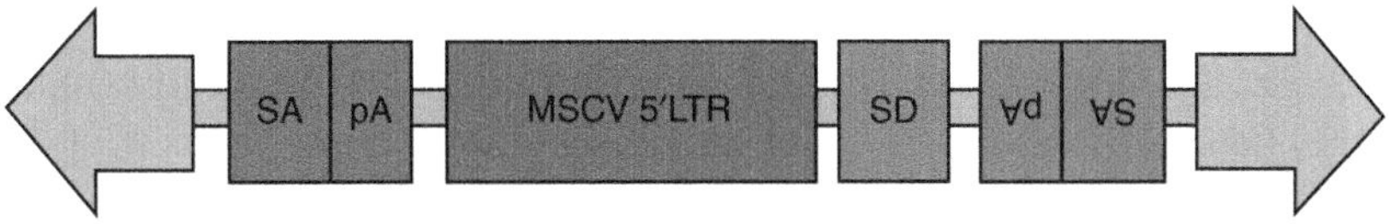

Fig. 1. T2/onc. T2/onc has SA/polyadenylation (pA) sequences in both orientations to generate loss-of-function mutations in genes in which it lands. Between the two SA are sequences from the long terminal repeat of the MSCV LTR, which contains promoter/enhancer elements to promote overexpression of genes near which it lands. Also present is a splice donor (SD) so that transcripts initiated in the MSCV LTR can splice into downstream endogenous exons.

formation ***(16,17)***. The transposon used for these experiments, T2/onc (**Fig. 1**), was introduced into mice by standard pronuclear injection techniques, resulting in a multicopy array, or concatomer, integrated into one place in the genome of the transgenic mice produced. T2/onc is designed with splice acceptors (SA)/polyadenylation sequences in both orientations. On mobilization by transposase, if T2/onc lands in a tumor suppressor gene these elements should disrupt splicing, and therefore generate a loss of function mutation. T2/onc also contains sequences from the murine stem cell virus (MSCV) LTR, which contains promoter/enhancer elements followed by a splice donor that facilitates splicing of transcripts initiated in the MSCV into downstream endogenous exons. Therefore, if T2/onc lands within or near a proto-oncogene it can promote its overexpression. Transgenic lines harboring approx 25 copies of T2/onc in their chromosomal concatomers have been generated and are referenced to as "low copy lines" ***(16)***. Dupuy et al. ***(17)*** modified T2/onc to contain a longer version of one SA and the resulting vector is named T2/onc2 to designate this difference. The transgenic lines that were generated using T2/onc2 have over 100 copies of transposons in their chromosomal concatomers and are therefore referred to as "high copy lines" ***(17)***.

1.1.2. Transposase Design

Two different sources of SB transposase have been used in published work on SB-induced tumors. A CAGGS-SB10 transgene ***(10)*** was used to accelerate sarcomagenesis in *p19Arf–/–*mice ***(16)***. This transgene is expressed from a multicopy concatomer created by standard transgenesis. Although the CAGGS promoter has been described as being active in all cells ***(21)***, we have recently found that the CAGGS-SB10 line expresses transposase primarily in mesenchymal cells. Furthermore, the expression is highly variegated and absent from most epithelial cell types (unpublished data). This may explain why CAGGS-SB10 was able to accelerate sarcoma development in T2/onc, *p19Arf–/–*mice ***(16)***. In a second paper, a transgene for the SB11 version of the transposase was introduced into the *Rosa26* locus ***(22)*** using homologous recombination in mouse embryonic cells ***(17)***. When mice carrying this *Rosa26*-SB11 allele were bred to high copy T2/onc2 transgenic mice, non-Mendelian inheritance was observed as most doubly transgenic mice died *in utero*. Those that survived to birth developed tumors, primarily lymphocytic leukemia, that was caused by T2/onc insertional mutagenesis ***(17)***.

In the future, modifications to the system will likely improve the utility of SB for cancer genetics studies in a wide variety of tissue types. Although beyond the scope of this chapter, we refer readers to several reviews that discuss how the system may be modified for future studies ***(1,23,24)***.

1.1.3. Considerations for Cloning and Sequence Analysis of Tumor-Associated Transposon Insertion Sites

Once the SB system has been used to create a cancer model the next step is to clone, sequence, and analyze a large number of transposon insertion sites. As in studies using slow transforming retroviruses, the goal is to identify CIS. The greater the number of transposon-induced or transposon-accelerated tumors available, the greater the power to identify CISs that are infrequently mutated by transposon insertion. In addition, the closer one can approach "saturation" cloning of all the transposon insertion sites within a tumor, the better the chances are that no CIS will be missed. These considerations apply to both retroviral mutagenesis using murine leukemia virus (MLV), or mammary cancer with mouse mammary tumor virus (MMTV) and to transposon mutagenesis for cancer gene discovery. In order to process a large number of tumor genomic DNAs and to approach saturation for insertion site recovery, we use polymerase chain reaction (PCR)-based methods for amplifying insertion sites. Several reviews discuss high-throughput proviral insertion site cloning and analysis using MLV ***(1,25–27)***.

Two general methods for PCR amplification have been used for cloning transposable element-genomic DNA junction fragments, namely inverse PCR and linker-mediated PCR (reviewed in **ref.** ***28***). For inverse PCR, a transposon/cellular genomic DNA junction fragment is first generated by restriction enzyme digestion, then is circularized by ligation of the restricted tumor DNA, and two rounds of PCR using outward facing primers from the transposon sequence are used to amplify the junction between the transposon and adjacent cellular DNA. The resulting PCR fragment is then cloned into a plasmid vector for sequencing. The circularization step may be inefficient and may bias against recovery of some insertions. Linker-mediated PCR is the most common technique currently used to clone viral or transposon integration sites from tumors. In this method, shown in **Fig. 2**, specially designed linkers are ligated onto restricted tumor genomic DNA, then subjected to two rounds of PCR using transposable element-specific and linker-specific primers before cloning into a plasmid vector for sequencing.

For either inverse PCR or linker-mediated PCR approaches, more complete saturation of insertion site cloning can be achieved by performing reactions that allow the cloning from both the left IRDR (5′ end) as well as the right IRDR (3′ end)

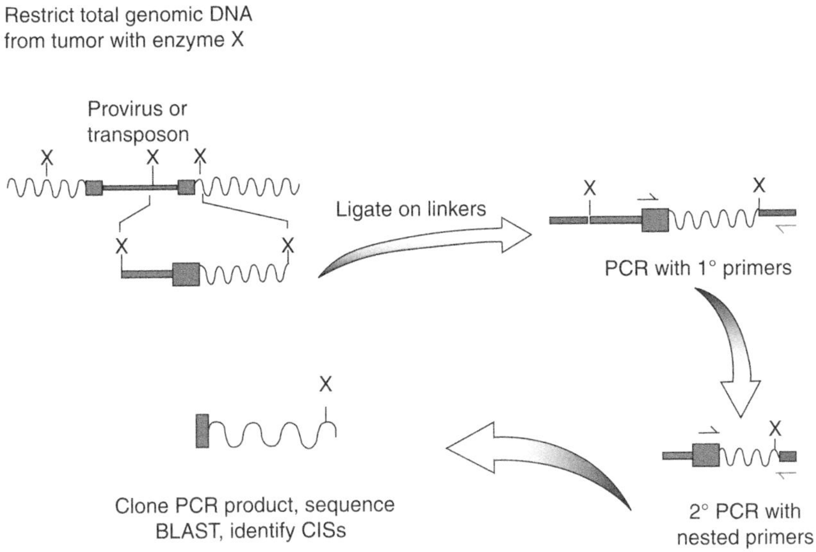

Fig. 2. Outline of linker-mediated PCR for cloning proviral- or transposon-genomic DNA junctions. Linker-mediated PCR involves the generation of transposon (or proviral)-genomic DNA fragments by restriction digest of tumor genomic DNA. Linkers with overhangs complementary to the restriction fragments used are then ligated onto the genomic DNA. Two rounds of PCR with linker and IRDR (or proviral) specific primers are used to amplify the transposon-genomic DNA junction.

of the transposable element. The protocol described here includes instructions for cloning from both ends of the transposon. Using multiple frequent-cutting restriction enzymes to digest the genomic DNA can also increase the number of transposon-genomic DNA junctions by assuring that a restriction enzyme site in adjacent cellular DNA creates a small enough junction fragment to be PCR amplified. Enzymes that frequently cut mammalian genomic DNA (e.g., enzymes that recognize 4 bp sites that lack CpG dinucleotides) are often used for this purpose. In addition, the restriction enzymes that are chosen for this approach must meet certain criteria including cutting close to the end of the transposon, cutting genomic DNA often enough to generate PCR-amplifiable junction fragments, and producing a sticky overhang that can be ligated to the designed double-stranded oligonucleotide linker efficiently. *Bfa*I (for cloning from the left IRDR) and *Nla*III (for cloning from the right IRDR) are two enzymes that we frequently use for this protocol.

Although the basic steps described above are common to methods for cloning MLV, MMTV, and transposon insertions, there are features unique to cloning SB transposon insertions as compared with MLV or MMTV. Most importantly, SB transposon insertions that cause cancer are derived by the mobilization of transposon vectors from a chromosomally resident multicopy concatomer. Therefore, in an SB-induced tumor there are tumor-specific, clonal transposon-insertion sites (anywhere from 5 to 30+ have been observed) and the remaining copies of the transposon vector that reside in the donor concatomer. A method to avoid repeatedly cloning the junction fragment between adjacent copies of the SB transposon vectors in the concatomers is needed in this situation. To achieve this, we have used "blocking" primers (with blocked 3′ ends that prevent polymerase extension) that are complementary to sequences from plasmid DNA that flanks each copy of the SB transposon in the concatomer. However, this method often has limited success. Alternately, after ligation of the linkers to the restricted DNA one can redigest with another restriction enzyme that cuts at least once in the plasmid DNA that flanks each copy of the SB transposon in the concatomers, but does not cut in the SB transposon sequence itself. Thus, the SB transposon junction fragments generated from within the concatomers cannot be amplified as the binding sites for the primers are now separated on two different DNA molecules. Using enzymes that cut more rarely in genomic DNA (e.g., restriction enzymes that have a 6-bp recognition sequence) reduce the possibility that the enzyme will also cut the genomic DNA between the end of the IRDR and the enzyme used for the first digest of genomic DNA. We currently use *Bam*HI (for cloning from the left IRDR) and *Xho*I (for cloning from the right IRDR) for this purpose. Although complex, the basic method described in **Subheadings 3.2.** to **3.4.** can be used to clone insertion sites from any application using a SB vector that uses "pT" or "pT2"-based IRDR. However, we recommend that the reader verify that the IRDR primer sequences provided are present in the transposon vector used. In addition, the restriction enzyme used for the second digestion (to prevent amplification of transposon junction fragments from within the donor concatomers) will depend on the sequence of the plasmid backbone sequences that were left linked to the SB transposon vector when the transgenic line was made. The protocol described below will work for tumors generated using the T2/onc ***(16)*** and T2/onc2 ***(17)*** transgenic lines. Whatever the method used to PCR amplify the junction fragments, by "shot-gun" cloning these PCR products and sequencing 96 or more clones per tumor, it is likely that even rare insertion sites will be identified, again contributing to saturation cloning.

2. Materials

2.1. DNA Preparation

1. 1 *M* Tris-HCl (pH 8.0).
2. 0.5 *M* Ethylenediamine tetra acetic acid (EDTA) (pH 8.0).
3. 10% Sodium dodecyl sulfate in distilled water (caution—dust is an irritant).
4. 5 *M* NaCl.
5. TE buffer: 4 mL of 1 *M* Tris-HCl (pH 8.0), 0.08 mL of 0.5 *M* EDTA (pH 8.0) to 400 total in distilled water.
6. Proteinase K: dissolved in TE at 10 mg/mL, incubated at 37°C for 1 h before storing in aliquots at –20°C.

Oligo name	Oligo sequence	Modifications	Notes
Bfa linker+	GTAATACGACTCACTATAG GGCTCCGCTTAAGGGAC	None	–
Bfa linker–	TAGTCCCTTAAGCGGAG	5′ Phosphate, 3′ amino group	*See* Note 12
NlaIII linker+	GTAATACGACTCACTATA GGGCTCCGCTTAAGGG ACCATG	None	–
NlaIII linker–	GTCCCTTAAGCGGAGCC	5′ Phosphate, 3′ amino group	*See* Note 12
IR/DR(R)KJC1	CCACTGGGAATGTGATGA AAGAAATAAAAGC	None	–
Long IR/DR (R)	GCTTGTGGAAGGCTACT CGAAATGTTTGACCC	None	–
Long IR/DR (L2)	CTGGAATTTTCCAAGCTG TTTAAAGGCACAGTCAAC	None	–
New L1	GACTTGTGTCATGCACAA AGTAGATGTCC	None	–
Linker primer	GTAATACGACTCACTATAG GGC	None	–
Linker nested primer	AGGGCTCCGCTTAAGGGAC	None	–

7. STE buffer: 1.6 mL of 0.5 *M* EDTA, 8 mL of 1 *M* Tris-HCl (pH 8.0), 16 mL of 5 *M* NaCl to 800 mL then autoclaved.
8. Tissue lysis buffer (for one tumor): 4.6 mL of STE, 100 μL of 0.5 *M* EDTA, 100 μL of proteinase K, and 200 μL of 10% sodium dodecyl sulfate made fresh before using.
9. RNase A: dissolved in TE at 2 mg/mL, incubated at 37°C for 1 h before storing in aliquots at –20°C.
10. pH 7.9-buffer saturated phenol (caution—hazardous).
11. Chloroform (caution—hazardous).
12. 95% Ethanol.
13. Glass rods for spooling DNA (we use Kimble melting point capillary tube part no. 34505–99).

2.2. DNA Digestion and Linker Annealing

1. Restriction enzymes: *Nla*III, *Xho*I, *Bfa*I, and *Bam*HI (New England Biolabs Ipswich, MA) (*see* **Note 1**). The appropriate 10X buffer and 100X bovine serum albumin, if required, for each enzyme is provided with the enzyme.
2. PCR reaction purification kit (QIAquick PCR Purification Kit by Qiagen, Maryland).
3. Linker oligonucleotides (*see* **Table 1** for sequences). Dissolve primers in Millipore purified or PCR grade water at 100 μ*M* and store at –20°C.
4. T4 DNA ligase supplied with 5X ligase buffer (Invitrogen, Carlsbad, CA) stored at –20°C.

2.3. PCR Reactions

1. Platinum *Taq* supplied with 10X PCR buffer and 50 m*M* $MgCl_2$ (Invitrogen).
2. dNTPs: 2.5 m*M* each for a total concentration of 10 m*M* in PCR grade water. Aliquot and store at –20°C.
3. PCR primers (*see* **Table 1**) dissolved in Millipore purified or PCR grade water at 25 μ*M*.
4. Agarose.
5. Ethidium bromide.
6. 50X TAE: 242.2 g of Tris Base, 100 mL of 0.5 *M* EDTA (pH 8.0), 57.1 mL of glacial acetic acid (caution—hazardous, use in fume hood) brought up to 1 L with Millipore water.
7. 100-bp DNA ladder.

2.4. Cloning of PCR Products

1. TA-based cloning vector kit (*see* **Note 2**).
2. Bacteria for transformation (store at –70 to –80°C) (*see* **Note 3**).
3. Luria Bertani (LB): 10 g of tryptone, 5 g of yeast extract, and 5 g of sodium chloride brought up to a liter in distilled water and autoclaved.
4. LB plates containing the appropriated antibiotic for selection for the TA cloning vector.

3. Methods

3.1. Preparation of Genomic DNA (See *Notes 4 and 5*)

1. Tumors isolated from mice are snap frozen in liquid nitrogen and stored at –70 to –80°C until ready to be processed.
2. Approximately 100–200 mg of tumor tissue is dissociated by douncing in 5 mL of tissue lysis buffer. Samples are incubated overnight in 15-mL conical polypropylene tubes (*see* **Note 6**) at 55°C with shaking to allow complete proteinase K digestion of samples. If smaller tumor samples are available, decrease digestion volume.
3. Allow tubes to cool slightly before addition of 50 μL of RNase A. Incubate at 37°C for 1 h to remove contaminating RNA. After this stage, samples may be stored at –20°C for several days.
4. Add an equal amount of buffered phenol (caution—hazardous, use in fume hood) to samples. Samples should be vortexed before centrifugation in a clinical centrifuge for 5 min.
5. The aqueous layer (top layer) is removed into a new polypropylene conical tube. Take care not to remove the fluffy precipitate that often forms at the junction of the aqueous and organic layers. An equal volume of 1:1 phenol:chloroform (caution—hazardous, use in fume hood) is then added, the sample is vortexed and centrifuged as earlier.
6. The aqueous layer should be essentially clear at this stage. If it is not, repeat **step 5**. If clear, remove aqueous layer to new tube taking care not to disturb the organic/aqueous interface. Add an equal volume of chloroform (caution—hazardous, use in fume hood), vortex and centrifuge.

7. Remove aqueous phase to a new tube and precipitate DNA by adding approx 3 vol 95% ethanol and mixing gently but well by inverting tube several times.
8. (Optional): store samples for a few hours at –20°C to enhance DNA precipitation.
9. For most samples a stringy or fluffy white pellet of precipitated DNA will be visible. This precipitate can be spooled on a glass rod and placed into a 1.5-mL centrifuge tube and allowed to air-dry. If no precipitate is visible, DNA can be precipitated by centrifugation in a clinical centrifuge. The supernatant should be removed carefully to prevent dislodging of the pellet and the pellet should be air-dried.
10. When dry, pellet should be resuspended in TE. For small or not easily visible pellets, use 50 μL of TE, large pellets may require 200–300 μL of TE.
11. Incubate at 37°C overnight to allow DNA to go into solution. If still viscous, additional TE should be added and sample should be reincubated. Vortexing is not recommended as it may shear DNA.
12. DNA concentration is determined using a spectrophotometer and DNA should be stored long-term at –20°C. Aliquoting is also recommended if the DNA is to be used for additional purposes such as Southern blotting.

3.2. DNA Digestion and Linker Ligation (See *Note 7*)

1. Linker annealing. Mix in an Eppendorf tube 50 μL each of *Bfa* linker+ and linker– (for left) or *Nla* linker+ and linker– (for right), then add 2 μL of 5 *M* NaCl. Incubate the tubes in a heating block at 95°C for 5 min, then turn the heating block off and allow them to anneal by slow cooling. Annealed linkers can be stored at –20°C.
2. First restriction digest. Approximately 2 μg of DNA is digested with *Bfa*I (for cloning off the left IRDR) or *Nla*III (for cloning off the right IRDR). Digests are performed at 37°C with 10 U of enzyme in the appropriate buffer in 50 μL reaction volume (to ensure appropriate dilution of the glycerol that the enzymes are stored in).
3. DNA purification. DNA is purified from the reaction buffer using the QIAquick PCR purification kit from Qiagen following the directions provided by the kit. There is one modification to the protocol—DNA is eluded from column using 30 μL of Millipore purified water.
4. Linker ligation. Linkers are ligated onto the DNA fragments in a 20 μL reaction that contains: 6 μL of annealed linkers (*Bfa* for left or *Nla*III for right), 8 μL of digested DNA (*Bfa* for left or *Nla*III for right), 4 μL of 5X ligase buffer, and 2 μL of T4 DNA ligase (Invitrogen). Incubate at 16°C for 6 h to overnight.
5. DNA purification is the same as **step 2** above.
6. Second restriction digest. Linker-ligated DNA fragments are digested with *Bam*H1 (left) or *Xho*I (right) in 50 μL of total reaction buffer. Use 22 μL of purified DNA from **step 5** (essentially all the eluate that is recovered), 10 U enzyme, and the appropriate buffers. Incubate at 37°C for 3 h to overnight (*see* **Note 8**).
7. DNA purification is the same as **step 2** above.

3.3. Polymerase Chain Reaction (See *Note 9*)

1. Primary PCR reaction conditions (*see* **Notes 10** and **11**):
   a. For one reaction:

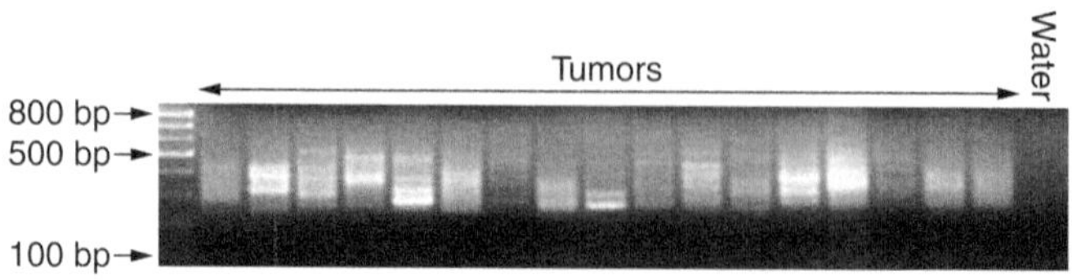

Fig. 3. A typical linker-mediated PCR result on leukemias induced by T2/onc insertional mutagenesis. Most successful linker-mediated PCR reactions produce a "smear" of PCR products, although some individual bands can be discerned.

10X PCR buffer	5 μL
50 m*M* $MgCl_2$	2 μL
10 m*M* dNTPs	1 μL
Primary IRDR primer (long IRDR [L2] for left, long IRDR [R] for right)	0.5 μL
Linker primer	0.5 μL
Platinum *Taq*	0.25 μL
PCR grade water	37.75 μL
Purified, linker-ligated DNA	3 μL

2. Primary PCR thermocycler conditions:
   a. 94°C for 2 min.
   b. 25 cycles of 94°C for 15 s, 60°C for 30 s, and 72°C for 90 s.
   c. 72°C for 5 min.
   d. 4°C hold.
3. For secondary PCR, the primary PCR product is first diluted 1:75 with water. In addition, the volume for secondary PCR is doubled compared with primary PCR to allow half the reaction to be visualized on an agarose gel and half to be saved for cloning.
4. Secondary PCR reaction conditions:

10X PCR buffer	10 μL
50 m*M* $MgCl_2$	4 μL
10 m*M* dNTPs	2 μL
Nested IRDR primer (new L1 for left, KJC1 for right)	1 μL
Linker nested primer	1 μL
Platinum *Taq*	0.5 μL
PCR grade water	75.5 μL
Diluted primary PCR product	6 μL

5. Primary PCR thermocycler conditions are the same as for primary PCR (**step 2** above).
6. Half of each PCR reaction is run on an approx 2% of agarose gel in 1X TAE containing ethidium bromide for visualization (caution ethidium bromide is a suspected carcinogen) using a 100-bp ladder for visualization. Linker-mediated PCR usually results in a smear or products, with occasional distinct bands visible (*see* **Fig. 3**).

7. The remaining half of each reaction is purified using QIAquick PCR columns for cloning. Again, purified DNA is eluted into 30 μL of Millipore purified water.

3.4. Cloning of PCR Products

1. Ligate purified PCR products into the appropriate TA cloning vector following manufacturers directions. We use the maximum volume of purified PCR in the ligation reaction as possible. For example, for pGEMT Teasy, the ligation reaction consists of 1 μL vector, 2 μL of 5X ligase buffer (Invitrogen), 6 μL of purified PCR product, and 1 μL of T4 DNA ligase (Invitrogen) and is allowed to incubate overnight at 16°C.
2. Transform bacteria with approximately half of the ligation (the rest can be stored at –20°C for later use if necessary) and plate onto LB plates containing the appropriate antibiotic for the cloning vector used.
3. For sequencing of a small number of PCR products, mini-preps can be performed by hand to isolate plasmid DNA for sequencing. For high-throughput sequencing, colonies are often grown in 96-well plates in preparation for automated plasmid purification.

4. Notes

1. The manufacturer recommends storage of *Nla*III and *Bfa*I at –70°C. We also recommend aliquoting these enzymes to prevent multiple rounds of freeze/thaw.
2. Any vector that allows the ligation of the 3′ A overhangs generated by *Taq* can be used. pCR®4-TOPO® (Invitrogen) is advantageous because self-ligated vectors restore a lethal *Escherichia coli* gene and therefore, cannot be recovered. We have also had success with cloning into pGEMT-easy (Promega, Madison, WI) followed by performing blue–white selection on 5-bromo-4-chloro-3-indolyl-β-D-galactopyranoside plates to detect plasmid-containing ligated PCR product inserts. In either case, ligation of insert and vector should be performed according to manufacturer's recommendations.
3. Some TA-cloning kits include bacteria for transformation. For example, TOPO cloning kits from Invitrogen include TOP10 competent cells. When not supplied, we use chemically competent DH5-α for routine cloning of PCR products. When large-scale sequencing of cloned products is required, we use electrocompetent cells such as ElectroMAX™ DH10B™ from Invitrogen as they have higher transformation efficiency. Follow the transformation directions supplied by the manufacturer.
4. Traditionally, our laboratory has used phenol/chloroform extraction followed by ethanol precipitation to purify genomic DNA from snap frozen tumor tissues digested with proteinase K as described here. We find that the protocol described here is the most consistent at isolating high-quality DNA. All steps utilizing phenol and chloroform must be carried out in an appropriately ventilated fume hood with appropriate safety precautions. Waste must be treated according to local regulations. However, for DNA extraction we have also had success using protocols that use "salting out" methods for protein precipitation followed by isopropanol precipitation of DNA from the resulting supernatant. We have not had great success using commercially available "spin column" type methods for preparing genomic DNA for linker-mediated PCR.
5. We take great care in handling reagents and samples to prevent cross-contamination. This involves using disposable single use plastics or glass pipets when possible, and using aerosol-resistant pipet tips (also known as "filter" tips).

6. Use polypropylene tubes and not polystyrene as polystyrene will not be resistant to the organic phenol used in later stages.
7. Cloning of junctions from both the right and left sides of the transposon requires two separate reactions for each tumor sample. Although the reactions are technically similar, they require different restriction enzymes, linkers, and PCR primers. Details are given for both left and right reactions.
8. It is very important that this digest goes to completion to prevent the amplification of the T2-vector junction that is present in every transposon remaining in the concatomer. Therefore, overnight incubations are recommended.
9. To prevent contamination of PCR with T2/onc plasmids from the laboratory, we perform PCR in a dedicated PCR hood equipped with an ultraviolet lamp. We have dedicated PCR pipets, filter tips, and water that do not leave the PCR hood. In addition, we expose our PCR tubes, PCR buffer, and $MgCl_2$ solutions to ultraviolet light for 30 min before PCR.
10. It is often convenient to make a PCR "master-mix" that contains everything but the DNA, which is then aliquoted into the individual PCR tubes before DNA addition.
11. In addition to the DNA samples, we always include a PCR reaction that contains only water to control that no plasmid contamination is occurred.
12. Linker-primers are 5′ phosphorylated and 3′ amino modified. The phosphate modification is necessary for ligation onto digested genomic DNA. The 3′ amino group prevents *Taq* from extending off the 3′ end of the linker and copying the region of nonhomology in linker+.

Acknowledgments

This work was supported by grants R21 CA118600 and R01 CA113636 from the National Cancer Institute (to DAL) and K01 CA122183 from the National Cancer Institute (to LSC).

References

1. Collier, L. S. and Largaespada, D. A. (2005) Hopping around the tumor genome: transposons for cancer gene discovery. *Cancer Res.* **65,** 9607–9610.
2. Ivics, Z. and Izsvak, Z. (2006) Transposons for gene therapy! *Curr. Gene Ther.* **6,** 593–607.
3. Miskey, C., Izsvak, Z., Kawakami, K., and Ivics, Z. (2005) DNA transposons in vertebrate functional genomics. *Cell. Mol. Life Sci.* **62,** 629–641.
4. Ivics, Z., Hackett, P. B., Plasterk, R. H., and Izsvak, Z. (1997) Molecular reconstruction of Sleeping Beauty, a Tc1-like transposon from fish, and its transposition in human cells. *Cell* **91,** 501–510.
5. Geurts, A. M., Yang, Y., Clark, K. J., et al. (2003) Gene transfer into genomes of human cells by the sleeping beauty transposon system. *Mol. Ther.* **8,** 108–117.
6. Geurts, A. M., Collier, L. S., Geurts, J. L., et al. (2006) Gene Mutations and Genomic Rearrangements in the Mouse as a Result of Transposon Mobilization from Chromosomal Concatemers. *PLoS Genet.* **2,** E156.

7. Keng, V. W., Yae, K., Hayakawa, T., et al. (2005) Region-specific saturation germline mutagenesis in mice using the Sleeping Beauty transposon system. *Nat. Methods* **2,** 763–769.
8. Balciunas, D., Davidson, A. E., Sivasubbu, S., et al. (2004) Enhancer trapping in zebrafish using the Sleeping Beauty transposon. *BMC Genomics* **5,** 62.
9. Davidson, A. E., Balciunas, D., Mohn, D., et al. (2003) Efficient gene delivery and gene expression in zebrafish using the Sleeping Beauty transposon. *Dev. Biol.* **263,** 191–202.
10. Dupuy, A. J., Fritz, S., and Largaespada, D. A. (2001) Transposition and gene disruption in the male germline of the mouse. *Genesis* **30,** 82–88.
11. Dupuy, A. J., Clark, K., Carlson, C. M., et al. (2002) Mammalian germ-line transgenesis by transposition. *Proc. Natl. Acad. Sci. USA* **99,** 4495–4499.
12. Carlson, C. M., Dupuy, A. J., Fritz, S., Roberg-Perez, K. J., Fletcher, C. F., and Largaespada, D. A. (2003) Transposon mutagenesis of the mouse germline. *Genetics* **165,** 243–256.
13. Horie, K., Kuroiwa, A., Ikawa, M., et al. (2001) Efficient chromosomal transposition of a Tc1/mariner-like transposon Sleeping Beauty in mice. *Proc. Natl. Acad. Sci. USA* **98,** 9191–9196.
14. Horie, K., Yusa, K., Yae, K., et al. (2003) Characterization of Sleeping Beauty transposition and its application to genetic screening in mice. *Mol. Cell. Biol.* **23,** 9189–9207.
15. Sivasubbu, S., Balciunas, D., Davidson, A. E., et al. (2006) Gene-breaking transposon mutagenesis reveals an essential role for histone H2afza in zebrafish larval development. *Mech. Dev.* **123,** 513–529.
16. Collier, L. S., Carlson, C. M., Ravimohan, S., Dupuy, A. J., and Largaespada, D. A. (2005) Cancer gene discovery in solid tumours using transposon-based somatic mutagenesis in the mouse. *Nature* **436,** 272–276.
17. Dupuy, A. J., Akagi, K., Largaespada, D. A., Copeland, N. G., and Jenkins, N. A. (2005) Mammalian mutagenesis using a highly mobile somatic Sleeping Beauty transposon system. *Nature* **436,** 221–226.
18. Mikkers, H., Allen, J., Knipscheer, P., et al. (2002) High-throughput retroviral tagging to identify components of specific signaling pathways in cancer. *Nat. Genet.* **32,** 153–159.
19. Wu, X., Luke, B. T., and Burgess, S. M. (2006) Redefining the common insertion site. *Virology* **344,** 292–295.
20. Wu, X., Li, Y., Crise, B., and Burgess, S. M. (2003) Transcription start regions in the human genome are favored targets for MLV integration. *Science* **300,** 1749–1751.
21. Okabe, M., Ikawa, M., Kominami, K., Nakanishi, T., and Nishimune, Y. (1997) 'Green mice' as a source of ubiquitous green cells. *FEBS Lett.* **407,** 313–339.
22. Soriano, P. (1999) Generalized lacZ expression with the ROSA26 Cre reporter strain. *Nat. Genet.* **21,** 70–71.
23. Dupuy, A. J., Jenkins, N. A., and Copeland, N. G. (2006) Sleeping beauty: a novel cancer gene discovery tool. *Hum. Mol. Genet.* **15(Spec No. 1),** R75–R79.
24. Starr, T. K. and Largaespada, D. A. (2005) Cancer gene discovery using the Sleeping Beauty transposon. *Cell Cycle* **4,** 1744–1748.

25. Uren, A. G., Kool, J., Berns, A., and van Lohuizen, M. (2005) Retroviral insertional mutagenesis: past, present and future. *Oncogene* **24,** 7656–7672.
26. Neil, J. C. and Cameron, E. R. (2002) Retroviral insertion sites and cancer: fountain of all knowledge? *Cancer Cell* **2,** 253–255.
27. Akagi, K., Suzuki, T., Stephens, R. M., Jenkins, N. A., and Copeland, N. G. (2004) RTCGD: retroviral tagged cancer gene database. *Nucleic Acids Res.* **32,** D523–D527.
28. Hui, E. K., Wang, P. C., and Lo, S. J. (1998) Strategies for cloning unknown cellular flanking DNA sequences from foreign integrants. *Cell. Mol. Life Sci.* **54,** 1403–1411.

8

Insertional Mutagenesis of the Mouse Germline With *Sleeping Beauty* Transposition

Junji Takeda, Zsuzsanna Izsvák, and Zoltán Ivics

Summary

Efficient linking of primary DNA sequence information to gene functions in vertebrate models requires that genetic modifications and their effects are analyzed in an efficacious, controlled, and scalable manner. Thus, to facilitate analysis of gene function, new genetic tools and strategies are currently under development. Transposable elements, by virtue of their inherent ability to insert into DNA, can be developed into useful tools for chromosomal manipulations. Mutagenesis screens based on transposable elements have numerous advantages as they can be applied in vivo and are therefore phenotype-driven, and molecular analysis of the mutations is straightforward. Current progress in this field indicates that transposable elements will serve as indispensable tools in the genetic toolkit of vertebrate models. Here, we provide experimental protocols for the construction, functional testing, and application of the *Sleeping Beauty* transposon for insertional mutagenesis of the mouse germline.

Key Words: Functional genomics; gene trap; insertional mutagenesis; poly A-trap; transgenesis; transposon.

1. Introduction

DNA transposons are natural, nonviral "vehicles" that are able to move a defined DNA segment from one genetic location to another. Transposons have been successfully used in lower metazoan model species and in plants for transgenesis and insertional mutagenesis, but until the reactivation of the *Sleeping Beauty* (SB) transposon system in 1997 *(1)*, there was no indication of DNA-based transposons sufficiently active for these purposes in vertebrates. SB exhibits high transpositional activity in a variety of vertebrate-cultured cell lines *(2)*, embryonic stem cells *(3)*, and in both somatic *(4)* and germline *(5–9)* cells of the mouse and rat *(10)* in vivo. Therefore, SB is a valuable tool for functional genomics in several model organisms (reviewed in **ref.** ***11***).

From: *Methods in Molecular Biology, vol. 435: Chromosomal Mutagenesis*
Edited by: G. Davis and K. J. Kayser © Humana Press Inc., Totowa, NJ

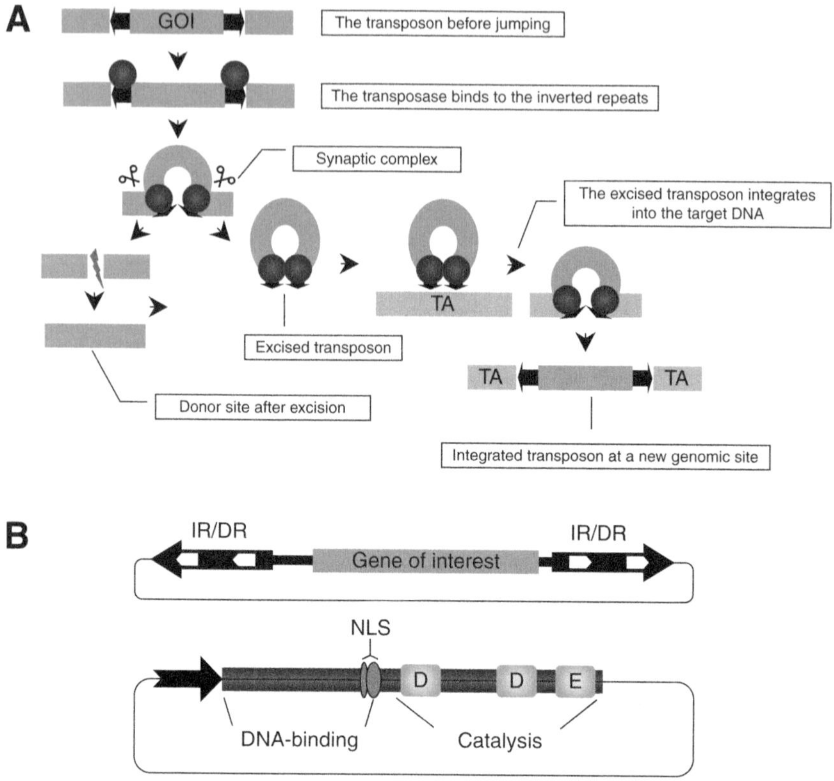

Fig. 1. The SB transposon system. **(A)** Mechanism of SB transposition. The transposable element carrying a GOI (orange box) is maintained and delivered as part of a DNA vector (blue DNA). The transposase (purple circle) binds to its sites within the transposon IRs (black arrows). Excision takes place in a synaptic complex. Excision separates the transposon from the donor DNA, and the double-strand DNA breaks that are generated during this process are repaired by host factors. The excised element integrates into a TA dinucleotides site in the target DNA (green DNA) that will be duplicated and will be flanking the newly integrated transposon. **(B)** Components and structure of a two-component gene transfer system based on SB. A GOI (orange box) to be mobilized is cloned between the terminal (IR/DR, black arrows) that contain binding sites for the transposase (white arrows). The transposase gene (purple box) is physically separated from the IR/DRs, and is expressed in cells from a suitable promoter (black arrow). The transposase consists of an N-terminal DNA-binding domain, a nuclear localization signal, and a catalytic domain characterized by the D aspartic acid and glutamic acid signature.

SB transposes through a conservative, cut-and-paste mechanism, during which the transposable element is excised from its original location by the transposase, and is integrated into a new location **(Fig. 1A)**. SB represents a two-component gene transfer vector system consisting of a transposase protein and a gene-of-interest (GOI) cloned between the terminal inverted repeats (IRs) containing binding sites for the transposase **(Fig. 1B)**. This enables the generation of transgenic stocks,

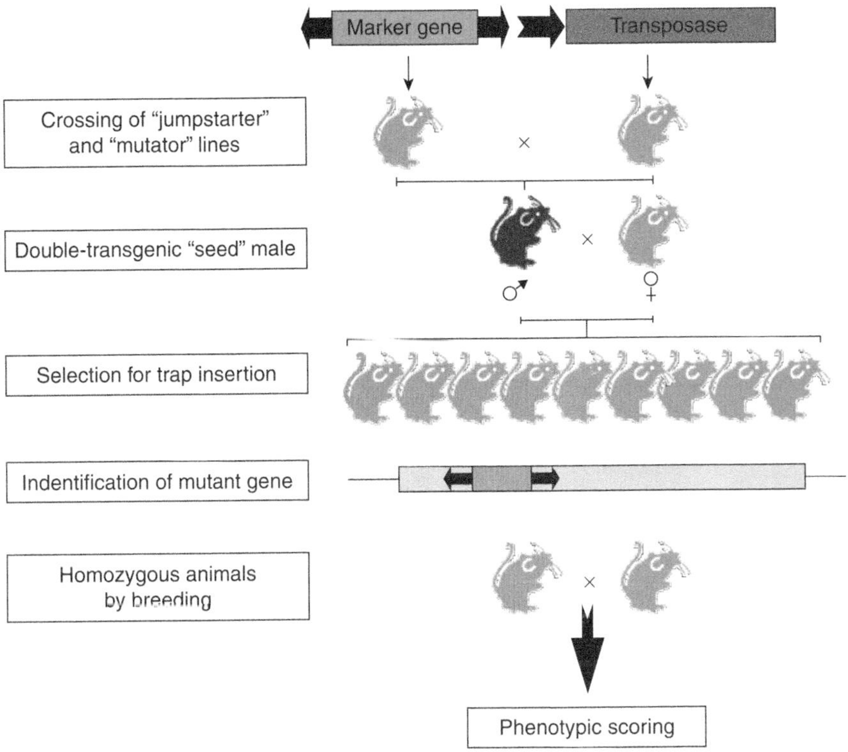

Fig. 2. In vivo germline mutagenesis of the mouse with transposable elements. Breeding of "jumpstarter" and "mutator" stocks induces transposition in the germline of double-transgenic "seed" males. The transposition events that take place in germ cells are segregated in the offspring. Animals with transposition events need to be bred to homozygosity in order to visualize the phenotypic effects of recessive mutations. Mutant genes can easily be cloned by different PCR methods making use of the inserted transposon as a unique sequence tag.

each containing a separate component of the binary transposon system in its genome: one component, encoding the transposase, is carried by the "jumpstarter" strain, which on intercrossing, efficiently mobilizes the second component, a nonautonomous transposon in the genome of the "mutator" strain **(Fig. 2)** *(12)*. Most transposon-based experimental strategies in vertebrates have been utilizing this two-component, binary approach in which transposition is controlled by *trans*-supplementation of the transposase. This experimental setup is especially useful for directing transposition events to particular tissues or organs by tissue-specific promoters driving transposase expression. Importantly, once integrated, transposase-deficient nonautonomous transposons are stable in the absence of the transposase.

Insertional mutagenesis using engineered transposable elements can be one of the most productive and versatile approaches toward disrupting and manipulating genes on a genome-wide scale. Transposon insertion into a gene can itself be mutagenic, if the insertion disrupts the transcriptional regulatory or coding region of a gene. However, most intronic insertions are not expected to be mutagenic (**Fig. 3A**). Thus, several features to enhance the mutagenicity as well to add reporting capabilities of insertional vectors by trapping transcription units were developed; these are summarized in **Fig. 3**. Gene trapping (often referred to as promoter or exon trapping) is based on the activation of a promoter-less reporter gene whose expression is dependent on splicing between the exons of the trapped gene and a splice acceptor (SA) site carried by the transposon (**Fig. 3B**). Thus, gene trap vectors both report the insertion of the transposon into an expressed gene, and have a mutagenic effect by truncating the transcript through imposed splicing.

More sophisticated vectors have been developed that contain a polyadenylation (polyA) trap cassette that reports insertion into a Pol II transcription unit (**Fig. 3C**). The marker gene lacks a polyA signal (pA), but contains a splice donor (SD) site. Thus, when integrating into an intron, a fusion transcript can be synthesized consisting the marker and the downstream exons of the trapped gene. Because polyA trap cassettes have their own promoters, they can report the insertion into genes irrespective of their expression status in a given cell type. The gene trap and polyA trap cassettes can be combined. Such dual tagging systems (**Fig. 3D**) allow the isolation of both upstream and downstream fusion transcripts of the trapped gene, and have been used in the mouse ***(13,14)***.

The mutagenicity of gene trap vectors is higher than that of simple insertional vectors, and they enable easy identification of the hit gene by reverse transcriptase-polymerase chain reaction (PCR) targeting the composite transcripts made up by sequences of the insertional vector and the endogenous gene. In cell culture, drug resistance markers are generally in use whereas in animal systems reporters offer the possibility to visualize spatial and temporal expression patterns of the mutated genes by using *LacZ* or fluorescent proteins (**Fig. 3**). SB can be equipped with gene trap cassettes ***(8,9,15)***, which significantly enhances its utility as a tool for functional genomics in vertebrate models. Furthermore, similar to the GAL4 transcriptional factor and its upstream activating sequences system in *Drosophila*, a conditional, tetracycline-regulated system has been shown to be applicable to transposon-mediated insertional mutagenesis in mice ***(16)***.

SB has been successfully used for forward genetics approaches in the mouse. Double transgenic mouse lines were generated bearing chromosomally present transposons and either an ubiquitously ***(6–9)*** or male germline-specifically ***(5)*** expressed transposase gene. Segregating the transposition events by mating the founder males to wild-type females revealed that up to 90% of the progeny can carry transposon insertions ***(7)***, and a single sperm of a founder can contain, on average, two insertion events ***(6)***. The germline of such a founder was estimated to harbor approximately 10,000 different mutations ***(8)***. Importantly, transposition of gene trap transposons identified mouse genes with ubiquitous and tissue-specific expression patterns, and mutant/lethal phenotypes were easily obtained by generating homozygous animals ***(8,9,17)***.

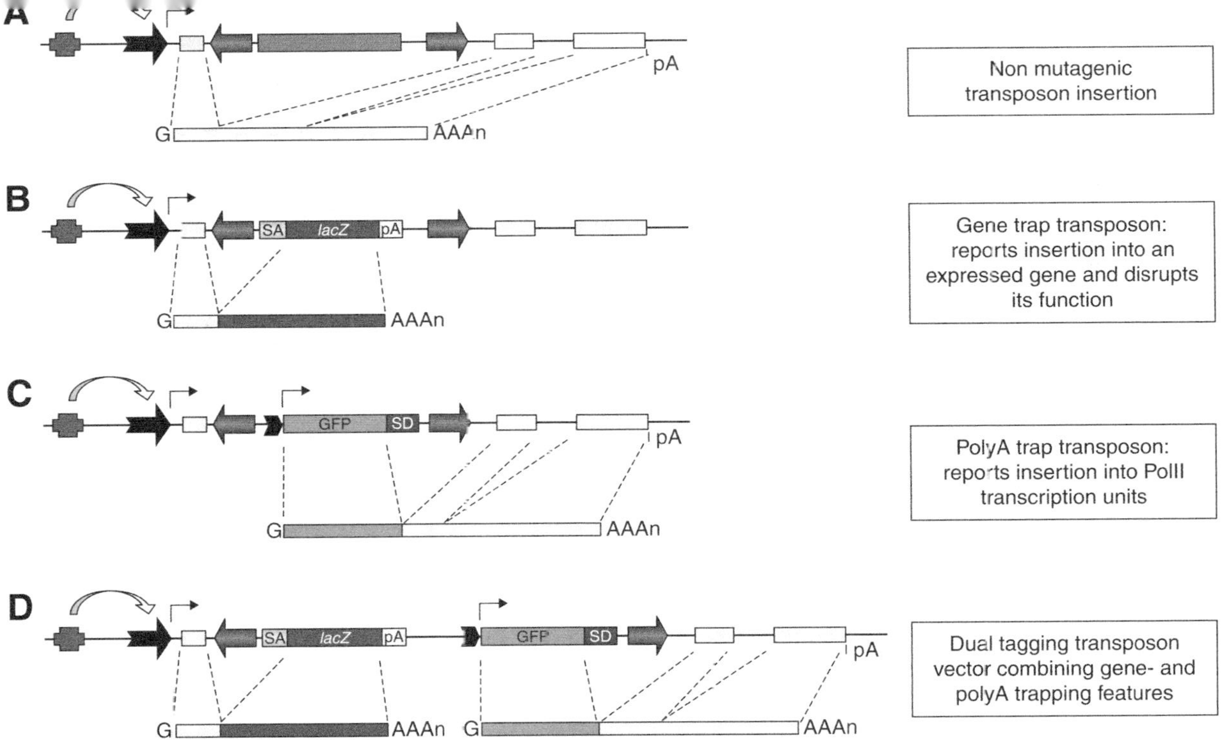

Fig. 3. Transposon-based vectors for insertional mutagenesis. **(A)** A hypothetical transcription unit is depicted with an upstream regulatory element (purple box), a promoter (black arrow), three exons (yellow boxes), and a pA. Transposon insertion into one of the introns is not expected to be mutagenic, because the element is spliced out of the pre-mRNA together with the intron sequences. Major classes of transposon-based trapping constructs and spliced transcripts are shown in the figure. Transposon IRs are indicated by gray arrows. **(B)** The conventional gene-trapping cassettes contain a SA followed by a reporter gene such as *lacZ* and a pA. The reporter is only expressed when transcription starts from the promoter of an endogenous transcription unit. Thus, the expression of the reporter follows the expression pattern of the trapped gene. **(C)** poly(A) traps contain a promoter followed by a reporter gene such as green fluorescent protein (GFP) and a SD site, but they lack a pA. Therefore, reporter gene expression depends on splicing to downstream exon/s of a Pol II transcription unit containing a pA. **(D)** The "dual tagging" vectors are based on both gene- and poly(A) trapping of a targeted transcription unit.

New transposon insertions tend to cluster around the original transposon donor locus, a phenomenon termed "local hopping." Keng et al. ***(18)*** took advantage of SB's local hopping behavior to provide proof-of-concept that transposon technology can be utilized to mutagenize mouse genes at a saturation level within a certain chromosomal interval.

The major advantage of transposon-mediated insertional mutagenesis in the mouse lies in the ability to generate and maintain whole libraries of insertional mutants in vivo, in the testes of founder animals. The phenotypic effects of these mutations can then be easily analyzed by simple breeding of the founders. In order to take full advantage of local hopping for saturation mutagenesis, libraries of transposon donors in chromosomal regions of interest (e.g., quantitative trait loci or syntenic regions of certain disease loci where genes of interest are located in clusters) can be generated.

2. Materials

2.1. Tissue Culture

1. Dulbecco's phosphate-buffered saline (PBS) (1X) without Mg^+ and Ca^+ (PAA Laboratories GmbH, Cölbe, Germany).
2. Dulbecco's modified Eagle's medium + GlutaMax (+4.5 g/L D-glucose; + pyruvate) (Gibco [Invitrogen Corporation, Carlsbad, CA, USA]).
3. JetPEI RGD transfection reagent (Biomol [BIOMOL GmbH, Hamburg, Germany]).
4. Antibiotic-antimycotic solution (100X) (Gibco).

2.2. PCR Assay for Transposon Excision in Cell Culture

1. Primer pUC1 5′-CAG TAA GAG AAT TAT GCA GTG CTG CC.
2. Primer pUC2 5′-GCG AAA GGG GGA TGT GCT GCA AGG.
3. Primer pUC3 5′-CGA TTA AGT TGG GTA ACG CCA GGG.
4. Primer pUC4 5′-CAG CTG GCA CGA CAG GTT TCC CG.
5. Primer pUC5 5′-TCT TTC CTG CGT TAT CCC CTG ATT C.
6. Primer pUC6 5′-CCA TTC GCC ATT CAG GCT GCG CAA C.
7. *Taq* polymerase (InviTek, Berlin, Germany).

2.3. PCR Assay for Transposon Excision In Vivo

1. HotStarTaq DNA polymerase (Qiagen GmbH, Hilden, Germany).
2. GeneAmp PCR system 9700 (PE Applied Biosystem, Forster City, CA USA).

2.4. Detection of poly(A) Trapped Events in Mice

1. Fluorescence microscope (WILD M10; Leica Geosystems AG, St. Gallen, Switzerland).

2.5. Detection of Promoter-Trapped Events in Mice

1. 4-Chloro-5-bromo-3-indolyl-β-D-galactopyranoside (X-Gal) (Nakalai tesque, Kyoto, Japan) is dissolved at 40 mg/mL in dimethyl sulfoxide and stored at –30°C.
2. 25% of Glutaraldehyde solution (Nakalai tesque) stored at 4°C.
3. 4% of Paraformaldehyde solution is dissolved in 0.001 *N* of NaOH.
4. 10% of Nonidet P-40 (NP-40) dissolved in H_2O.

2.6. Ligation-Mediated PCR (LM-PCR) to Determine Integration Sites of the SB Transposon

1. HotStarTaq DNA polymerase (Qiagen).
2. GeneAmp PCR system 9700 (PE Applied Biosystem).
3. Splinkerette linkers:
   a. Spl-top (molecular weight 19361.8) 5′-CGA ATC GTA ACC GTT CGT ACG AGA ATT CGT ACG AGA ATC GCT GTC CTC TCC AAC GAG CCA AGG.
   b. SplB-Sau (molecular weight 9154.1) 5′-GAT CCC TTG GCT CGT TTT TTT TTG CAA AAA.
   c. SplB-BLT (molecular weight 7918.2) 5′-CCT TGG CTC GTT TTT TTT TGC AAA AA.
   d. Linker for blunt end: 10.6 μg of Spl-top and 4.4 μg of SplB-BLT are combined in 50 μL of a solution containing 10 m*M* of Tris-HCl pH 7.5 and 5 m*M* of $MgCl_2$. The mixture is soaked in 95°C followed by gradual cooling to room temperature.
   e. Linker for cohesive end: 10.2 μg of Spl-top and 4.8 μg of SplB-Sau are combined in 50 μL of a solution containing 10 m*M* of Tris-HCl pH 7.5 and 5 m*M* of $MgCl_2$. The mixture is soaked in 95°C followed by gradual cooling to room temperature. The prepared linkers are kept at –30°C until use.
4. Oligonucleotides for PCR:
   a. T/direct repeat (DR) 5′-CTGGAATTGTGATACAGTGAATTATAAGTG.
   b. T/BAL 5′-CTTGTGTCATGCACAAAGTAGATGTCC.
   c. Spl-P1 5′-CGAATCGTAACCGTTCGTACGAGAA.
   d. Spl-P2 5′ TCGTACGAGAATCGCTGTCCTCTCC.
5. TaKaRa Ligation Kit version 1 (TAKARA Bio, Shiga, Japan).

3. Methods

3.1. Construction of Mutagenic Transposon Vectors

To generate pTrans-SA-IRES*LacZ*-CAG-GFP-SD:*Neo* construct **(Fig. 4)**, the following steps are performed.

1. *Unique restriction enzyme (RE) sites are introduced just outside of the transposon vector*: multicloning sites of pBluescript II were replaced with *Asc*I, *Xho*I, *Not*I, and *Swa*I sites by PCR amplification with primers 5′ GCCGCTCGAGGGCGCGCCAGATT-TAAATC AGCTTTTGTTCCCTTTAGTGAG 3′ and 5′ CGCAGCGGCCGCATT-TAAATGAGG CGCGCCGCTCCAATTCGCCCTATAGTG 3′ using pBluescriptII as a template. A 2.9-kb *Xho*I-*Not*I fragment of the PCR product was ligated to a 0.8-kb *Xho*I-*Not*I fragment of IR/DR(R,L) from pBS-IR/DR(R,L), resulting in pBS-IR/DR-AS, which contains *Asc*I and *Swa*I sites flanking the IRs and DRs.
2. *Introduction of unique enzyme sites into the old transposon vector*: linkers containing *Asc*I-*Kpn*I-*Swa*I sites and *Pme*I-*Pac*I sites were created by annealing oligonucleotides 5′ GTACGGCGCGCCGGTACCATTTAAAT 3′ and 5′ GTACATTTAAATGGTAC-CGGCGCGCC 3′ and oligonucleotides 5′ CGTTTAAACTTAATTAAGAGCT 3′ and 5′ CTTAATTAAGTTTAAACGAGCT 3′, respectively. Each linker was inserted into

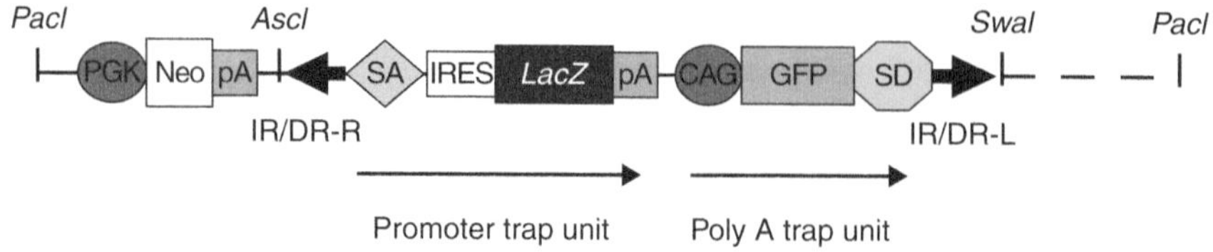

Fig. 4. Schematic representation of a "dual tagging" gene trap transposon construct. The orientations of transcription are shown by an arrow. Some of unique RE sites (*Pac*I, *Swa*I, and *Asc*I) are also shown. Dashed line indicates the vector backbone. SD, splice donor; SA, splice acceptor; pA, polyA signal.

the unique *Kpn*I and *Sac*I sites, respectively, of pTransCX-*GFP*:*Neo* after the removal of the TransCX-*GFP* fragment, resulting in pAKS:*Neo*:PP.

3. *Construction of GFP-SD unit for poly(A) trap*: the *Sal*I-*Bam*HI fragment of pCX-enhanced green fluorescent protein (*EGFP*) *PigA*, containing CAG-*EGFP*, and the 256-bp fragment of the *Neo* cassette, consisting of SD sequences from the mouse *hprt* gene exon 8/intron 8 region and the mRNA instability signal derived from the 3′ untranslated region of the human granulocyte-macrophage colony-stimulating factor cDNA were inserted into *Sal*I-blunted *Not*I sites of pBluescript II, resulting in pCAG-*GFP*-SD.
4. *Addition of SA-LacZ-poly(A) unit for promoter trap*: an *Xba*I-blunted-*Hin*dIII fragment of the rabbit β-globin poly(A) addition signal and a *Sac*II-*Not*I fragment of the *lacZ* gene, containing the nuclear localizing signal were isolated and were inserted at *Xba*I and *Sma*I sites and *Sac*II and *Not*I sites of pBluescriptII, respectively, resulting in p*LacZ*-BS. A *Sal*I-*Xho*I fragment of CAG-*GFP*-SD from pCAG-*GFP*-SD was inserted at the *Xho*I site of p*LacZ*-BS, resulting in p*LacZ*-CAG-*GFP*-SD. The human *bcl-2* intron 2/exon 3 SA sequence was amplified by using primers 5′ CGGCAAGCTTCTCGAGCTGTATCTCTAAGATGGCTGG 3′ and 5′ GCCACGGTCGACGCCTGCATATTATTTCTACTGC 3′, with the removable exon trap ***(19)*** (RET) vector as a template. The internal ribosome entry site (IRES) sequence was amplified with primers 5′ GGAGCGTCGACTACGTAAATTCCGCCCCTCTCCCTC 3′ and 5′ GGAGCGTCGACTACGTAAATTCTCCCTCCCC 3′, with the RET vector as a template. A *Hin*dIII-*Sal*I SA-containing fragment and a *Sal*I-*Bam*HI IRES-containing fragment were simultaneously cloned into the *Hin*dIII and *Bam*HI sites of p*LacZ*-CAG-*GFP*-SD, resulting in pSA-IRES*LacZ*-CAG-*GFP*-SD.
5. *Generation of the final construct*: the *Xho*I fragment of pSA-IRES-*LacZ*-CAG-*GFP*-SD containing SA, IRES, *lacZ*, poly(A), and CAG-*GFP*-SD was blunted and inserted at the *Eco*RI and *Bam*HI sites of the pBS-IR/DR-AS after both sites were blunted, resulting in pTrans-SA-IRES*LacZ*-CAG-*GFP*-SD. The *Asc*I-*Swa*I fragment of pTrans-SA-IRES*LacZ* CAG-*GFP*-SD was inserted at the *Asc*I and *Swa*I site of pAKS:*Neo*:PP, resulting in pTrans-SA-IRES*LacZ*-CAG-*GFP*-SD:*Neo* (*see* **Note 1**).

3.2. Testing Transposon Excision In Vitro

The main steps of the procedure are illustrated in **Fig. 5**. The following procedure is applicable for any transposon that is cloned into the multiple cloning regions of pUC and pBluescript-derived plasmid vectors.

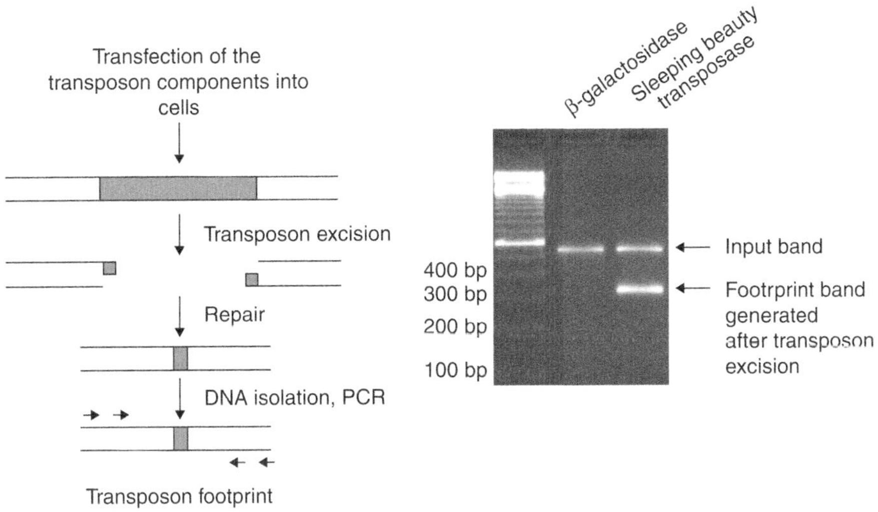

Fig. 5. Transposon excision assay in transfected cells. HeLa cells were cotransfected with a *neo*-marked transposon plasmid and vectors expressing the proteins indicated. Transposon excision is assayed with PCR that amplifies a footprint product. PCR-amplification of the *neo* marker inside the transfected transposon donor serves as a loading control.

3.2.1. Transfection of Human HeLa Cells

1. Trypsinize a 10-cm HeLa plate when the plate is 80–85% confluent.
2. Resuspend the cells in 4 mL of medium.
3. Plate out 50 µL in a 6-well plate (TPPAG, Trasadingen, Switzerland) or 1–1.5 × 10^5 cells/well (*see* **Note 2**).
4. Let the cells grow for 24 h.
5. Transfect cells using jetPEI RGD as transfection reagent with 100 µg of transposase-expressing helper plasmid DNA and 1 µg of transposon donor plasmid.
6. Put together the plasmid DNAs and then add 150 m*M* of NaCl to 50 µL.
7. Prepare a mastermix with the transfection reagent: 2 µL of jetPEI with 48 µL of 150 m*M* NaCl; add this to the prepared plasmid mix.
8. Incubate for 30 min and then pipet the mixture put it to the cells.
9. Incubate the cells in the presence of the transfection reagent for 2 d.

3.2.2. Harvest of the Cells

1. Aspirate medium and wash the cells with PBS.
2. Trypsinize the cells.
3. Resuspend the cells in 1 mL of serum-containing medium and transfer them into an 2-mL Eppendorf tube.
4. Pellet cells for 3 min at 800*g*.
5. Aspirate medium and wash with 1 mL of PBS.
6. Repeat centrifugation step and aspirate PBS.
7. Pellet is used for DNA preparation or stored at –80°C for later use.

3.2.3. Preparation of Plasmids From Transfected Cells

This method is based on the Qiagen Spin Miniprep Kit (Qiagen).

1. Resuspend harvested cells in 300 μL buffer P1.
2. Add 300 μL 1.2% sodium dodecyl sulfate and 5 μL of proteinase K (10 mg/mL), mix well but do not vortex.
3. Incubation at 55°C for 30 min.
4. Add 400 μL of buffer N3, mix well but do not vortex.
5. Incubate on ice for 30 min.
6. Centrifuge for 10 min at 16,000*g*.
7. Pipet the supernatant into the spin column.
8. Follow the subsequent steps (washing and elution) exactly as specified in the Qiagen protocol for Spin Prep.
9. Measure DNA concentration.

3.2.4. Excision PCR

3.2.4.1. PCR-I

DNA	1.5 μL (50 ng)
10X PCR buffer	2.0 μL
$MgCl_2$ (25 m*M*)	1.2 μL
dNTPs (10 m*M*)	0.4 μL
Primer pUC (10 μ*M*)	1.0 μL
Primer pUC6 (10 μ*M*)	1.0 μL
Taq DNA polymerase	0.25 μL
H_2O	To a final volume of 20 μL

PCR-cycle:

95°C—5′

95°C—30″ }
65°C—30″ } 30X
72°C—1′ }

72°C—5′

4°C

3.2.4.2. PCR-II

Dilute products 1:100 for PCR-II:

DNA	3 μL diluted PCR product
10X PCR buffer	5 μL
$MgCl_2$ (25 m*M*)	2.5 μL
dNTPs (10 m*M*)	1 μL
Primer pUC5 (10 μ*M*)	2.5 μL
Primer pUC2 (10 μ*M*)	2.5 μL
Taq DNA polymerase	0.5 μL
H_2O	To a final volume of 50 μL

- PCR cycle as above.
- Run PCR products on a 1.5% agarose gel.
- If your product is to weak you can perform a third PCR as followed.
- Dilute products of PCR-II with sterile distilled water (DW) 1:100.

3.2.4.3. PCR-III

DNA	3 μL diluted PCR product
10X PCR buffer	5 μL
$MgCl_2$ (25 m*M*)	2.5 μL
dNTPs (10 m*M*)	1.0 μL
puc3 Primer (10 μ*M*)	2.5 μL
puc4 Primer (10 μ*M*)	2.5 μL
Taq DNA polymerase	0.5 μL
H_2O	To a final volume of 50 μL

- PCR cycle as in PCR-1.
- Run PCR products on a 1.5% agarose gel.

3.3. Generation of Transgenic Mice: SB Transposase-Expressing (SB Transgenic) Mice and Transposon-Containing (GFP Transgenic) Mice

To generate pCAG-SB construct, a blunt-ended *Sac*II SB fragment pSB10 was inserted at the blunt end of *Eco*RI site of pCX-EGFP, after removal of an *Eco*RI EGFP fragment. The *Sal*I-*Bam*HI fragment of pCAG-SB was gel purified and injected into fertilized eggs obtained from the mating of BCF1(C57BL/6 × C3H) × BCF1 mice to generate SB transgenic mice (*see* **Note 3**). The SB transgenic line was established by mating the founder mice with C57BL/6 mice (*see* **Note 4**). pTrans-SA-IRES*LacZ*-CAG-*GFP*-SD:*Neo* was linearized with *Pac*I and injected into BDF1 (C57BL/6 × DBA) × BDF1 fertilized eggs to generate GFP mice (*see* **Note 5**).

3.4. Selection of GFP Transgenic Mice

As most transpositions occur locally, close to the donor sites in GFP-transgenic mice, mutant mice homozygous for a new transposon insertion often contain the donor site at both alleles. Selection of GFP transgenic mice in which the donor site does not affect phenotype when homozygous is important for phenotype screening by the SB transposon system.

3.5. Breeding, Generation of Double-Transgenic "Seed" Mice

As illustrated in **Fig. 2**, SB transgenic mice are mated with GFP-transgenic mice to generate double-transgenic male "seed" mice (*see* **Note 6**).

3.6. Testing Transposon Excision In Vivo

Excision of SB transposon was examined by PCR with following primers: TgTP-2L, 5′-ACACAGGAAACAGCTATGACCATGATTACG-3′ and TGTP-1U, 5′ GACCGCTTCCTCGTGCTTTACGGTATC-3′. Each primer is located just

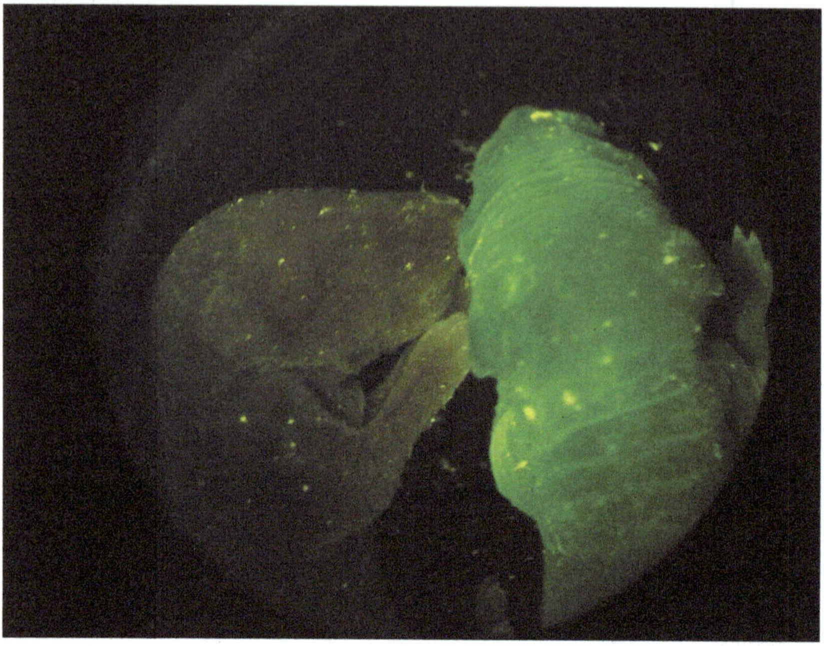

Fig. 6. Examination of polyA trap events using fluorescence stereomicroscopy in newborn mice. The mouse at the left is GFP-negative and the mouse at the right is GFP-positive.

outside of the IR/DR-R and IR/DR-L of pTrans-SA-IRES*LacZ*-CAG-*GFP*-SD:*Neo*. PCR conditions were 95°C for 15 min, 50 cycles 94°C for 1 min, 59°C for 1 min, and 72°C for 1 min, followed by 72°C for 10 min. As this PCR condition detected approximately one excision event, particular GFP transgenic line could be evaluated by frequency of excision in "seed" mice generated by mating with SB mice.

3.7. Detection of poly(A)-Trap Events in Mice

Newborn mice are checked with a fluorescence microscope with GFP-specific filter before the appearance of hair. GFP-positive mice (**Fig. 6**) are candidates of gene-trapped mice.

3.8. Detection of Gene-Trap Events in Mice

To examine the expression patterns of trapped genes, tissues or embryos are fixed with 1% paraformaldehyde, 0.2% glutaraldehyde, and 0.02% NP-40 in PBS (pH 7.3) for 30 min at room temperature, washed with PBS containing 0.02% NP-40 for three times, and then stained in a solution of 1 mg of X-Gal/mL, 2 m*M* $MgCl_2$, 4 m*M* $K_3Fe(CN)_6$, and 4 m*M* $K_4Fe(CN)_6$ in PBS (pH 7.3). A β-galactosidase-expressing tissue sample is illustrated in **Fig. 7**.

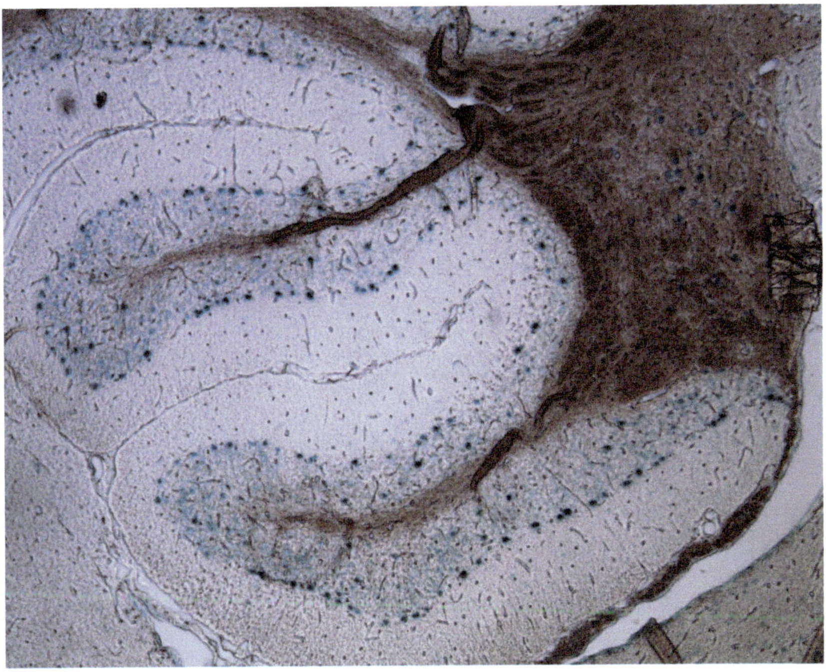

Fig. 7. Detection of promoter trap events using X-Gal staining. The cross-section of the cerebellum derived from one of the SB transposon-inserted lines is shown. Purkinje cells are positive for X-Gal staining.

3.9. LM-PCR to Determine Integration Sites of SB Transposon

1. Isolate genomic DNA from mouse tail using 500 μL of DNA extraction buffer (1 m*M* EDTA, 1X SSC, 1% sodium dodecyl sulfate, and 10 m*M* Tris-HCl pH 7.4) with 10 μL of proteinase K (10 mg/mL) and incubate at 56°C overnight.
2. Centrifuge at 15,000 rpm for 5 min at 4°C to separate undissolved tissue and transfer liquid phase to fresh Eppendorf tube using blue pipet tips cut at the end to avoid genomic DNA shearing.
3. Add equal volume (500 μL) of phenol:chloroform and mix by rotation for 15–30 min.
4. Centrifuge at 15,000 rpm for 5 min at 4°C and transfer aqueous phase to fresh Eppendorf tube using blue pipet tips cut at the end to avoid genomic DNA shearing.
5. Add 0.7 vol (350 μL) of isopropanol and mix by gentle inversion.
6. Centrifuge at 18,000*g* rpm for 10 min at 4°C to pellet genomic DNA.
7. Discard supernatant and wash with 500 μL of 80% ethanol.
8. Centrifuge at 15,000 rpm for 5 min at 4°C to pellet genomic DNA.
9. Discard supernatant and dissolve genomic DNA in 50 μL TE.
10. Incubate at 56–60°C for 15–30 min to dissolve genomic DNA. *Note*: **steps 1–10** can be done using automated DNA isolation equipment.
11. Measure DNA concentration using Nano-Drop spectrophotometer (NanoDrop Technologies, Wilmington, DE USA).

12. Dilute genomic DNA to a 10 ng/μL concentration using DW.
13. Proceed to digest 100 ng of diluted genomic DNA at 37°C for 3 h using one of the following RE (4-base cutters) in a final volume of 50 μL.
    a. *Alu*I
    b. *Mbo*I
    c. *Hae*III
    d. *Rsa*I;

    For IF or IR vectors, use *Mbo*I or *Alu*I, respectively, as the first choice RE for initial screening, followed by remaining enzymes. If sample quantity is small, use all three enzymes (*Alu*I, *Mbo*I, and *Hae*III). Use *Rsa*I only if all other enzymes fail.
14. Heat inactivation for 20 min after incubation:
    a. *Alu*I, *Mbo*I, or *Rsa*I at 65°C.
    b. *Hae*III at 80°C.
15. Linker ligation at 16°C for at least 2 h:
    a. *Alu*I, *Rsa*I, and *Hae*III—use Spl-top/blunt.
    b. *Mbo*I—use Spl-top/Sau.

Linker ligation reaction:

RE digested genomic DNA	2 μL
Appropriate linker	1 μL
Takara ligation buffer A	12 μL
Takara ligation buffer B	3 μL
Total	18 μL

16. Purify using Qiagen PCR purification kit (using manufacturer's instructions)—resuspend in 38 μL DW. For each sample, proceed to digest at 37°C for 3 h using *Kpn*I.

RE digest reaction:

Linker-ligated/purified genomic DNA	38 μL
Buffer (10X)	5 μL
BSA (10X)	5 μL
*Kpn*I (in excess)	2 μL
Total	50 μL

17. Purify using Qiagen PCR purification kit (using manufacturer's instructions) and resuspend in 50 μL DW (may omit if sample quantity is large—proceed directly to nested-PCR).
18. Nested-PCR using 1 μL template with the following primer sets (first PCR):

First PCR:

DNA (from **step 17**)	1 μL
Buffer (10X)	5 μL
dNTP (10 m*M*)	1 μL
Primer 1 T/DR (10 μ*M*)	1 μL
Primer 2 Sp1-P1 (10 μ*M*)	1 μL
Hot start *Taq*	0.25 μL
DW	40.75 μL
Total	50 μL

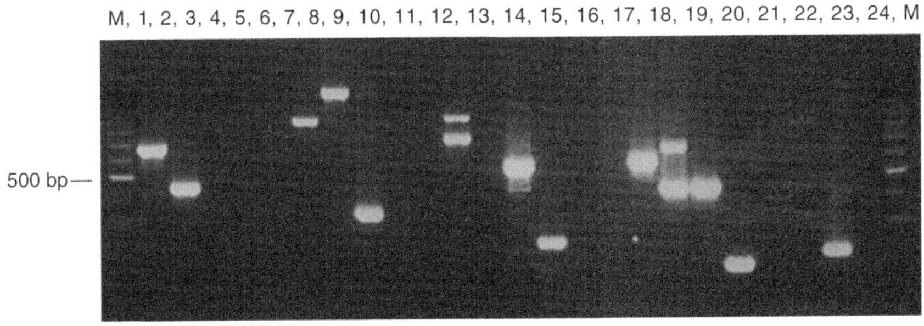

Fig. 8. LM-PCR analysis of transposon insertions. The agarose gel shows products of LM-PCR on DNA samples prepared from mice with transposon insertions. M indicates a marker of a 100-bp DNA ladder.

PCR condition:

95°C	15 min	
94°C	1 min	
55°C	1 min	30X
72°C	1 min	
72°C	7 min	

Cool to 25°C

Second PCR using the following primer sets:

First PCR	1 μL
Buffer (10X)	5 μL
dNTP (10 m*M*)	1 μL
Second primer 1 T/BAL (10 μ*M*)	1 μL
Second primer 2 Spl-P2 (10 μ*M*)	1 μL
Hot start *Taq*	0.25 μL
DW	40.75 μL
Total	50 μL

PCR condition:

Similar to first PCR.

19. Check PCR product by running 5 μL of each sample in a 2% agarose gel (ethidium bromide added) **(Fig. 8)**. *Note*: for samples with no visible PCR product, proceed to repeat using other 4-base RE and start from **step 13** again.
20. Proceed to run preparative 2% agarose gel (ethidium bromide added).
21. Gel extraction of bands under ultraviolet and proceed to purification using Qiagen gel extraction kit (using manufacturer's instructions)—resuspend in either 30 μL or 50 μL DW, depending on the intensity of the PCR band.
22. Recheck for purity by running 5 μL of resuspended PCR-band product in a 2% agarose gel (ethidium bromide added).
23. Proceed to cycle sequencing using 1 μL of the purified PCR-band product with the appropriate primer.

4. Notes

1. When the blunted SA-IRES-*LacZ*-CAG-GFP-SD was inserted into the pBS-IR/DR-AS, both orientations were obtained. IF was that, transcriptional orientation of *LacZ*, GFP, and *Neo* was same in the final construct. On the other hand, IR was that *Neo* and other units were reverse.
2. Because there is always some variation from one transfection to another, it is good practice to transfect two wells with the same plasmid combination. The transfected cells harvested from the two wells can be combined for subsequent DNA preparation.
3. SB10 transposase could be replaced by hyperactive versions of the SB transposase ***(20–23)***. C57BL/6 fertilized eggs can be used for generation of SB transgenic mice.
4. In our experience, it took a long time (~6 mo) for the SB transgene in the founder mouse to be transmitted to the next generation.
5. The vector backbone was included for suppression of *LacZ* and GFP expression, because the SD of GFP and downsteam of SA of *LacZ* within tandem array of transgenes may allow expression of GFP. As predicted, GFP signal was not detected in most founder mice (seven out of eight).
6. Transposition efficiencies in male or female germ cell from double-positive mice were examined by comparing GFP-positive progeny. Male double-positive (seed) mice could generate much higher percentage of GFP-positive mice, suggesting higher transpositions in male germ cells.

References

1. Ivics, Z., Hackett, P. B., Plasterk, R. H., and Izsvak, Z. (1997) Molecular reconstruction of Sleeping Beauty, a Tc1-like transposon from fish, and its transposition in human cells. *Cell* **91,** 501–510.
2. Izsvak, Z., Ivics, Z., and Plasterk, R. H. (2000) Sleeping Beauty, a wide host-range transposon vector for genetic transformation in vertebrates. *J. Mol. Biol.* **302,** 93–102.
3. Luo, G., Ivics, Z., Izsvak, Z., and Bradley, A. (1998) Chromosomal transposition of a Tc1/mariner-like element in mouse embryonic stem cells. *Proc. Natl. Acad. Sci. USA* **95,** 10,769–10,773.
4. Yant, S. R., Meuse, L., Chiu, W., Ivics, Z., Izsvak, Z., and Kay, M. A. (2000) Somatic integration and long-term transgene expression in normal and haemophilic mice using a DNA transposon system. *Nat. Genet.* **25,** 35–41.
5. Fischer, S. E., Wienholds, E., and Plasterk, R. H. (2001) Regulated transposition of a fish transposon in the mouse germ line. *Proc. Natl. Acad. Sci. USA* **98,** 6759–6764.
6. Dupuy, A. J., Fritz, S., and Largaespada, D. A. (2001) Transposition and gene disruption in the male germline of the mouse. *Genesis* **30,** 82–88.
7. Horie, K., Kuroiwa, A., Ikawa, M., Okabe, M., Kondoh, G., Matsuda, Y., and Takeda, J. (2001) Efficient chromosomal transposition of a Tc1/mariner-like transposon Sleeping Beauty in mice. *Proc. Natl. Acad. Sci. USA* **98,** 9191–9196.
8. Horie, K., Yusa, K., Yae, K., et al. (2003) Characterization of Sleeping Beauty transposition and its application to genetic screening in mice. *Mol. Cell. Biol.* **23,** 9189–9207.
9. Carlson, C. M., Dupuy, A. J., Fritz, S., Roberg-Perez, K. J., Fletcher, C. F., and Largaespada, D. A. (2003) Transposon mutagenesis of the mouse germline. *Genetics* **165,** 243–256.

10. Kitada, K., Ishishita, S., Tosaka, K., et al. (2007) Transposon-tagged mutagenesis in the rat. *Nat. Methods* **4,** 131–133.
11. Miskey, C., Izsvak, Z., Kawakami, K., and Ivics, Z. (2005) DNA transposons in vertebrate functional genomics. *Cell. Mol. Life Sci.* **62,** 629–641.
12. Cooley, L., Kelley, R., and Spradling, A. (1988) Insertional mutagenesis of the *Drosophila* genome with single P elements. *Science* **239,** 1121–1128.
13. Friedrich, G. and Soriano, P. (1991) Promoter traps in embryonic stem cells: a genetic screen to identify and mutate developmental genes in mice. *Genes Dev.* **5,** 1513–1523.
14. Zambrowicz, B. P., Friedrich, G. A., Buxton, E. C., Lilleberg, S. L., Person, C., and Sands, A. T. (1998) Disruption and sequence identification of 2,000 genes in mouse embryonic stem cells. *Nature* **392,** 608–611.
15. Clark, K. J., Geurts, A. M., Bell, J. B., and Hackett, P. B. (2004) Transposon vectors for gene-trap insertional mutagenesis in vertebrates. *Genesis* **39,** 225–233.
16. Geurts, A. M., Wilber, A., Carlson, C. M., et al. (2006) Conditional gene expression in the mouse using a Sleeping Beauty gene-trap transposon. *BMC Biotechnol.* **6,** 30.
17. Yae, K., Keng, V. W., Koike, M., et al. (2006) Sleeping beauty transposon-based phenotypic analysis of mice: lack of Arpc3 results in defective trophoblast outgrowth. *Mol. Cell. Biol.* **26,** 6185–6196.
18. Keng, V. W., Yae, K., Hayakawa, T., et al. (2005) Region-specific saturation germline mutagenesis in mice using the Sleeping Beauty transposon system. *Nat. Methods.* **2,** 763–769.
19. Ishida, Y. and Leder, P. (1999) RET: a poly A-trap retrovirus vector for reversible disruption and expression monitoring of genes in living cells. *Nucleic Acids Res.* **27,** E35
20. Baus, J., Liu, L., Heggestad, A. D., Sanz, S., and Fletcher, B. S. (2005) Hyperactive transposase mutants of the Sleeping Beauty transposon. *Mol. Ther.* **12,** 1148–1156.
21. Geurts, A. M., Yang, Y., Clark, K. J., et al. (2003) Gene transfer into genomes of human cells by the sleeping beauty transposon system. *Mol. Ther.* **8,** 108–117.
22. Zayed, H., Izsvak, Z., Walisko, O., and Ivics, Z. (2004) Development of hyperactive sleeping beauty transposon vectors by mutational analysis. *Mol. Ther.* **9,** 292–304.
23. Yant, S. R., Park, J., Huang, Y., Mikkelsen, J. G., and Kay, M. A. (2004) Mutational analysis of the N-terminal DNA-binding domain of sleeping beauty transposase: critical residues for DNA binding and hyperactivity in mammalian cells. *Mol. Cell. Biol.* **24,** 9239–9247.

9

Conditional Gene Trapping Using the FLEx System

Thomas Floss and Frank Schnütgen

Summary

The knowledge about the complete genome sequences of mouse, human, and other organisms is only the first step toward the functional annotation of all genes. It facilitates the recognition of sequence conservation, which helps to distinguish between important and not important and also coding from noncoding sequence. Nevertheless, approximately only 50% of all mouse genes have been entirely annotated to date. In the postgenomic era, large-scale projects have been initiated to describe also the expression (Emap, Eurexpress) and the function (International Gene Trap Consortium, Eucomm, Norcomm, Komp) of all mouse genes. By building up on these resources, the average amount of time starting from a gene-coding sequence to finally studying its function in a living organism or embryo, has shortened significantly within the last decade. Several recent developments, namely, in bioinformatics and gene synthesis but also in targeted and random mutagenesis have contributed to the current status. This chapter will highlight the milestones that have been undertaken in order to saturate the mouse genome with gene trap mutations. We have no intention to cover the entire field but will instead focus on most recent vectors and protocols, which have turned out to be most useful in order to promote the technology. Therefore, we apologize upfront to the many studies that could not be mentioned here solely owing to space limitations but which nevertheless made significant contributions to our current understanding. This chapter will finally provide guidance on possible uses of conditional gene trap alleles as well as detailed protocols for the application of this recent technology.

Key Words: Conditional; gene trap; mutagenesis; Cre; Flpe; recombinase; FlEx; inversior.

1. Introduction

The first documented effort of a strategically placed gift in order to achieve a specific goal is described in Homers report about the ancient greek Trojan Horse, which was donated only in order to defeat Agamemnon, King of Mykene and has led to the roman proverb: "timeo danaos et dona ferentes," or: "I am afraid of the greeks, even if they are bringing presents." A similar powerful idea has been the gene-trapping technique, which is based on the introduction of small artificial DNA cassettes in embryonic stem (ES) cells in order to simultaneously identify

From: *Methods in Molecular Biology, vol. 435: Chromosomal Mutagenesis*
Edited by: G. Davis and K. J. Kayser © Humana Press Inc., Totowa, NJ

and mutate genes and in general also report their in vivo expression pattern. Gene trapping has been first applied successfully to the mouse ES cell system in the late 1980's by Joachim Gossler and colleagues *(1)*. At the time, the promising feature of the technique has mainly been the gene discovery aspect. Many studies were applied to study gene expression patterns in embryos and to retrieve new interesting genes that function during embryonic development. Soon afterwards, with newly developed linker-mediated polymerase chain reaction (PCR) technologies it has become feasible to determine also the site of the vector integration in a straightforward manner (for a review *see* **ref.** ***2***). Once the mouse and human genomes had been completely sequenced, the gene discovery aspect of gene trapping is not anymore a priority of the community. Gene trapping today has the goal of mutating as many mouse genes as possible using comparatively inexpensive techniques. Since the founding members of the International Gene Trap Consortium first got together in 2000, it has been demonstrated that different groups with different goals can work together cooperatively and complementarily, but at the same time compete with each other for international grants and publications.

1.1. Possibilities and Limitations

The real value of gene trapping lies in the thousands of individually frozen ES cell clones, which all represent potential mouse models, available to the whole scientific community at prices which typically cover only shipping and handling. However, although we have been talking about the *possibilities* of random mutagenesis in the past, we have to talk about its *limitations* today. Most of the big gene trap libraries today have been built up on vectors that rely on the expression of genes in ES cells. Splice-donor/polyA trap vectors, which initially promised to circumvent this restriction, have been discussed controversially, because theoretically, they could cause a variety of unwanted effects as integrations mostly 3′ in genes and trapping of polyA-like sequences, which are in fact not related to genes. And finally, polyA trap vectors may have unwanted influences of the integrated promoter on neighboring loci *(3–5)*. The use of an Internal Ribosomal Entry Site (IRES) after the stop sequence of these cassettes elegantly circumvents the nonsense-mediated decay and thereby also selects integrations of splice-donor traps 5′ in genes *(6)*. However, a drawback could be that exons downstream of the promoter-IRES cassette may be even overexpressed or expressed ectopically from the included ubiquitous promoter/IRES element and could lead to unexpected phenotypes. Therefore, the promoter element should be removed using Cre recombinase before generation of a mouse. There may be better solutions to this limitation in the future; however currently, the most widely applied technique remains splice acceptor trapping using promoter-less constructs *(1,7–9)*. Based on the number of genes in public and private databases, it has been estimated that the number of genes expressed in ES cells is at least 60% of all genes *(10)*. Initially, gene trapping is inexpensive and large numbers of new genes can be retrieved in a screen. However, as the screen progresses, vectors tend to integrate in large, expressed multiexon genes. After trapping the estimated number of 40% of all mouse genes, the effort to select new genes raises the cost per new mutation for random mutagenesis over the cost for gene targeting. A more recent technique, which overcomes this limitation, uses the introduction of promoter-less gene trap vectors into larger genomic fragments, which serve as homologous arms for gene targeting. This technique yields considerably higher homologous recombination frequencies as compared with classical gene targeting with promoter-containing selection markers *(11)*. The application of this so-called "targeted trapping" technique will allow the mutation of all expressed genes, which are underrepresented or missing in the public libraries. These are namely smaller genes, containing only 1–3 exons.

1.2. Secretory Traps

The use of more specialized vectors allows to select for specific proteins, for example, secreted molecules, transmembrane receptors, or mitochondrially imported proteins. This improved trapping technique has initially been demonstrated by Skarnes and colleagues *(8)*, but suffered from the use of the transmembrane part of the rat CD4 molecule, which traps also proteins, which do not necessarily contain leader peptides. We conclude that the CD4 transmembrane part is oriented as a type 2 transmembrane molecule by default (for an overview on what determines the orientation of transmembrane molecules see also **ref. *12***). The use of a human CD2 transmembrane peptide/neo fusion (Ceo; **Fig. 2**) has been shown to be advantageous in trapping only secreted molecules and transmembrane receptors (80% of all traps without any preselection, as compared with 20% for CD4 after preselection of clones by means of *lacZ* patterning; *[13]*). From these data we conclude that the default orientation of the human CD2 transmembrane part is a type 2 orientation with the COOH terminus, and thereby the neomycin selection cassette outside of the cell or inside the ER lumen *(14)*.

1.3. Conditional Gene Traps

Until recently, the major drawback of gene trapping has been that, the "null"-mutations generated by the integration of a gene disruption cassette often result in embryonic lethal phenotypes, which precludes the analysis of gene function in the adult animal. Therefore, we generated a novel generation of gene trap vectors *(15)*, which allow two subsequent inversions of the trapping cassette mediated by the recombinase systems Cre/loxP and FLPe/frt using the recently developed flip-excision system (FlEx) for unidirectional recombinase-mediated inversions *(16)*. Thereby, the mutation brought in by the gene trap can be inactivated by FLPe-mediated inversion. Reactivation of the trap using tissue- and time-specific inducible Cre-mediated reinversion allows the investigation of specific gene functions in defined tissues at defined time-points of development (**Fig. 1**). This

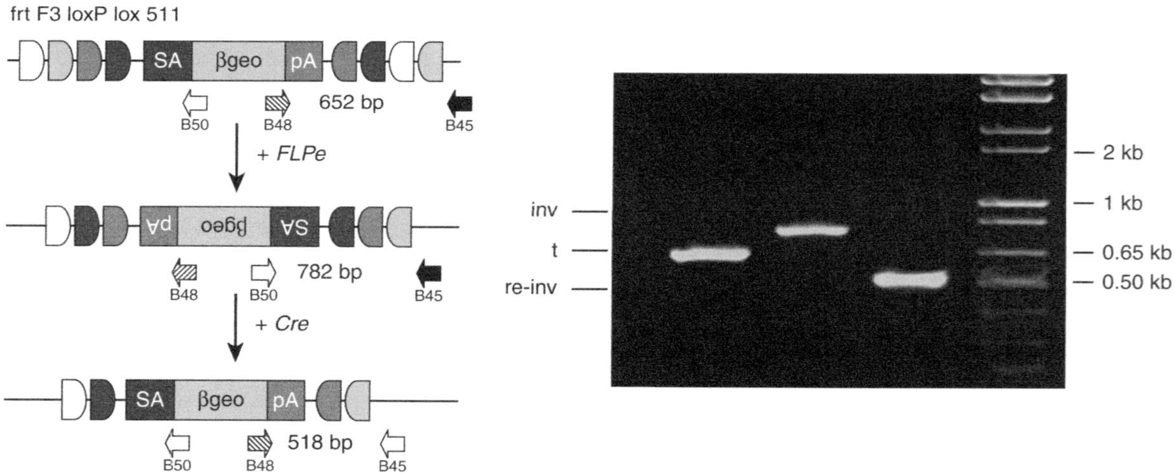

Fig. 1. Multiplex PCR for detection of the inversion status of FlipROSAβ-geo trapped ES cell lines. DNA was extracted from an ES cell clone containing an FLPe-inverted FlipROSAβ-geo gene trap and from an ES cell clone that underwent subsequent FLPe and Cre inversions of FlipROSAβ-geo. These sublines were subjected to a multiplex PCR to identify inversions. Arrows indicate primer positions within FlipROSAβ-geo; specific amplification products are visualized on ethidium bromide stained gels.

concept has been successfully adapted to the trapping cassettes β-galactosidase/neomycinphosphotransferase (β-geo, *[17]*) and CD2 antigen/neomycinphosphotransferase (Ceo, *[13]*). Additionally to the invertability, clones carrying conditional gene traps can be postinsertionally modified by recombinase-mediated cassette exchange ***(18)***. Examples include replacing the gene trap cassettes with inducible Cre recombinase genes to expand the Cre-zoo, or point-mutated minigenes to mimic human genetic diseases. Further options are the induction of gain-of-function mutations or the ablation of specific cell lineages by inserting gain-of-function cassettes or toxin genes ***(19)***, respectively. These gene traps are now in use of the German Gene Trap Consortium and the European Conditional Mouse Mutagenesis Project ***(20)***. All these publicly available gene trap lines have been recently centralized on the International Gene Trap Consortium's Web page (http://www.igtc.ac.org/; *[21]*). In the following, we will provide protocols about the molecular characterization of conditional gene trap clones.

2. Materials

1. 1X Trypsin (Invitrogen, Karlsruhe).
2. Electroporation cuvet (4 mm gap, BioRad, München).
3. Gene pulser X-cell; BioRad.
4. Puromycin (Sigma, München, P-8833).
5. Proteinase K (Qiagen, Hilden, Germany).
6. RNAse A (Sigma R61-48).
7. Oligos (Invitrogen).
8. *Taq* polymerase (Invitrogen).

3. Methods

3.1. Integrity of loxP and Frt Sites

Before extensive modifications of the ES cell lines as well as carrying out expensive microinjections into blastocysts one might wish to test the integrity of all loxP and frt sites present in the gene trap cassette. Therefore, PCR reactions have to be carried out on either side of the gene disruption cassette and sequenced.

1. FlipROSAβ-geo:
   a. For detection of the 5′ loxP and frt sites, pipet a mastermix containing per PCR sample 2.5 μL 10X PCR buffer, 0.75 mL 2 μ*M* $MgCl_2$, 0.5 μL dNTPs (10 μ*M* each), 1 μL SR2 primer, 1 μL B050 primer (**Table 1**; 10 pmol/μL each), 0.5 μL Dimethyl sulfoxide, and 0.2 μL *Taq* polymerase (Invitrogen). For the 3′ loxP and frt sites pipet the same mastermix using the primers B045 and B048 (**Table 1**).
   b. Aliquot into PCR tubes, add 2 μL of DNA, and run the following PCR program: 94°C for 4 min initial denaturation step. 30 cycles of 94°C for 45 s, 61°C for 45 s, and 72°C for 60 s, final elongation step 72°C for 7 min.
   c. Gel purify the resulting DNA fragments and sequence using the PCR primers.
2. FlipROSACEO.

Table 1
PCR Primers Used to Identify the Status of FlipROSA Genetrap Inversions

Primer name	Sequence
B045	5′-CTC CGC CTC CTC TTC CTC CAT C-3′
B048	5′-TCC CAC TGT CCT TTC CTA ATA A-3′
B050	5′-TTT GAG GGG ACG ACG ACA GTA T-3′
G001	5′-CAA GTT GAT GTC CTG ACC CAA G-3′
G013	5′-TAT CAG GAC ATA GCG TTG GCT ACC CG-3′
I001	5′-CGC CTC CTC TTC CTC CAT CC-3′
SR2	5′-GCC AAA CCT ACA GGT GGG GTC TTT-3′

Follow the protocol described earlier for the FlipROSAβ-geo, but use for the 5′ lox and frt sites the primers SR2 and G01 and for the 3′ sites the primers I01 and G13 (**Table 1**). The optimal annealing temperature is 57°C.

3.2. In Vitro Inversion of Gene Disruption Cassette

To generate conditional alleles from FlipROSAβ-geo or FlipROSACeo-trapped ES cell lines either in vitro inversions in the ES cell line or in vivo inversions by mating mice carrying the trapped allele with FLPe-expressing deleter *(22)* strains is possible. For both strategies arguments can be found (*see* **Note 1**). Here we describe the protocols for in vitro inversions, but we recommend considering the possible risk involved.

3.2.1. Transfection of Recombinase-Expressing Plasmids

This protocol is used for in vitro inversion of the gene trap cassette using site-specific Flpe- or Cre-recombinases, respectively (*see* **Note 2**).

1. Grow cells to reach 1×10^7 cells in log phase of growth.
2. After 24 h, wash cells with phosphate buffered saline (PBS) and trypsinize with 1 mL trypsin (Invitrogen) for 5 min at 37°C (*see* **Note 3**).
3. Inactivate trypsin with 9 mL ES medium and pipet the suspension using a 10 mL pipet (8–10 times) in order to get single cells.
4. Take an aliquot of 10 μL and count cells in the haematocytometer (Roth, Karlsruhe).
5. Centrifuge in 15-mL Falcon tube at 200 g for 5 min at room temperature (RT).
6. Resuspend the pellet in PBS in a concentration of 1×10^7 cells in 900 μL PBS.
7. Mix 900 μL cell suspension with the DNA (40 μg, circular) and transfer to an electroporation cuvet (4 mm gap; BioRad).
8. Leave on ice for 5–10 min.
9. Electroporate using the following conditions: 250 V, 500 μF (Gene pulser X-cell; BioRad).
10. After electroporation immediately flick the cuvet to stabilize the pH.

11. Place the cuvet on ice.
12. Transfer the electroporated suspension using 1 mL pipet in a 15-mL Falcon tube containing 9.5 mL ES medium. Total volume is around 10.4 mL.
13. Distribute the cells in 10-cm culture dishes (1 mL per dish—10 dishes) that are already filled with 10 mL ES medium each.
14. After 24 h replace the ES medium with one that contains 1 μg/mL puromycin (Sigma P-8833).
15. Keep the selection for 48 h, without changing the medium.
16. Remove the selection medium and replace with normal ES medium and change every 3 d. 10–12 d after electroporation colonies become visible.

3.2.2. Low-Density Seeding of ES Cells

In case selection is not possible or colonies exhibit mosaic genotype, cells have to be diluted and seeded onto feeder layers for growing clonal cell colonies.

1. Proceed as described earlier until **step 13**.
2. Let cells grow for 48 h.
3. Wash cells with PBS and trypsinize with 1 mL trypsin for 5 min at 37°C (*see* **Note 3**).
4. Inactivate trypsin with 9 mL ES medium and pipet the suspension using a 10 mL pipet (8–10 times) in order to get single cells.
5. Take an aliquot of 10 μL and count cells in the haematocytometer.
6. Seed 500 cells onto gelatinized 6-cm dishes containing feeder layers.

Grow cells for 10–12 d with daily change of medium until colonies reach an appropriate size for picking.

3.2.3. Picking of Colonies

1. Prepare feeder coated 96-well plates the day before picking
2. Picking is carried out in PBS to allow trypsination. Therefore, aspirate culture medium and add prewarmed PBS. If colonies from more than one plate have to be picked, process them one by one. If possible colonies should be picked under a laminar flow hood to minimize contaminations (*see* **Note 5**). Scrape the colonies from the surface of the culture dish using a Pipetman (P200) and sterile tips, and aspirate the cell clumps in about 30 μL of PBS. Transfer the cells to a fresh 96-well plate and repeat until all colonies are picked.
3. Add 30 μL of 2X trypsin solution to each well and incubate until cells can be easily dispersed by tapping the plate (*see* **Note 3**). Add 100 μL of medium to each well and transfer cells onto 96-well plates containing feeder cells.
4. When cells approach confluence, trypsinize and expand for freezing and further analysis (*see* **Note 4**).

3.3. Determination of Gene Trap Configuration

Conditional gene trap colonies contain invertible cassettes flanked by two FlEx systems using Cre and Flpe recombinases, respectively ***(15)***. In vitro inversions of these cassettes can easily be monitored using a multiplex PCR strategy.

3.3.1. Isolation of DNA From ES Cell Colonies

1. Aspirate off medium from confluent 24-well plates containing the expanded cell clones.
2. Wash carefully with PBS.
3. Add 400 µL lysis buffer (50 m*M* Tris/HCl, 1 m*M* EDTA, 0.1 *M* NaCl, 0.5% sodium dodecyl sulfate, pH 7.5) and incubate for 10 min at room temperature. At this step, cells may be frozen and kept on –20°C for up to a month.
4. Use a pipetman (P1000) and cut pipet tips, scrape off the cells from the wells, and pipet immediately into 1.5-mL microcentrifuge tubes.
5. Add 5 µL proteinase K solution (20 mg/mL, Qiagen) and incubate overnight at 55–60°C.
6. Add 75 µL 8 *M* KAc and 600 µL chloroform and mix carefully without vortexing.
7. Leave on ice for 30 min.
8. Centrifuge at 4°C, 13,000*g* for 5 min.
9. Transfer the upper aqueous phase to a new tube. Be careful not to touch the white precipitate at the interphase.
10. Add 1 mL ice-cold ethanol to precipitate the DNA and centrifuge at room temperature for 10 min 13,000*g*.
11. Remove the supernatant without disturbing the DNA pellet.
12. Wash with 200 µL ice-cold 70% ethanol and centrifuge again at room temperature 13,000*g* for 10 min.
13. Remove the supernatant and dry for 10 min in the bench.
14. Dissolve the DNA in 50–100 µL of purified water containing 100 µg/mL RNAse A (Sigma R61-48).

3.3.2. Multiplex PCR

3.3.2.1. FLIPROSAβ-GEO

Monitoring the inversion status of gene trap cassettes requires three primers. One primer anneals to a region located downstream of the gene disruption cassette outside of the recombinase target sites in antisense direction (B045). The second primer binds to the 3′ part of the G418 resistance cassette (B048) in sense orientation. A PCR using these two primers will result in a PCR fragment for the trapped and the reinverted configuration of the gene trap cassette. The removal of recombinase target sites results in differences in sizes of the PCR product. The third primer binds to the 5′ of the β-galactosidase cDNA in antisense orientation (B050). A PCR reaction of B045 and B050 only gives a PCR product in the inverted gene trap status (**Fig. 1**). The same procedure can be used to detect the gene trap configuration on the 5′ side of the gene trap using the primers SR/B048/B050.

1. Pipet a mastermix containing per PCR sample 2.5 µL 10X PCR buffer, 0.75 mL 2 µ*M* $MgCl_2$, 0.5 µL dNTPs (10 µ*M* each), 1 µL B045 primer, 1 µL B048 primer, 1 µL B050 primer (10 pmol/µL each), 0.5 µL dimethyl sulfoxide, and 0.2 µL *Taq* polymerase (Invitrogen). Put 23 µL per 0.2-mL PCR tube.

2. Add 2 μL of DNA, and run the following PCR program: 94°C for 4 min initial denaturation step. 30 cycles of 94°C for 45 s, 61°C for 45 s, and 72°C for 60 s, final elongation step 72°C for 7 min.

Load samples onto 1.5% agarose gels. Expected sizes are: 652 bp for the trapped configuration, 782 bp inverted and 518 bp for reinverted gene trap cassettes.

3.3.2.2. FLIPROSACEO

Essentially, the procedure is the same as described earlier for FlipROSAβ-geo, except that different primers have to be used and the PCR protocol differs slightly. The primer I001 binds to the region downstream of the gene disruption cassette in antisense orientation, G013 binds to the 3′ part of the neomycin resistance cassette in sense orientation, and G001 binds to the 5′ part of the CD2 sequence in antisense orientation (**Fig. 2**).

All these PCR reactions can be carried out as multiplex PCR under the same reaction conditions. Thereby, the detection of mixed clones can be facilitated.

4. Notes

1. The in vitro inversion of the gene trap cassette requires additional manipulation of the ES cell line and thereby might affect germline transmission capacity but in cases of essential X-chromosomal genes or mutations leading to haploinsufficiency, the generation of mouse lines containing the mutation may be otherwise impossible. In vivo inversion by mating to deleter strains is time consuming, as it requires two more generations of mouse breeding and may further dilute possible phenotypes on undesired genetic backgrounds.
2. For higher efficiency of recombination (or reducing work load) the recombinases are bicistronically expressed with puromycin-resistance gene (*pac*) under the control of the CAGGs promoter *(23)*. Thereby, transient selection for 48 h can be used to reduce background level *(24)*. A similar vector for Cre recombinase (pCAGGs-Cre-IRES-puro-pA) is available.
3. Extensive trypsination is detrimental to ES cells. Cells should be watched carefully during trypsination, and trypsin activity should be blocked immediately after cells become detached from the plastic surface.
4. In cases where a selection cassette has been inverted, screening colonies for antibiotic sensitivity is possible. For this reason make duplicates. One plate will grow in normal medium and the second in medium containing antibiotic for which the cells should have the selection marker inactivated. Colonies that grow in normal medium but die in antibiotic-containing medium have been efficiently recombined. We do not advise screening for loss of β-gal activity, as false-negative clones could be selected owing to low dosage of *lacZ* expression.
5. Colonies normally differ in size, probably owing to the level of expression of the resistance cassette. In order to obtain homogenously grown cell cultures during expansion, colonies of similar sizes should be picked into one 96-well plate. But to reach a larger target of trapped genes, colonies of different sizes should be picked but processed separately.

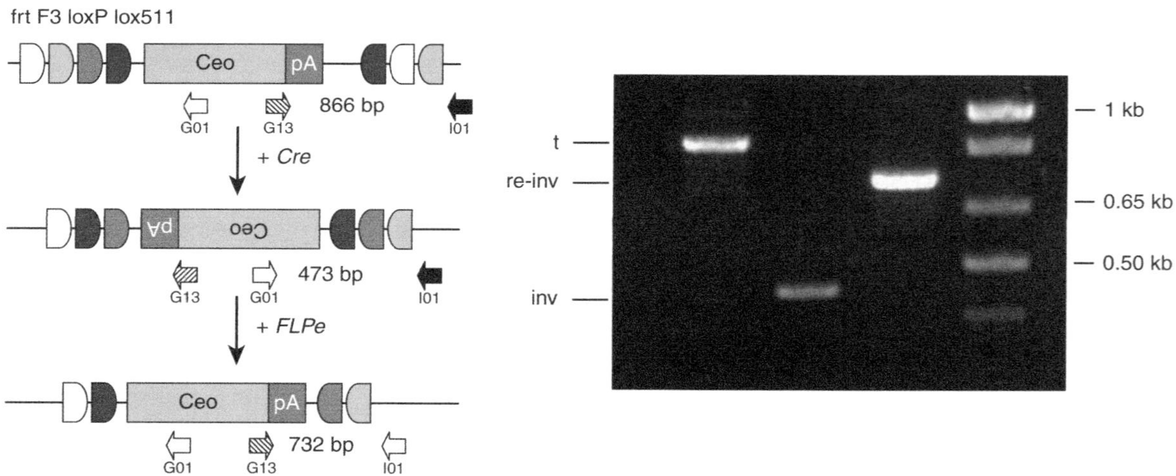

Fig. 2. Multiplex PCR for detection of the inversion status of FlipROSACeo-trapped ES cell lines. DNA was extracted from as ES cell clone containing an FLPe-inverted FlipROSACeo gene trap and from as ES cell clone that underwent subsequent FLPe and Cre inversions of FlipROSACeo. These sublines were subjected to a multiplex PCR to identify inversions. Arrows indicate primer positions within FlipROSACeo; specific amplification products are visualized on ethidium bromide stained gels.

References

1. Gossler, A., Joyner, A. L., Rossant, J., and Skarnes, W. C. (1989) Mouse embryonic stem cells and reporter constructs to detect developmentally regulated genes. *Science* **244,** 463–465.
2. Hui, E. K., Wang, P. C., and Lo, S. J. (1998) Strategies for cloning unknown cellular flanking DNA sequences from foreign integrants. *Cell Mol. Life Sci.* **54,** 1403–1411.
3. Fiering, S., Epner, S., Robinson, K., et al. (1995) Targeted deletion of 5′HS2 of the murine β-globin LCR reveals that it is not essential for proper regulation of the β-globin locus. *Genes Dev.* **9,** 2203–2213.
4. Floss, T., Arnold, H. H., and Braun, T. (1996) Myf-5(m1)/Myf-6(m1) compound heterozygous mouse mutants down-regulate Myf-5 expression and exert rib defects: evidence for long-range cis effects on Myf-5 transcription. *Dev. Biol.* **174,** 140–147.
5. Olsen, E. N., Arnold, H.-H., Rigby, P. W. J., and Wold, B. J. (1996) Know your neighbors: three phenotypes in null mutants of the myogenic bHLH gene MRF4. *Cell* **85,** 1–4.
6. Shigeoka, T., Kawaichi, M., and Ishida, Y. (2005) Suppression of nonsense-mediated mRNA decay permits unbiased gene trapping in mouse embryonic stem cells. *Nucleic Acids Res.* **33,** E20.
7. Chen, W. V. and Soriano, P. (2003) Gene trap mutagenesis in embryonic stem cells. *Methods Enzymol.* **365,** 367–386.
8. Skarnes, W. C., Moss, J. E., Hurtley, S. M., and Beddington, R. S. (1995) Capturing genes encoding membrane and secreted proteins important for mouse development. *Proc. Natl. Acad. Sci. USA* **92,** 6592–6596.
9. Wiles, M. V., Vauti, F., Otte, J., et al. (2000) Establishment of a gene-trap sequence tag library to generate mutant mice from embryonic stem cells. *Nat. Genet.* **24,** 13–14.
10. Skarnes, W. C., von Melchner, H., Wurst, W., et al. (2004) A public gene trap resource for mouse functional genomics. *Nat. Genet.* **36,** 543–544.
11. Friedel, R. H., Plump, A., Lu, X., et al. (2005) Gene targeting using a promoterless gene trap vector (targeted trapping) is an efficient method to mutate a large fraction of genes. *Proc. Natl. Acad. Sci. USA* **102,** 13,188–13,193.
12. Spiess, M. (1995). Heads or tails—what determines the orientation of proteins in the membrane. *FEBS Lett.* **369,** 76–79.
13. Gebauer, M., von Melchner, H., and Beckers, T. (2001) Genomewide trapping of genes that encode secreted and transmembrane proteins repressed by oncogenic signaling. *Genome Res.* **11,** 1871–1877.
14. De-Zolt, S., Schnütgen, F., Seisenberger, C., et al. (2006) High-throughput trapping of secretory pathway genes in mouse embryonic stem cells. *Nucleic Acids Res.* **13,** E25.
15. Schnütgen, F., De-Zolt, S., Van Sloun, P., et al. (2005) Genomewide production of multipurpose alleles for the functional analysis of the mouse genome. *Proc. Natl. Acad. Sci. USA* **102,** 7221–7226.
16. Schnütgen, F., Doerflinger, N., Calleja, C., Wendling, O., Chambon, P., and Ghyselinck, N. B. (2003) A directional strategy for monitoring Cre-mediated recombination at the cellular level in the mouse. *Nat. Biotechnol.* **21,** 562–565.

17. Friedrich, G. and Soriano, P. (1991) Promoter traps in embryonic stem cells: a genetic screen to identify and mutate developmental genes in mice. *Genes Dev.* **5,** 1513–1523.
18. Baer, A. and Bode, J. (2001) Coping with kinetic and thermodynamic barriers: RMCE, an efficient strategy for the targeted integration of transgenes. *Curr. Opin. Biotechnol.* **12,** 473–80.
19. Buch, T., Heppner, F. L., Tertilt, C., et al. (2005) A Cre inducible diphtheria toxin receptor mediates cell lineage ablation after toxin administration. *Nat. Methods* **2,** 419–426.
20. Auwerx, J., Avner, P., Baldock, R., et al. (2004) The European dimension for the mouse genome mutagenesis program. *Nat. Genet.* **36,** 925–927.
21. Nord, A. S., Chang, P. J., Conklin, B. R., et al. (2006) The International Gene Trap Consortium Website: a portal to all publicly available gene trap cell lines in mouse. *Nucleic Acids Res.* **34,** D642–D648.
22. Rodriguez, C. I., Buchholz, F., Galloway, J., et al. (2000) High-efficiency deleter mice show that FLPe is an alternative to Cre-loxP. *Nat. Genet.* **25,** 139–140.
23. Schaft, J., Ashery-Padan, R., van der Hoeven, F., Gruss, P., and Stewart, A. F. (2001) Efficient FLP recombination in mouse ES cells and oocytes. *Genesis* **31,** 6–10.
24. Taniguchi, M., Sanbo, M., Watanabe, S., Naruse, I., Mishina, M., and Yagi, T. (1998) Efficient production of Cre-mediated site-directed recombinants through the utilization of the puromycin resistance gene, pac: a transient gene-integration marker for ES cells. *Nucleic Acids Res.* **26,** 679–680.
25. Wurst, W., Rossant, J., Prideaux, V., et al. (1995) A large-scale gene-trap screen for insertional mutations in developmentally regulated genes in mice. *Genetics* **139,** 889–899.

10

Steps Toward Targeted Insertional Mutagenesis With Class II Transposable Elements

Sareina Chiung-Yuan Wu, Kommineni J. Maragathavally, Craig J. Coates, and Joseph M. Kaminski

Summary

Insertional mutagenesis can be achieved by a variety of approaches, including both random and targeted methods. In contrast to chemical mutagenesis, insertional mutagens provide a molecular tag, thereby allowing rapid identification of the mutated genomic region. Integration into defined genomic locations has great utility for both gene insertion and mutagenesis. Our laboratories have explored targeted integration through the use of transposases coupled to defined DNA-binding domains. This technology holds great promise for targeted insertional mutagenesis by biasing integration events to regions recognized by the chosen DNA-binding domain. Herein, we provide a brief background on targeted transposon integration and detailed protocols for testing chimeric transposases in both mammalian cell culture and insect embryos.

Key Words: Cell culture; chimeric transposases; DNA-binding domain; Gal4-UAS; *piggyBac*; insertional mutagenesis.

1. Introduction

Class II transposable elements have been utilized to modify the genomes of a variety of organisms. These transposons move by a simple cut and paste mechanism, rather than copy and paste as in the case of retrotransposons. For example, the *Tc1/mariner* elements have been successfully used as transgene vectors in bacteria, protozoa, flies, mosquitoes, and vertebrates. Additionally, transgenic animals have been produced with both prokaryotic and eukaryotic transposons. For example, Moisyadi et al. were able to produce transgenic mice with *Tn5*, a prokaryotic transposon, through intracytoplasmic sperm injection *(1)*. We have recently demonstrated that *piggyBac* integrates in a variety of mammalian cell lines with efficiencies far superior than the hyperactive version of *Sleeping Beauty (2)*. However, transposons integrate nonspecifically, thus potentially resulting in untoward

From: *Methods in Molecular Biology, vol. 435: Chromosomal Mutagenesis*
Edited by: G. Davis and K. J. Kayser © Humana Press Inc., Totowa, NJ

sequelae such as decreased fitness of the organism. This nonspecific integration is problematic for gene therapy and for transgenesis; however, it is advantageous for applications such as insertional mutagenesis.

Transposons mobilized from the genome preferentially integrate (50–80%) at loci closely linked to the donor site, thus resulting in region-specific mutagenesis *(3)*. For other applications, it would be useful to target integration to allow for site-directed insertional mutagenesis and also to lessen the frequency of regional integration if the transposon is mobilized from the genome. Over the past several years, we have developed methods that may allow the use of Class II transposons as tools for targeted insertional mutagenesis.

By coupling a DNA-binding domain (DBD) to the transposase, it was suggested that integration could be directed to specific regions of the genome *(4)*. Wilson et al. *(5)* demonstrated that a zinc finger DBD could be fused to the *Sleeping Beauty* transposase; however, this addition unfortunately resulted in a marked reduction in transposition activity. In contrast, we have demonstrated that the use of the *piggyBac* transposase coupled to the GAL4 DBD (GAL4-*piggyBac*) resulted in augmented and site-directed integration as compared with controls in plasmid-based assays in mosquito embryos *(6)*. We have since verified that the GAL4- *piggyBac* transposase is also active in human embryonic kidney cells (HEK293), whereas maintaining transpositional activity as compared with the wild-type, i.e., nonmodified transposase *(2)*. It is also plausible that there will be increased transpositional activity when using a chimeric transposase as compared with the wild-type if the cell contains the DNA sequence recognized by the DBD of the chimera, as demonstrated in the plasmid assay system in mosquito embryos *(6)*.

Herein, we provide protocols for targeted transposon integration in both mammalian cell lines and insects. These protocols can be easily adapted for the purposes of targeted insertional mutagenesis.

2. Materials

2.1. Plasmid Construction

1. Oligonucleotide primers (Operon Biotechnologies, Huntsville AL) are dissolved in water to a final concentration of 100 μ*M* and stored at –20°C. The working concentration is 10 μ*M*.
2. *Taq* DNA polymerase (Stratagene, La Jolla, CA) is supplied with a 10X buffer (*see* **Note 1**).
3. dNTP mixture (2.5 m*M* for each nucleotide) (Roche Diagnostics, Indianapolis, Indiana).
4. Polymerase chain reaction (PCR) purification kit (Qiagen, Valencia, CA).
5. Transposase template (10 ng/μL, *see* **Note 2**).
6. Expression vector (*see* **Note 3**).
7. Restriction enzymes, supplied with 10X buffer (NEB, Beverly, MA).
8. T4 DNA ligase (NEB).
9. Competant cells (Stratagene).

10. *Escherichia coli* culture media (Luria-Bertani [LB] medium).
11. Maxi-prep DNA kit (Promega, Madison, WI) (*see* **Note 4**).

2.2. Excision Assay

1. Hirt lysis solution (1X): 10 m*M* Tris-HCl, pH 7.5, 10 m*M* EDTA, pH 7.5, 0.6% (w/v) sodium dodecyl sulfate (SDS). Store at room temperature.
2. 5 *M* NaCl (Fisher Scientific, Suwanee, Georgia).
3. Rubber policeman (GSC International, West Melbourne, FL).
4. Proteinase K (20 mg/mL) (Ambion, Austin, TX). Store at –20°C. The working concentration is 100 μg/mL.
5. Saturated phenol and phenol/chloroform/isoamyl alcohol (Invitrogen, Carlsbad, CA).

2.3. Transposition Assay

1. Hygromycin (Mediatech, Herndon, Virginia).
2. Modified eagle's medium (MEM)-α medium, (HyClone, Logan, Utah) supplemented with 5% fetal bovine serum ([FBS], HyClone).
3. FuGene 6 transfection reagent (Roche Diagnostics).
4. Phosphate-buffered saline (PBS) (1X): 137 m*M* NaCl, 2.7 m*M* KCl, 10 m*M* Na_2HPO_4, 2 m*M* KH_2PO_4. Store at room temperature.
   Dissolve 8 g of NaCl, 0.2 g of KCl, 1.44 g of Na_2HPO_4, and 0.24 g of KH_2PO_4 in 800 mL of water. Adjust pH to 7.4 with HCl. Add water to 1 L. Autoclave and store in aliquots at room temperature.
5. Solution of trypsin (0.05%) (HyClone). Store at –20°C.
6. Solution of paraformaldehyde (Fisher Scientific): 4% (w/v) in 1X PBS.
   Dissolve 2 g of paraformaldehyde in 40 mL of 1X PBS (pH 7.4). Add 0.5 mL of 1 *N* NaOH. Heat the solution in 80°C water bath with agitation until the paraformaldehyde is completely dissolved. Add 0.5 mL of 1 *N* HCl and add to 50 mL with 1X PBS (pH 7.4). Immediately freeze in single use (50 mL) aliquots at –20°C.
7. Methylene blue (ACROS/Fisher Scientific) solution: 0.2% (w/v) of methylene blue in water. Store at room temperature.

2.4. Plasmid Rescue

1. Cloning cylinders (Bel-art Products, Pequannock, NJ).
2. DNeasy kit (Qiagen).
3. Phenol/chloroform/isoamylalcohol solution (Invitrogen).
4. 100% Ethanol (Fisher Scientific).
5. 3 *M* Sodium acetate (pH 5.2) (Fisher Scientific).

Dissolve 408.3 g of sodium acetate in 800 mL of water. Adjust the pH to 5.2 with glacial acetic acid. Adjust the volume to 1 L with water. Autoclave and store at room temperature.

1. Mini-prep DNA Kit (Promega).
2. DNA sequencing.

ABI 3730 XL 96-capillary sequencer (Applied Biosystem, Foster City, CA).
ABI's Big Dye Terminator 3.1. kit (Applied Biosystem).
The following programs were used to edit/view/print the sequence profiles: Edit View for Mac: http://www.appliedbiosystems.com/support/software/dnaseq/installs.cfm or Finch Trace View for Windows or Mac OS X and Redhat/SuSE Linux: http://www.geospiza.com/finchtv/. For PC, use Chromas, this can be downloaded from: http://www.technelysium.com.au/chromas14x.htmlS.

2.5. Embryo Injections and Transposition Assays

1. 100% Ethanol.
2. 3 *M* Sodium acetate pH 5.2.
3. Plasmid injection buffer: 5 m*M* KCl, 0.1 m*M* sodium phosphate pH 6.8.
4. Water saturated halocarbon oil 700 (Halocarbon Products Corp, River Edge, NJ).
5. Grind buffer: 0.5% SDS, 0.08 *M* NaCl, 0.16 *M* sucrose, 0.06 *M* EDTA, 0.12 *M* Tris-HCl, pH 9.0.
6. Disposable plastic pestles for microfuge tubes.
7. 8 *M* Potassium acetate.
8. 95% and 70% ethanol.
9. TE buffer: 10 m*M* Tris-HCl pH 8.0, 1 m*M* EDTA.
10. ElectroMAX DH10β cells (Invitrogen).
11. Super optimal broth with catabolite repression (SOC) medium: 0.5% yeast extract, 2% tryptone, 10 m*M* NaCl, 2.5 m*M* KCl, 10 m*M* $MgCl_2$, 10 m*M* $MgSO_4$; sterlize by autloclaving; add filter-sterilized glucose to 20 m*M*.
12. LB (per litre): 10 g tryptone, 5 g yeast extract, 5 g NaCl, 1 mL 1 *N* NaOH, 15 g agar; sterilize by autoclaving, + ampicillin (100 μg/mL) plates.
13. LB (per litre): 10 g tryptone, 5 g yeast extract, 5 g NaCl, 1 mL 1 *N* NaOH, 15 g agar, sterilize by autoclaving, (kanamycin (25 μg/mL) + chloramphenicol (10 μg/mL) plates.
14. Mini-prep plasmid DNA preparation kit (Promega).

3. Methods

Mammalian cell culture and insect embryo assays provide rapid means to test the transpositional activities and insertional preferences of modified transposons. In both cases multiple plasmid constructs can be introduced and transposase activity measured. Excision assays can be utilized to ensure that the transposase is being expressed at suitable levels and is catalyzing the first step of the "cut and paste" process; excision of the transposon construct from the donor plasmid. Transposition assays can be utilized to quantify transpositional activity and can either be used as plasmid-to-plasmid or plasmid-to-chromosome mobilization assays. The assays described in **Subheadings 3.2.** and **3.3.** are plasmid-to-chromosome assays in mammalian cell culture and plasmid-to-plasmid assays in insect embryos. The main advantages of the plasmid-to-plasmid assays are that they can be performed rapidly and importantly, they can also be utilized to confirm DBD-target sequence interactions without a need to have the specific target already located on a chromosome. The plasmid-to-plasmid assays can also be performed in mammalian cell culture.

3.1. Plasmid Construction

1. The Gal4 DNA-binding domain was PCR amplified (example amplification conditions shown below) from the pAS1-2 plasmid (Clontech, CA) using primers that incorporated *Sac*II restriction enzyme recognition sites, a nuclear localization signal (NLS), and a flexible linker sequence. A consensus NLS (TPPKKKRKVED) *(7)* was incorporated upstream of the Gal4 DBD as part of the PCR forward primer (5′-CCGCGGATGACCCCCCCCAAGAAGAAGCGCAAGGTGGAGGACggaatgaagct actgtcttctatc-3′). The *Sac*II site is underlined and the Gal4 DBD—specific sequences—are shown in lower case. A flexible linker (KLGGGAPAVGGGPK) *(7)* was inserted between the Gal4 DBD and the transposase by incorporation of the linker sequence in the reverse PCR primer (5′-CCGCGGCCTTGGGGCCGCCG CCCACGGCGGGGGCGCCGCCGCCCAACTTcgatacagtcaactgtctttg-3′). The *Sac*II site is underlined and the Gal4 DBD-specific sequences are shown in lower case (*see* **Note 1**). The amplified Gal4 DBD was cloned into a *piggyBac* helper plasmid (pIE1-3-pB-ORF) at the *Sac*II site between the hr5-IE1 enhancer-promoter and the *piggyBac* transposase to form pIE1-Gal4-pB **(Fig. 1)** *(6)*. This transposase plasmid was used in all insect embryo assays.
2. A transposase expression plasmid for mammalian cell culture experiments **(Fig. 2)** was produced using the pIE1-Gal4-pB plasmid as a template for a PCR amplification using the following primers—(1) gatcgaattcaccATGACCCCCCCCAAGAAGAAGC and (2) CTCTAATAGTCCTCTGTGGC, with the amplified product being cloned into the expression vector, pcDNA3.1Δneo *piggyBac* (*see* **Notes 2** and **3**).
3. The PCR amplification conditions were as follows:

Template (10 ng/μL)	1 μL
Primer1 (10 μ*M*)	1 μL
Primer2 (10 μ*M*)	1 μL
Taq polymerase	0.5 μL
10X buffer	5 μL
dNTP (2.5 m*M*)	1 μL
dH_2O	41 μL
Total	50 mL

Step 1	94°C for 2 min
Step 2	94°C for 30 s
Step 3	55°C for 45 s
Step 4	72°C for 1 min
Step 5	Go to step 2 for 29 times
Step 6	72°C for 20 min
Step 7	Maintain reaction at 4°C

4. The PCR product was purified using the Qiagen PCR purification kit and then subjected to restriction enzyme digestion as follows:

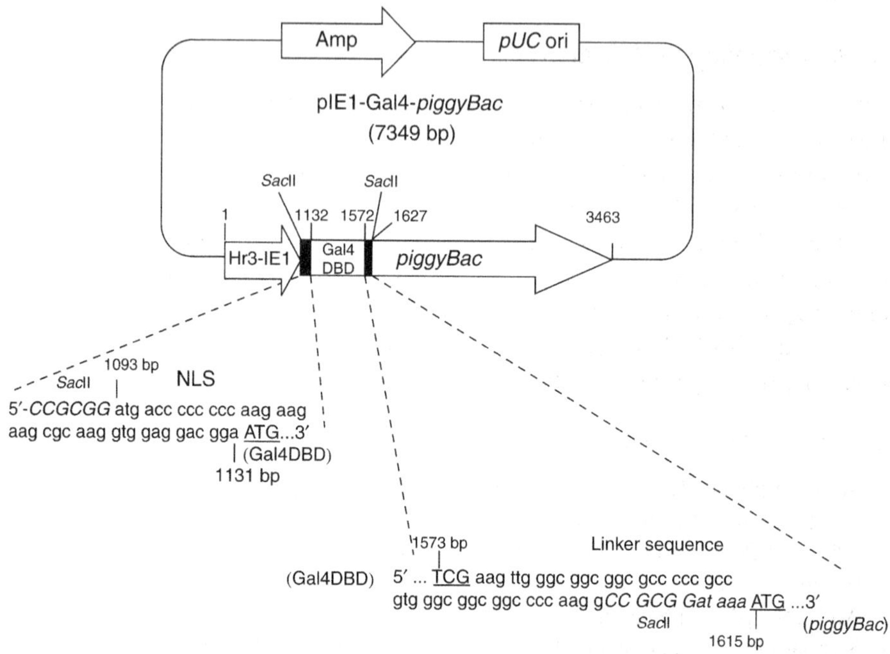

Fig. 1. Gal4 DNA-binding domain (DBD)—*piggyBac* transposase helper plasmid for insect cells. This plasmid provides a source of the chimeric transposase by utilizing the Hr3-enhancer-IE1-promoter regulatory sequences, which result in high levels of transcription in insect cells. A nuclear localization signal (NLS) was included to maximize transport of the transposase to the nucleus of the cells and a flexible linker sequence was incorporated to allow correct folding of both the Gal4 DBD and the *piggyBac* transposase.

DNA (1 μg)	X μL
Enzyme	2 μL
Buffer (10X)	5 μL
ddH_2O	Add to 50 μL
Total	50 μL

Incubate at 37°C for 2 h. Run the digested DNA on a 1% agarose gel (1X TAE). Purify the DNA fragment by following the provided protocol from the DNA purification Kit (Mol Bio lab, Inc; Carlsbad, CA).

5. The ligation reaction was performed as follows:

Digested vector (50 ng/μL)	1 μL
PCR product (10 ng/μL)	1 μL
10X buffer	1 μL
T4 DNA ligase	1 μL
ddH_2O	6 μL
Total	10 μL

Incubate at 16°C overnight.

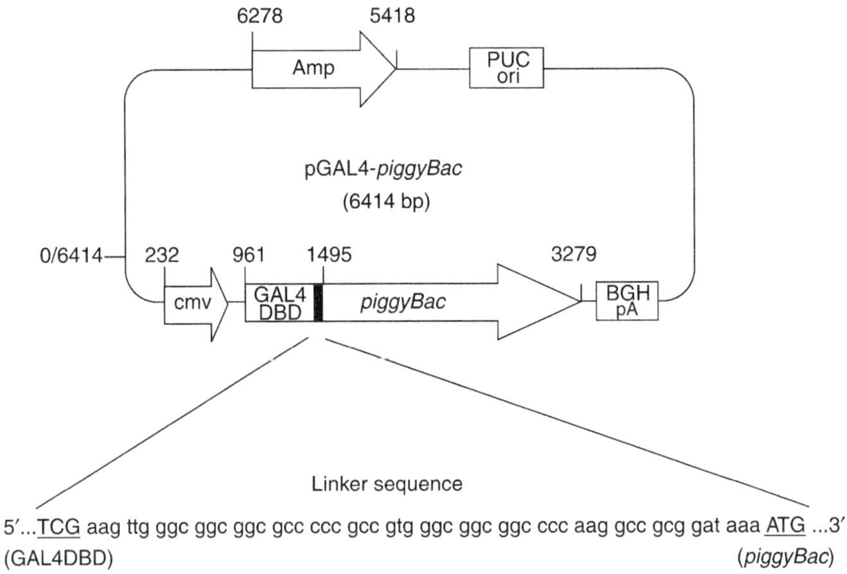

Fig. 2. Gal4 DNA-binding domain (DBD)—*piggyBac* transposase helper plasmid for mammalian cells. This plasmid provides a source of the chimeric transposase by utilizing the cytomegalovirus (CMV) promoter, which results in high levels of transcription in mammalian cells. A flexible linker sequence was incorporated to allow correct folding of both the Gal4 DBD and the *piggyBac* transposase.

6. Bacterial competent cell transformation was performed as follows:
   a. Add the ligation reaction to 50 μL of competent cell and incubate on ice for 30 min.
   b. Heat shock at 42°C for 1 min and stand on ice for 1 min.
   c. Add LB and culture at 37°C for 1 h.
   d. Spread the transformed bacteria onto selection plates.
   e. Incubate at 37°C overnight.
7. Plasmid DNA was isolated from individual clones and the construct was verified by automatic DNA sequencing.
8. Large-scale preparation of plasmid DNA constructs was performed using the Endotoxin Free Maxi-prep kit (Promega).

3.2. Excision Assay

1. Day 1: preparation of cells for transfection.
   Seed 1 × 10^6 of HEK293 cells onto the 60-mm plate with 5 mL of MEM containing 10% FBS around 2:00 PM.
2. Day 2: start the transfection procedure at 9:00 AM the next day.
3. 6 μL of FuGene 6 (Roche) and 194 μL of MEM are added to a sterile 1.5-mL Eppendorf tube.
4. Flick tube to mix and let it stand at room temperature for 5 min.

5. The 200 μL is then added to each DNA sample (1 μg of donor + 1 μg of helper).
6. Flick tube to mix and let it stand at room temperature for 20 min.
7. Remove 4 mL of medium from each well; therefore, each well should contain about 1 mL of medium.
8. Dropwise add the complex mixture from **step 3** (i.e., 200 μL of DNA/MEM/FuGene) to each well and swirl to ensure even dispersal.
9. Return cells to a 37°C humidified incubator with an atmosphere of 5% CO_2.
10. Day 5: plasmid extraction using Hirt method.
11. Rinse the 60-mm plates once with ice-cold PBS.
12. Cells are lysed by adding 200 μL of Hirt solution consisting of 10 m*M* Tris, 10 m*M* EDTA (pH 7.5), and 0.6% SDS.
13. Following incubation for 10 min at room temperature, 10 μL of 5 *M* NaCl are added and the plates rocked gently for 2 min.
14. The cell lysate is scraped from the plate with a rubber policeman, transferred to a 1.5-mL of Eppendorf tube, and then incubated overnight at 4°C.
15. After centrifugation at 4°C for 40 min, the pellet is discarded and 25 μg of proteinase K (34 U/mg protein) is added to the supernatant and incubated at 37°C for 1 h.
16. LMW DNA is extracted twice with saturate phenol and once with chloroform/isoamyl alcohol.
17. Plasmids are precipitated at –80°C for 30 min by adding 2 vol of ethanol and centrifuged at 13,000 rpm for 15 min at 4°C.
18. The pellet is washed with 70% ethanol and centrifuged for 5 min at 13,000 rpm.
19. The pellet is dissolved in 20 μL of water.
20. Design two pairs of primers positioned at approx 500 bp outside the putative excision site (e.g., primer pair 1 and 2 and primer pair 3 and 4).
21. PCR reaction conditions.

Template (from Hirt extraction)	1 μL
Primer 1 (10 μ*M*)	1 μL
Primer 2 (10 μ*M*)	1 μL
Taq polymerase (Stratagene; no. 600132-81)	0.5 μL
10X buffer (supplied with *Taq*)	5 μL
dNTP (2.5 m*M*)	1
dH_2O	41 μL
Total	50 μL

Step 1	94°C for 2 min
Step 2	94°C for 30 s
Step 3	55°C for 45 s
Step 4	72°C for 1 min
Step 5	Go to step 2 for 29 times
Step 6	72°C for 20 min
Step 7	Maintain reaction at 4°C

22. Perform a 1–50 dilution of the 1st round of PCR product with water and use 1 μL to perform the 2nd round of PCR with the same reaction mixture (except primer 3 and 4) and PCR condition.
23. Second round PCR reaction conditions.

Template (1:50 dilution from 1st PCR)	1 μL
Primer 3 (10 μ*M*)	1 μL
Primer 4 (10 μ*M*)	1 μL
Taq polymerase (Stratagene; no. 600132-81)	0.5 μL
10X buffer (supplied with *Taq*)	5 μL
dNTP (2.5 m*M*)	1
dH_2O	41 μL
Total	50 μL

Step 1	94°C for 2 min
Step 2	94°C for 30 s
Step 3	55°C for 45 s
Step 4	72°C for 1 min
Step 5	Go to step 2 for 29 times
Step 6	72°C for 20 min
Step 7	Maintain reaction at 4°C

24. Run the PCR products on the 1% agarose gel in 1X TAE buffer.

3.3. Transposition Assay and Plasmid Rescue

1. Mammalian cells lines are maintained in MEM-α medium (Hyclone) containing 5% FBS (Hyclone) at 37°C in a humidified incubator with an atmosphere of 5% CO_2.
2. Day 1: seed 1 × 10^5 cells/well into individual wells of a (24-well plate) containing 500 μL medium.
3. Day 2: performing transfection using a total 400 ng of pcDNA3.1ΔNeo, helper and donor. The cells should be approx. 75% confluent at the time of transfection.
4. 1 μL of FuGene 6 (Roche) and 19 μL of MEM are added to a sterile 1.5-mL Eppendorf tube.
5. Flick tube to mix and let stand at room temperature for 5 min.
6. The 20 μL is then added to each DNA sample.
7. Flick tube to mix and let stand at room temperature for 20 min.
8. Remove 300 μL of medium from each well; therefore, each well should contain about 200 μL of medium.
9. Dropwise add the complex mixture from **step 3** (i.e., 20 μL of DNA/MEM/FuGene) to each well and swirl to ensure even dispersal.
10. Return cells to a 37°C humidified incubator with an atmosphere of 5% CO_2.
11. Day 3: plating cells on hygromycin plates 18 h after transfection.
12. Aspirate off spent media and discard.
13. Add 1 mL of PBS, being careful not to disturb the monolayer and gently rock back and forth. Remove PBS and discard.

14. Add 300 μL of trypsin to each well and rock the plate to ensure the entire monolayer is covered with trypsin solution.
15. Incubate until the cells begin to detach.
16. Add cells from each well to individual 1.5-mL Eppendorf tubes containing 700 μL of MEM-α medium containing 10% FCS for a total of 1 mL.
17. Take 100 μL from **step 5**, i.e., 1/10th of the solution containing the cells, and add it to 100-mm plates (MEM/5%FCS + hygromycin), and evenly disperse by pipeting. The concentrations of hygromycin B used in HeLa, HEK293, H1299, and Chinese hamster ovary cells are 200, 100, 400, and 400 μg/mL, respectively.
18. Day 17: counting clones.
19. To count the clones, cells are fixed with PBS containing 4% paraformaldehyde for 10 min.
20. Stain fixed cells with 0.2% methylene blue for 1 h.
21. After 14 d of hygromycin selection, only colonies larger than 0.5 mm in diameter are counted.

3.4. Plasmid Rescue Procedure

1. Eighteen hours after transfection, one fiftieth of cells from individual wells (24-wells plate) are seeded onto each 150-mm plate with 20 mL of α-MEM medium containing 5% FBS and hygromycin. The concentrations of hygromycin B used in HeLa, HEK293, H1299, and Chinese hamster ovary cells are 200, 100, 400, and 400 μg/mL, respectively.
2. Day 15: isolate individual clones.
3. Fourteen days after transfection, remove the medium and rinse each plate once with 20 mL of PBS and then remove PBS by aspiration. Individual hygromycin-resistant clones are isolated using cloning cylinder (Bel-art products no. 378470200) by placing each clone within the cylinder.
4. Add 20 μL of trypsin to each cylinder and incubate for 2 min.
5. Cells from each cylinder are transferred into individual wells (96-well plate) containing 180 μL of MEM/5% FBS/hygromycin.
6. Expand cells by gradually transferring to a larger well (i.e., 48-well, 24-well, 12-well, 6-well, 60 mm, and 100-mm plate) once cells reach 100% confluency on plates.
7. One to two months: isolate genomic DNA.
8. Harvest cells once the cells reach 100% of confluency on 100-mm plate. It takes about 1–2 mo. to reach this point from a single colony.
9. DNA of individual hygromycin-resistant clones is isolated using DNeasy kit from Qiagen. Follow detailed instructions described in DNeasy Tissue Handbook.
10. 5 μg of genomic DNA is digested overnight with appropriate restriction enzymes. *Xho*I (100 U per digestion) is used for retrieving genomic information adjacent to one end of the *piggyBac* inverted repeat in 100 μL of reaction solution.
11. Stop the enzyme reaction with phenol/chloroform exaction (100 μL of phenol/ chloroform) once.
12. Precipitate DNA with 250 μL of 100% ETOH and 10 μL of 3 *M* NaOAC (incubate at –20°C for 2 h or at –80°C for 20 min). Make sure to mix the mixture well before incubation.

13. After centrifuging at 13,000 rpm for 20 min, the pellet is dissolved in 25 μL of water.
14. 2.5 μL of ligase (NEB M0202) and 2.5 μL of 10X ligation buffer are added to the 25 μL of DNA from **step 5**. Incubate the reaction at 16°C overnight.
15. 30 μL of ligation reaction is transformed into 200 μL of ultracompetent cells (Stratagene 200150) following the Stratagene protocol.
16. For sequencing, individual plasmids rescued are isolated using plasmid purification kit (Promega A7640). Primer for sequencing: GTCCTGTCGGGTTTCGC.

3.5. Embryo Injections and Transposition Assays

1. Combine the pIE1-Gal4-*pB* helper (0.25 μg/μL), pB(KOα) donor (0.25 μg/μL), and pGDV1-UAS target (0.5 μg/μL) plasmids *(6)* and precipitate with 1/10th vol of 3 *M* sodium acetate pH 5.2 and 2 vol 100% ethanol.
2. Wash DNA pellet with 70% ethanol and air-dry.
3. Resuspend DNA pellet in injection buffer at a concentration up to 1 mg/mL (*see* **Note 5**).
4. In a cotton plugged vial, place absorbent cotton, add water and a round egg collecting paper (Whatman 3 MM) on top of the wet absorbent cotton.
5. Add six blood-fed females to the vial and keep in the dark for 40 min.
6. Keep looking for white eggs every 5 min after 40 min and remove light-gray eggs and align on filter paper with posterior (thin) ends facing one direction.
7. Once the eggs are all aligned, remove all moisture from the paper by placing dry paper around eggs and rubbing with finger to wick moisture away.
8. Use a cover slip holding a slice of double-sided sticky tape to press down on the eggs and lift off paper.
9. Desiccate in air until a couple of "dimples" appear on the egg's surface (*see* **Note 6**).
10. Immediately cover eggs with halocarbon oil, inject as soon as possible (*see* **Note 7**).
11. Allow eggs to recover for 20 min and then remove excess oil, gently remove eggs onto wet filter paper, maintain moisture, and keep overnight at 25°C.
12. Transfer embryos to a microfuge tube, add 100 μL of grind buffer, and homogenize with disposable plastic pestle.
13. Wash pestle into tube with an additional 20 μL and incubate at 65°C for 30 min.
14. Add 14 μL of 8 *M* potassium acetate and place on ice for 30 min.
15. Centrifuge at full speed for 10 min at room temperature.
16. Carefully remove supernatant and transfer to a fresh microfuge tube, taking note of the approximate volume.
17. Add 2 vol of 95% ethanol, mix by gently inverting tube and place at room temperature for 2 min.
18. Centrifuge at full speed for 10 min at room temperature.
19. Discard supernatant and wash pellet in 70% ethanol.
20. Spin as above, carefully remove supernatant and air-dry pellet.
21. Resuspend pellet in an appropriate volume of TE buffer. 3 μL of buffer is used per 30 embryos.
22. Electroporate 1 μL of plasmid rescue DNA into DH10B cells (*see* **Note 8**).
23. Add 1 mL of SOC media and recover with shaking at 250 rpm for 1 h.

24. Plate 5 μL on a LB + ampicillin plate to determine the recovery of donor plasmids.
25. Spin down the remaining cells at 5000 rpm and discard all but 100 μL of SOC media.
26. Resuspend the cells in the remaining SOC and plate on a LB + kanamycin + chloramphenicol plate to recover transposition products.
27. Plasmid DNA is prepared by the Promega mini-prep kit and sequencing performed with the following primer: GTCCTGTCGGGTTTCGC.
28. Transformation frequency is calculated as the number of transpositions recovered per donor plasmid recovered.

4. Notes

1. Because of the long lengths of these oligonucleotide primers, they were purified by polyacrylamide gel electrophoresis to ensure that full-length primers were being utilized, and thus, maintaining the correct reading frame for the DBD-transposase fusions, including the NLS and linker.
2. The Gal4 DNA-binding domain was PCR amplified from pIE1-Gal4-*pB* ***(6)*** followed by *Eco*RI and *Sex*AI digestion. The resulting fragment was then cloned into the *Eco*RI and *Sex*AI sites of pcDNA3.1ΔNeo-*piggyBac* ***(2)***.
3. pcDNA3.1ΔNeo was obtained by deleting the *Nsi*I and *Bsm*I fragment of pcDNA3.1(+) (Invitrogen) to remove the most of the *Neo* coding sequence. The ends of the resulting fragment were blunted with T4 DNA polymerase followed by self-ligation using T4 DNA ligase. The wild-type *piggyBac* expression plasmid, pcDNA3ΔNeo-piggyBac, for mammalian cell culture was produced by cloning the *piggyBac* coding sequence, obtained from the *Sac*II and *Bam*HI fragment of PIE3-pigORF, into the *Eco*RV site of pcDNA3.1ΔNeo ***(2)***.
4. Plasmids prepared for microinjection into insect embryos have traditionally been prepared by utilizing two rounds of CsCl centrifugation and banding to ensure the highest purity. However, several laboratories have reported successful insect transformation experiments utilizing plasmid DNA that was prepared using commercial kits that produce endotoxin-free DNA.
5. DNA concentrations above 1 mg/mL can result in more frequent clogging of the microinjection needle and another important consideration is that it has been demonstrated that some transposases suffer from overproduction inhibition and as such may actually be more active at lower concentrations.
6. Dessication time and conditions will be species-specific. Some insect embryos will survive once dechorionated (manual removal on double-sided sticky tape or mild bleach treatment), which makes for easier injections whereas other species like mosquitoes will not survive dechorionation. Some species can be dessicated in air on the bench whereas others will require placement in a dessication chamber with a moisture-absorbing compound.
7. Injection techniques and volumes are insect-specific. Manual injections with a large syringe can be used to generate the pressure to drive DNA solutions out of a needle or automated microinjection systems are often used. The corresponding pressure and time of injection vary for each species and even with the degree of dessication.
8. Electorporation parameters equivalent to 2.5 kV, 200 Ω with a 2-mm cuvet gap.

Acknowledgment

This work was supported in part by National Institute of Health grant no. AI047303 to Craig J. Coates.

References

1. Suganuma, R., Pelczar, P., Spetz, J. F., Hohn, B., Yanagimachi, R., and Moisyadi, S. (2005) Tn5 transposase-mediated mouse transgenesis. *Biol. Reprod.* **73,** 1157–1163.
2. Wu, C. -Y., Meir, Y. -J., Coates, C. J., et al. (2006) *piggyBac* is a flexible and highly active transposon as compared to Sleeping Beauty, Tol2, and Mos1 in mammalian cells. *Proc. Natl. Acad. Sci. USA* **103,** 15,008–15,013.
3. Carlson, C. M. and Largaespada, D. A. (2005) Insertional mutagenesis in mice: new perspectives and tools. *Nat. Rev. Genet.* **6,** 568–580.
4. Kaminski, J. M., Huber, M. R., Summers, J. B., and Ward, M. B. (2002) The design of a non-viral vector for site-selective, efficient integration into the human genome. *FASEB J.* **16,** 1242–1247.
5. Wilson, M. H., Kaminski, J. M., and George, A. L. Jr. (2005) Functional zinc finger/Sleeping Beauty transposase chimeras exhibit loss of overproduction inhibition. *FEBS Lett.* **579,** 6205–6209.
6. Maragathavally, K. J., Kaminski, J. M., and Coates, C. J. (2006) Chimeric *Mos1* and *piggyBac* chimeric transposases result in site-directed integration. *FASEB J.* **20,** 1880–1882.
7. Szuts, D. and Bienz, M. (2000) LexA chimeras reveal the function of *Drosophila* Fos as a context-dependent transcriptional activator. *Proc. Natl. Acad. Sci.* **97,** 5351–5356.

11

Targeting Integration of the *Saccharomyces* Ty5 Retrotransposon

Troy L. Brady, Clarice L. Schmidt, and Daniel F. Voytas

Summary

Many retrotransposons and retroviruses display integration site specificity. Increasingly, this specificity is found to result from recognition by the retroelement of specific chromatin states or DNA-bound protein complexes. A well-studied example of such a targeted retroelement is the *Saccharomyces* Ty5 retrotransposon, which integrates into heterochromatin at the telomeres and silent mating loci. Targeting is mediated by an interaction between Ty5 integrase (IN) and the heterochromatin protein silent information regulator 4 (Sir4). A small motif of IN, called the targeting domain, is responsible for this interaction. Ty5 integration can be directed to DNA sites outside of heterochromatin by tethering Sir4 to ectopic locations using fusion proteins between Sir4 and a DNA-binding domain. Alternatively, the targeting domain of Ty5 can be swapped with peptides that recognize other protein partners, thereby generating Ty5 elements with new target specificities. The mechanism of Ty5 target site choice suggests that integration specificity of other retrotransposons and retroviruses can be altered by engineering integrases to recognize DNA-bound protein partners. Retroelements can also be used to probe chromatin dynamics and the distribution of protein complexes on chromosomes. Here, we describe the basic assay by which Ty5 integration is monitored to sites of tethered Sir4.

Key Words: Chromatin; integration; integrase; retrotransposon; retrovirus; specificity.

1. Introduction

Retrotransposons and retroviruses are abundant components of eukaryotic genomes *(1)*. Retroelements replicate by copying mRNA into cDNA and then inserting the cDNA into the host genome using an element-encoded integrase (IN). Insertion sites for many retroelements are not distributed randomly along chromosomes *(2)*. Although numerous forces, including selection and recombination, can influence retroelement distribution patterns, targeted integration is emerging as an important underlying force *(3)*. Retroelement target site choice can be influenced by specific types of chromatin or DNA-bound protein complexes. For example,

From: *Methods in Molecular Biology, vol. 435: Chromosomal Mutagenesis*
Edited by: G. Davis and K. J. Kayser © Humana Press Inc., Totowa, NJ

the *Saccharomyces cerevisiae* Ty1 and Ty3 retrotransposons integrate upstream of sites of RNA polymerase III transcription ***(4,5)***. For both of these elements, the occupancy of target sites by RNA polymerase III is required for integration specificity, suggesting they recognize polymerase directly, as appears to be the case for Ty3 ***(6)***, or some chromatin feature associated with RNA polymerase III transcription ***(7)***. Chromatin or DNA-bound protein complexes also appear to play a role in retroviral target specificity. HIV, for example, integrates preferentially near actively transcribed genes, and murine leukemia virus (MLV) insertions are biased toward promoter regions ***(8,9)***. HIV IN interacts with a specific DNA-bound protein—namely, lens epithelium-derived growth factor (LEDGF)—and this interaction is critical for determining HIV integration sites. ***(10,11)***.

An important model for understanding mechanisms of retroelement target site choice is the *Saccharomyces* retrotransposon Ty5. Ty5's preferred target site is heterochromatin: about 90% of *de novo* Ty5 transposition events occur in heterochromatin at the telomeres or silent mating loci (*HML* and *HMR*) ***(12–14)***. Ty5 IN determines target site choice through a small motif at the C-terminus (the targeting domain, TD), and single amino acid substitutions within TD randomize Ty5 integration patterns ***(15,16)***. TD interacts with a protein component of heterochromatin, namely, silent information regulator 4 (Sir4) ***(15,17)***. This interaction tethers the integration complex to target sites, and the TD/Sir4 interaction is necessary and sufficient for target specificity ***(17)***. Ty5 target site choice can be altered by swapping TD with short peptides that recognize other protein partners. Such engineered Ty5 elements exhibit new target specificities and integrate at sites to which the protein partner is tethered. Surprisingly, the efficiency with which the engineered Ty5 elements integrate in the vicinity of their new protein partners is comparable with the efficiency with which wild-type Ty5 integrates near sites of tethered Sir4 ***(17)***.

Understanding the mechanism of Ty5 target specificity suggests ways in which other retrotransposons and retroviruses can be harnessed for genome modification. For example, the *Schizosaccharomyces pombe* Tf1 retrotransposon normally integrates in intergenic regions ***(18)***, yet, we have engineered a Tf1 element to integrate in heterochromatin by adding to IN a chromodomain that recognizes heterochromatin-specific histone modifications (Gao and Voytas, unpublished). Although targeting to heterochromatin occurred at low frequencies, the results suggest that this approach may be generally applicable. In the case of retroviruses, there is considerable interest in developing vectors for gene therapy with enhanced integration specificity both to minimize deleterious mutations caused by retrovirus integration and to better control the expression of virally delivered therapeutic transgenes.

A more direct exploitation of Ty5-targeting specificity is to use Ty5 to probe chromosomes for sites occupied by specific DNA-bound proteins ***(19)***. For example, Sir4 can be fused to transcription factors or other DNA-binding proteins, and Ty5 would be predicted to integrate at chromosomal sites occupied by the transcription factor-Sir4 fusion. The utility of this approach has recently been demonstrated in

Saccharomyces to identify promoters bound by seven different transcription factor-Sir4 fusions ***(20)***. This technique complements approaches such as ChIP-Chip assays, which are often used to identify chromosomal sites occupied by DNA-binding proteins ***(21)***. Here, we provide a detailed description of the assay used to measure targeting of Ty5 to sites of tethered Sir4 ***(17)***. This assay serves as the cornerstone for understanding Ty5 IN/Sir4 interactions and for developing other applications that derive from the ability to control target specificity of this retroelement.

2. Materials

2.1. Plasmids and Yeast Strains

1. Yeast strain YPH499 (*MATa ura3-52 lys2-801 ade2-101 trp1Δ63 his1Δ200 leu2Δ1*) ***(22)***.
2. *Escherichia coli* strain eDW335. (*recA1 endA1 gyrA96 thi-1 hsdR17 supE44 relA1 lac hisB463[del] zef-3128::Tn10 Tet*^{r}). This strain is a derivative of XL1 Blue (Stratagene, San Diego, CA) that has been cured of its F' plasmid. P1 phage transduction was performed to introduce a nonreverting *hisB* allele from the *E. coli* strain CGSC no.7397 (Wright and Voytas, is still unpublished).
3. Plasmid pDR14. This plasmid carries a galactose-inducible Ty5 element with a *his3AI* marker gene to select for Ty5 transposition events. A *URA3* marker gene is present for selection in yeast, and the plasmid confers ampicillin resistance to *E. coli* ***(23)***.
4. Plasmid pYZ316. This plasmid serves as a target for Ty5 integration. Four copies of LexA operators are imbedded within 7 kb of plant DNA. The plant DNA ensures that most Ty5 insertions do not compromise plasmid function. A *TRP1* marker gene enables selection in yeast, and the plasmid confers chloramphenicol resistance to *E. coli* ***(17)***.
5. Plasmid pYZ317. This plasmid is similar to pYZ316, except no LexA operators are present, and so it serves as a control. *TRP1*, Cam^r ***(17)***.
6. Plasmid pYZ276. This plasmid expresses a LexA-Sir4 fusion protein (amino acids 951–1358). A *LEU2* marker is present for selection in yeast, and the plasmid confers ampicillin resistance to *E. coli* ***(17)***.

2.2. Media for Growth of Yeast

1. Yeast extract/peptone/dextrose (YPD) medium: dissolve 1% yeast extract and 2% peptone (United States Biological, Swampscott, MA) in water and autoclave. Add dextrose from a sterile (10X) stock solution to 2% final concentration. For solid media, add 2% bacto agar ([BD], Franklin Lakes, NJ) before autoclaving.
2. SC-Ura-Leu-Trp-His medium: dissolve 0.17% yeast nitrogen base without amino acids (United States Biological), 0.06% CSM-His-Leu-Trp-Ura (MP Biomedicals, Solon, OH), and 0.002% L-histidine hydrochloride (Sigma, St. Louis, MO) in water and autoclave. Add carbon source (dextrose or galactose) from sterile (10X) stock solution to 2% final concentration. For solid media, add 2% BD before autoclaving.
    a. SC-Ura medium: prepare SC-Ura-Leu-Trp-His as described earlier and supplement with 0.003% L-leucine, 0.002% L-tryptophan, and 0.002% L-histidine hydrochloride.

b. SC-Ura-Leu medium: prepare SC-Ura-Leu-Trp-His as described earlier and supplement with 0.002% L-tryptophan and 0.002% L-histidine hydrochloride.
c. SC-Ura-Leu-Trp medium: prepare SC-Ura-Leu-Trp-His as described earlier and supplement with 0.002% L-histidine hydrochloride.
d. SC-Trp-His medium: prepare SC-Ura-Leu-Trp-His as described earlier and supplement with 0.002% uracil and 0.003% L-leucine.

2.3. Media for Growth of E. coli

1. Luria Bertani (LB): dissolve 1% tryptone, 0.5% yeast extract, and 0.5% sodium chloride (United States Biological) in water, adjust the pH to 7.0 and autoclave. For solid media, add 2% BD before autoclaving.
2. Super optimal broth plus catabolite (SOC) dissolve 0.5% yeast extract, 2% tryptone, 20 m*M* dextrose, 10 m*M* sodium chloride (United States Biological), 2.5 m*M* potassium chloride, 10 m*M* magnesium chloride, and 10 m*M* magnesium sulfate (Fisher Scientific, Pittsburg, PA) in water and autoclave.
3. Chloramphenicol: dissolve 2% chloramphenicol (Sigma) in ethanol. Aliquot and store at –20°C.
4. LB + Cam: prepare LB media as described earlier, cool to 50°C and add chloramphenicol to 20 mg/L. Store at 4°C.
5. M9 + Cam: dissolve 0.6% sodium phosphate dibasic, 0.3% potassium phosphate monobasic, 0.1% ammonium chloride, (Fisher Scientific), 0.05% sodium chloride, and 2% BD in water and autoclave. Add 1 m*M* magnesium sulfate, 0.2% dextrose and 0.01% thiamine (Sigma) from filter sterilized (10X) stock solutions. Cool media to 50°C and add chloramphenicol to 20 mg/L, pour plates, and store at 4°C.

2.4. Preparation of the Yeast Strain

1. Lithium acetate: 0.1 *M* lithium acetate (Sigma) in water, autoclaved.
2. ssDNA: dissolve 2 mg/mL DNA sodium salt from salmon testes (Sigma) in water and stored in aliquots at –20°C. Boil for 5 min to denature and chill in ice/water just before use.
3. Polyethylene glycol (PEG) solution: dissolve 50% PEG 3350 (Sigma) in water and filter sterilize.
4. Sterile distilled water.
5. 30% glycerol: dissolve 30% (v/v) glycerol in water and autoclave.

2.5. Induction of Transposition

1. Sterile flat toothpicks.
2. Supplies required for replica plating: velveteens, 15 cm^2 cotton velveteen or cheesecloth squares sterilized in autoclave; replica plating block (Cora Styles Needles 'N Blocks, Hendersonville, NC); sterile forceps for positioning velveteens on replica block.

2.6. Preparation of DNA From Yeast Cells

1. Sterile flat toothpicks.
2. Breaking buffer: dissolve 2% (v/v) Triton X-100, 1% sodium dodecyl sulfate, 100 m*M* sodium chloride, 10 m*M* Tris-Cl (from 1 *M* stock solution, pH 8.0) and 1 m*M* ethylenediaminetetraacetic acid (from 0.5 *M* stock solution, pH 8.0) in water.

3. Screw-cap centrifuge tubes: screw-cap microcentrifuge tubes with sealing O-ring (cat no. 05-669-43 from Fisher Scientific) should be used to prevent leaking of contents during vortexing.
4. Acid-washed glass beads: treat glass beads (425–600 μm or 30–40 US sieve) with 1 *M* nitric acid overnight under agitation. Rinse several times with water. Dry and store at –20°C.
5. Phenol/chloroform mixture: phenol/chloroform/isoamyl alcohol (25:24:1 mixture, pH 6.7) (Fisher Scientific).
6. Vortex: strong vortex apparatus with minimum of 3000 rpm speed and multitube attachment is recommended.

2.7. Electroporation of E. coli

1. 500 mL ice cold sterile water.
2. 10% glycerol: sterile solution of 10% (v/v) glycerol in water.
3. Electroporator: gene pulser/*E. coli* pulser (Bio-rad Laboratories, Hercules, CA) settings are: 2.5 kV, 25 μF, 200 Ω.
4. Cuvets: 2-mm gap electroporation cuvets.

3. Methods

This protocol measures the frequency with which the yeast Ty5 retrotransposons integrates at DNA sites to which the Sir4 protein is tethered ***(17)***. As illustrated in **Fig. 1**, Sir4 is expressed as a fusion to the LexA DNA-binding domain and tethered to DNA sites that have LexA operators. Ty5 typically integrates within a 100-bp window on either side of the operators.

Executing this protocol requires a yeast strain with three separate plasmids. One plasmid carries a Ty5 element under control of a galactose-inducible promoter (pDR14). The Ty5 element has a *his3AI* selectable marker for detecting transposition events ***(24)***. This marker is a *HIS3* gene with an intron that renders it nonfunctional. However, when the intron is spliced from Ty5 mRNA and reverse transcribed into cDNA, the *HIS3* gene becomes functional and transposition events can be selected as His^+ yeast cells. A second plasmid in the yeast strain serves as a target for Ty5 integration (pYZ316). This plasmid has LexA operators that bind the LexA-Sir4 fusion protein. Typically, a target plasmid without LexA operators is used as a control (pYZ317). The third plasmid expresses the LexA-Sir4 fusion protein (pYZ275). As Ty5 IN interacts with the Sir4 C-terminus ***(15,17)***, typically only LexA-Sir4 C-terminal fusions are expressed (e.g., aa 951–1358 in the case of pYZ275).

Each plasmid encodes a selectable marker that confers growth on yeast minimal medium lacking a particular nutrient supplement (uracil, tryptophan, or leucine). To select for these plasmids, the recipient yeast strain must carry mutations that prevent the synthesis of each of these nutrients, as well as histidine so that transposition events can be selected. The yeast strain YPH499 has the requisite auxotrophies for these selections.

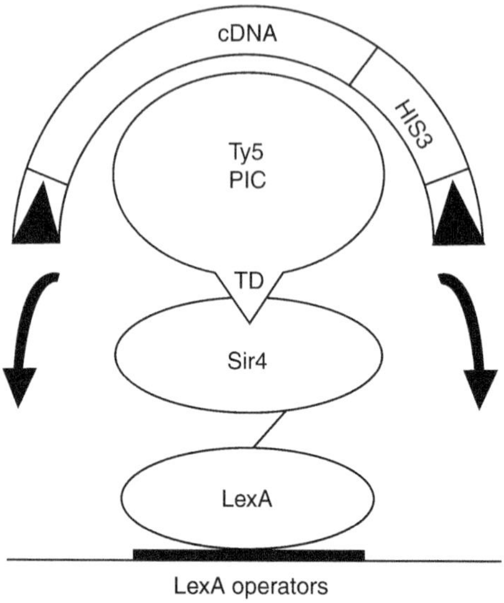

Fig. 1. The tethered integration assay. The Ty5 preintegration complex (PIC) is shown complexed to Ty5 cDNA. The TD of Ty5 IN interacts with Sir4. The target plasmid carries LexA operators that bind LexA-Sir4. *Arabidopsis* DNA flanks the operators to create a Ty5 landing pad. The target plasmid also has selectable markers and replication origins that function in yeast (*TRP1*, 2 μm) and *E. coli* (chloramphenicol-resistance gene, *ColE1*). After induction of Ty5 transposition, target plasmids are recovered in *E. coli* by selecting either for chloramphenicol resistance (Chl^r), or for chloramphenicol resistance and histidine prototrophy (Chl^rHis^+). The latter selection determines the number of target plasmids that carry Ty5 with its *HIS3* marker. The ratio of Chl^rHis^+ to Chl^r indicates the efficiency with which Ty5 integrates into the plasmid.

3.1. Preparation of the Yeast Strain

1. Prepare YPH499 cells for transformation. Inoculate 5 mL of liquid YPD medium with a colony of YPH499. Grow to early log phase (OD_{600} = 0.5). Centrifuge 1 mL of the culture in a 1.5-mL microcentrifuge tube for 10 s at 5000*g*. Discard the supernatant and resuspend the pellet in 1 mL of sterile water. Centrifuge again and resuspend the cell pellet in 1 mL 0.1 *M* LiOAc. Repeat the centrifugation once again and resuspend the pellet to 50 μL of 0.1 *M* LiOAc.
2. Add 0.1–0.5 μg of pDR14 DNA and 5 μL of ssDNA to the yeast cells. Mix by tapping. Add 350 μL PEG solution and mix again. Incubate at 30°C for 30 min before transferring to a 42°C water bath for 15–20 min. Centrifuge the cells for 10 s at 5000*g*. Discard the supernatant and resuspend the cell pellet in 100 μL sterile water. Plate the cells onto a SC-Ura plate and incubate at 30°C for 2–3 d. Pick a well-isolated colony and streak out for single colonies on a fresh SC-Ura plate for use in the next step (*see* **Note 1**).
3. Repeat **step 2**, except, grow the yeast cells on liquid SC-Ura medium. Repeat **step 3**, except, transform with pYZ276 and plate the cells onto an SC-Ura-Leu plate. Select an isolated colony for use in the next step.

4. Repeat **step 2**, except, grow the yeast cells on liquid SC-Ura-Leu medium. Repeat **step 3**, except, transform individual aliquots of cells with either pYZ316 (the target plasmid with four LexA operators) or pYZ317 (a control target plasmid with no LexA operators). Plate the cells onto an SC-Ura-Leu-Trp plate.
5. Pick at least three well-isolated colonies from both the pYZ316 (test) and pYZ317 (control) transformations for subsequent experiments. These strains can be stored indefinitely by growing 5 mL cultures to early log phase (OD_{600} = 0.5) in SC-Ura-Leu-Trp medium. Add 0.5 mL of 305 glycerol to 0.5 mL of the culture. Mix and freeze at –80°C.

3.2. Induction of Transposition

1. Using sterile toothpicks, inoculate three cultures each of the test and control strains into individual wells of a sterile 96-well plate containing 200 µL of liquid SC-Ura-Leu-Trp medium per well. Incubate the plate at 30°C for 24 h without shaking. (*see* **Note 2**).
2. After 24 h, the yeast cells will have grown across the bottom of the well and can be suspended by pipeting up and down with a micropipet. Spot 80 µL of the cell suspension onto a SC-Ura-Leu-Trp plate that has 2% galactose and 0.5% raffinose. The galactose induces the expression of the Ty5 element. Incubate the plate at room temperature (22–20°C) for 3 d.
3. Replica plate the induced yeast cells onto an SC-Trp-His plate. This medium selects for transposition of Ty5 elements, because transposition through a RNA intermediate results in loss of an inactivating intron in the *his3AI* marker gene and confers histidine prototrophy *(24)*. The absence of tryptophan in the media ensures retention of the target plasmid, as it carries the *TRP1* gene. Incubate the plates at 30°C for 3–4 d.
4. For each test and control strain, scrape the yeast cells from the SC-Trp-His plate and resuspend them in 5 mL SC-Trp-His liquid medium. Incubate the culture overnight at 30°C with shaking (*see* **Notes 3** and **4**).

3.3. Preparation of DNA From Yeast Cells

1. Using the overnight cultures generated in **Subheading 3.2., step 4**, centrifuge the cells for 2 min at 5000*g* in a clinical centrifuge. Discard the supernatant and resuspend the cell pellet in 1 mL water. Transfer the resuspended cells to a screw-cap microcentrifuge tube. Centrifuge for 10 s at 5000*g*. Discard the supernatant. Cells can be stored at –20°C for future analysis, if desired.
2. Add 200 µL of DNA-breaking buffer to the cell pellets. Vortex until the cells are resuspended. Add approx 200 µL (~0.3 g) glass beads and 200 µL phenol/chloroform/isoamyl alcohol (25:24:1). Vortex for 20–30 min. Centrifuge the broken cells at 16*g* for 5 min (*see* **Note 5**).
3. Remove the supernatant, be careful not to contaminate the supernatant with cellular debris that has accumulated at the interface of the aqueous and organic layers. Typically, only 100 µL of the supernatant can be removed safely. The supernatant can be stored at –20°C for future use.

3.4. Preparation of E. coli *Cells for Transformation by Electroporation*

1. The day before the preparation of the electro-competent cells, inoculate a 5 mL culture of LB medium with a single colony of eDW335. Grow overnight at 37°C with shaking.

2. Use the 5 mL starter culture to inoculate 500 mL of LB medium in a sterile 1-L flask. Incubate with shaking at 37°C until OD_{600} = 0.4–0.6. Divide the culture into two, sterile 250-mL centrifuge tubes and centrifuge for 8 min at 5000–6000*g* at 4°C.
3. Discard the supernatant and resuspend each cell pellet in 50 mL ice cold, sterile water.
4. Combine the resuspended cells and centrifuge for 8 min at 5000–6000*g* at 4°C. Repeat **steps 3** and **4** two more times using 100 mL of ice cold sterile water.
5. After the final wash, resuspend the pellet in 25 mL of ice cold, sterile 10% glycerol. Pellet cells by centrifuging 5000–6000*g* at 4°C. Pour off supernatant and resuspend the pellet in 3 mL ice cold, sterile 10% glycerol.
6. Aliquot 100 μL of cells to individual, sterile microcentrifuge tubes. Use immediately for electroporation or freeze the cells at –80°C for future use.

3.5. Transformation of E. coli and Recovery of Target Plasmids With Integrated Ty5 Elements

1. The electro-competent eDW335 *E. coli* cells are transformed with DNA present in the yeast cell supernatants prepared in **Subheading 3.3., step 3**. If frozen electro-competent cells are used for the transformation, allow them to first thaw on ice. To each 100 μL aliquot of cells, add 2 μL of the yeast supernatant prepared from either test or control yeast cells. Mix by tapping.
2. Transfer the DNA/cell mixture to an ice cold electroporation cuvet with a 2 mm gap. Electroporate at 2.5 kV, 200 Ω, 25 μF.
3. Immediately after electroporation, add 1.5 mL SOC to the cuvet. Transfer the cells to a culture tube and grow for 45 min at 37°C.
4. Add to each culture, 0.5 mL of LB with 100 μg/mL chloramphenicol. This yields a 2 mL culture volume with 25 μg/mL chloramphenicol. The chloramphenicol selects for the target plasmid. Grow cells for an additional 3 h at 37°C with shaking.
5. Harvest cells by centrifuging the culture in a 2-mL microcentrifuge tube for 1 min at 8000*g*. Discard the supernatant and resuspend the cells in 1 mL of sterile water. Centrifuge again for 1 min at 8000*g* and resuspend the cell pellet in 400 μL of sterile water (*see* **Note 6**).
6. To determine the fraction of target plasmids that have acquired a Ty5 insertion, 200 μL of the cell suspension is plated on an M9 + Cam plate. Plates are typically incubated at 37°C for 3 d. The minimal medium with the antibiotic ensures growth of only those cells with target plasmids with an integrated Ty5 element and its *HIS3* marker gene. The average colony yield is between 100 and 300 colonies per plate. To obtain adequate colony numbers, the amount of DNA used to electroporate and the volume of cells plated may be adjusted accordingly.
7. The frequency of targeting is calibrated relative to the total population of target plasmids. This is determined by combining 20 μL of the resuspended cells (1/10th the volume used to measure targeting) with 180 μL of sterile water in a microcentrifuge tube and plating the entire volume on a LB + Cam plate. Plates are incubated overnight at 37°C.
8. Colonies on the LB + Cam and M9 + Cam plates are counted. The frequency of targeting is considered the number of colonies that appear on the M9 + Cam plates divided by the normalized number of colonies that grow on the LB + Cam plates (i.e., 10 times the number of actual colonies). The targeting efficiency is also compared relative with efficiencies determined for the control target plasmid (*see* **Note 7**).

4. Notes

1. In **Subheading 3.1., steps 2–4**, the plasmids are added sequentially to yeast strain YPH499. To save time, it is possible to introduce two plasmids simultaneously. However, attempts to introduce three plasmids at once, have met with little success.
2. Instead of using a 96-well plate, sterile toothpicks can be used to inoculate 1-mL aliquots of liquid SC-U-L-T medium with each of the three test and control strains. Incubate at 30°C with shaking for 24 h.
3. The principal difference between this protocol and the published version is the addition of **Subheading 3.2., step 4**. This step was implemented to improve recovery of target plasmids by extending the period of growth under selection. This additional growth period does not significantly alter the targeting efficiency of wild-type Ty5.
4. One note of caution, which pertains to the original as well as this revised protocol, concerns discerning whether or not transposition events are independent of each other. In the original protocol, cells that sustained a transposition event were amplified by growth under selection on Petri plates (**Subheading 3.2., step 3**). In this protocol, a second period of growth under selection in liquid medium is invoked (**Subheading 3.2., step 4**). In either assay, the relative frequencies of targeted transposition are meaningful, as long as appropriate controls are used that provide a basis for comparison, such as a target plasmid without LexA operators. However, if integration events are analyzed at the DNA sequence level (*see* **Note 7**), and two plasmids are recovered with Ty5 elements at the same nucleotide site, it is not possible to determine whether these are independent events or siblings.
5. Although a multitube vortexer may be used for breaking the yeast cells, it should not be used to resuspend the cell pellet, as this results in excessive foaming and poor resuspension (**Subheading 3.3., step 2**). The time required to break the yeast cells by vortexing can be reduced; however, fewer target plasmids will be recovered. Whereas extensive vortexing can shear chromosomal DNA, it does not affect recovery of plasmid DNA.
6. It is important to carefully wash the *E. coli* cells with water in **Subheading 3.4., step 4**. Any contaminating histidine can result in background growth of the eDW335 on M9 medium, giving rise to very small colonies that appear after the first day of incubation at 37°C. An additional wash step will reduce this background growth, but often result in a slight decrease in cell survival. If background colonies do appear, they can be distinguished from cells with a transposed Ty5 element after continued incubation, because only the cells with the Ty5 *HIS3* marker produce large colonies.
7. Target plasmids can be recovered from colonies that grow on M9 + Cam plates (**Subheading 3.5., step 6**). DNA sequencing with a Ty5-specific primer can be used to determine precisely where on the target plasmid the Ty5 element has integrated.
8. In this protocol, the LexA-Sir4 fusion protein is tethered to the target plasmid by LexA operators. It should be possible to integrate LexA operators at any chromosomal site and direct Ty5 integration to these target sequences.
9. The domain of Sir4 that interacts with Ty5 IN can be fused to other DNA-binding proteins, such as transcription factors. On interaction with the transcription factor-Sir4 fusion, Ty5 will integrate at chromosomal sites bound by the transcription factor. This technique holds much promise for identifying promoters bound by specific transcription factors, and the technique can complement approaches such as the widely

used ChIP-Chip method for identifying chromosomal sites occupied by DNA-binding proteins ***(21)***. A proof-of-concept experiment that demonstrates the utility of this approach has been undertaken in *Saccharomyces* using seven different transcription factor-Sir4 fusions ***(20)***.

10. Ty5 target specificity can be altered by replacing the IN TD with other peptide motifs that interact with known protein partners. Integration occurs at high efficiency and in close proximity to DNA sites where the protein partners are tethered. This approach can also be used to engineer Ty5 elements to probe the chromosomal occupancy of certain DNA-binding proteins.
11. The general approach for modifying Ty5 target specificity can be applied to other retrotransposons to control their sites of integration. For example, the *S. pombe* Tf1 retrotransposon has been engineered to integrate in heterochromatin through the addition of a chromodomain to Tf1 IN, which recognizes histone modifications that are enriched in heterochromatin (Gao and Voytas, unpublished).
12. The general approach for modifying Ty5 target specificity may also prove useful for controlling sites of retroviral integration. Such engineered retroviruses would have particular utility in human gene therapy where it is particularly important to control the site of integration.

Acknowledgment

This work was supported by RO1 GM61657.

References

1. Kazazian, H. H., Jr. (2004) Mobile elements: drivers of genome evolution. *Science* **303,** 1626–1632.
2. Hua-Van, A., Le Rouzic, A., Maisonhaute, C., and Capy, P. (2005) Abundance, distribution and dynamics of retrotransposable elements and transposons: similarities and differences. *Cytogenet. Genome Res.* **110,** 426–440.
3. Bushman, F. D. (2003) Targeting survival: integration site selection by retroviruses and LTR-retrotransposons. *Cell* **115,** 135–138.
4. Chalker, D. L. and Sandmeyer, S. B. (1992) Ty3 integrates within the region of RNA polymerase III transcription initiation. *Genes Dev.* **6,** 117–128.
5. Devine, S. E. and Boeke, J. D. (1996) Integration of the yeast retrotransposon Ty1 is targeted to regions upstream of genes transcribed by RNA polymerase III. *Genes Dev.* **10,** 620–633.
6. Yieh, L., Kassavetis, G., Geiduschek, E. P., and Sandmeyer, S. B. (2000) The Brf and TATA-binding protein subunits of the RNA polymerase III transcription factor IIIB mediate position-specific integration of the *gypsy*-like element, Ty3. *J. Biol. Chem.* **275,** 29,800–29,807.
7. Bachman, N., Gelbart, M. E., Tsukiyama, T., and Boeke, J. D. (2005) TFIIIB subunit Bdp1p is required for periodic integration of the Ty1 retrotransposon and targeting of Isw2p to *S. cerevisiae* tDNAs. *Genes Dev.* **19,** 955–964.
8. Schroder, A., Shinn, P., Chen, H., Berry, C., Ecker, J., and Bushman, F. (2002) HIV-1 integration in the human genome favors active genes and local hotspots. *Cell* **110,** 521.
9. Wu, X., Li, Y., Crise, B., and Burgess, S. M. (2003) Transcription start regions in the human genome are favored targets for MLV integration. *Science* **300,** 1749–1751.

10. Ciuffi, A., Llano, M., Poeschla, E., et al. (2005) A role for LEDGF/p75 in targeting HIV DNA integration. *Nat. Med.* **11,** 1287–1289.
11. Cherepanov, P., Sun, Z. Y., Rahman, S., Maertens, G., Wagner, G., and Engelman, A. (2005) Solution structure of the HIV-1 integrase-binding domain in LEDGF/p75. *Nat. Struct. Mol. Biol.* **12,** 526–532.
12. Zou, S. and Voytas, D. F. (1997) Silent chromatin determines target preference of the *Saccharomyces* retrotransposon Ty5. *Proc. Natl. Acad. Sci. USA* **94,** 7412–7416.
13. Zou, S., Kim, J. M., and Voytas, D. F. (1996) The *Saccharomyces* retrotransposon Ty5 influences the organization of chromosome ends. *Nucleic Acids Res.* **24,** 4825–4831.
14. Zou, S., Ke, N., Kim, J. M., and Voytas, D. F. (1996) The *Saccharomyces* retrotransposon Ty5 integrates preferentially into regions of silent chromatin at the telomeres and mating loci. *Genes Dev.* **10,** 634–645.
15. Xie, W., Gai, X., Zhu, Y., Zappulla, D. C., Sternglanz, R., and Voytas, D. F. (2001) Targeting of the yeast Ty5 retrotransposon to silent chromatin is mediated by interactions between integrase and Sir4p. *Mol. Cell Biol.* **21,** 6606–6614.
16. Gai, X. and Voytas, D. F. (1998) A single amino acid change in the yeast retrotransposon Ty5 abolishes targeting to silent chromatin. *Mol. Cell* **1,** 1051–1055.
17. Zhu, Y., Dai, J., Fuerst, P. G., and Voytas, D. F. (2003) Controlling integration specificity of a yeast retrotransposon. *Proc. Natl. Acad. Sci. USA* **100,** 5891–5895.
18. Bowen, N. J., Jordan, I. K., Epstein, J. A., Wood, V., and Levin, H. L. (2003) Retrotransposons and their recognition of pol II promoters: a comprehensive survey of the transposable elements from the complete genome sequence of *Schizosaccharomyces pombe*. *Genome Res.* **13,** 1984–1997.
19. Zhu, Y., Zou, S., Wright, D. A., and Voytas, D. F. (1999) Tagging chromatin with retrotransposons: target specificity of the *Saccharomyces* Ty5 retrotransposon changes with the chromosomal localization of Sir3p and Sir4p. *Genes Dev.* **13,** 2738–2749.
20. Wang, H., Mitra, R., and Johnston, M. (2006) Development of 'calling cards' for DNA binding proteins. Abstract, *Yeast Genetics Meeting*, 505A.
21. Horak, C. E. and Snyder, M. (2002) ChIP-chip: a genomic approach for identifying transcription factor binding sites. *Methods Enzymol.* **350,** 469–483.
22. Sikorski, R. S. and Hieter, P. (1989) A system of shuttle vectors and yeast host strains designed for efficient manipulation of DNA in *Saccharomyces cerevisiae*. *Genetics* **122,** 19–27.
23. Gao, X., Rowley, D. J., Gai, X., and Voytas, D. F. (2002) Ty5 *gag* mutations increase retrotransposition and suggest a role for hydrogen bonding in the function of the nucleocapsid zinc finger. *J. Virol.* **76,** 3240–3247.
24. Curcio, M. J. and Garfinkel, D. J. (1991) Single step selection for Ty1 element retrotransposition. *Proc. Natl. Acad. Sci. USA* **88,** 936–940.

12

Site-Specific Chromosomal Integration Mediated by ϕC31 Integrase

Annahita Keravala and Michele P. Calos

Summary

ϕC31 integrase is a site-specific recombinase from a bacteriophage that has become a useful tool in mammalian cells. The enzyme normally performs precise, unidirectional recombination between two attachment or *att* sites called *attB* and *attP*. We have shown that an *attP* site preintegrated into a mammalian chromosome can serve as a target for integration of an introduced plasmid carrying an *attB* site. Recombination leads to precise integration of the plasmid into the chromosome at the *attP* site. This reaction is useful for placing introduced genes into the same chromosomal environment, in order to minimize position effects associated with random integration. Because ϕC31 integrase can also mediate integration at endogenous sequences that resemble *attP*, called pseudo *attP* sites, a selection system is used that yields integration only at the desired preintegrated *attP* site. This chapter provides a protocol that features a simple antibiotic selection to isolate cell lines in which the introduced plasmid has integrated at the desired *attP* site. A polymerase chain reaction assay is also presented to verify correct chromosomal placement of the introduced plasmid. This integration system based on ϕC31 integrase supplies a simple method to obtain repeated integration at the same chromosomal site in mammalian cells.

Key Words: *attB*; *attP*; ϕC31 integrase; phage integrase; position effects; site-specific integration.

1. Introduction

A commonplace need in genetic studies is to insert novel genetic material into the chromosomes. Chromosomal integration of introduced DNA occurs spontaneously in mammalian cells, but the process is rare and such integration is random. Random integration subjects the integrated DNA to position effects in which the chromosomal context influences expression of the integrated DNA. This influence is usually undesirable. Integration can be directed to specific locations by homologous recombination, but spontaneous homologous recombination is also an infrequent

From: *Methods in Molecular Biology, vol. 435: Chromosomal Mutagenesis*
Edited by: G. Davis and K. J. Kayser © Humana Press Inc., Totowa, NJ

process. To catalyze a higher frequency of integration, enzymes may be used. In mammalian cells, several classes of enzymes have been used for this purpose. Transposases derived from transposable elements such as *Sleeping Beauty* can be used to integrate segments of DNA that have been cloned between inverted repeat sequences recognized by the transposase ***(1)***. This process brings about fairly efficient, but nearly random, integration into the genome ***(2)***. As well, retroviral integrases, which are distantly related to transposases, mediate efficient, nearly random integration ***(3)***.

Although transposases and retroviral integrases are efficient at chromosomal integration, they have several limitations, including an absence of control over the location where the DNA will be integrated in the genome and size limits for the DNA to be integrated. In order to remove size limits and gain control over the site of integration, site-specific recombinases, which require recognition of specific DNA sequences at the target site, have been used. These enzymes derive from single-celled organisms and typically involve the precise recombination of two sequences recognized by the recombinase. Cre was the first such enzyme to find widespread utility in mammalian cells ***(4)***. Cre requires no host-specific cofactors and is catalytically robust in the mammalian environment. It recombines two identical 34-bp long *loxP* sites. The similar flippase (FLP) can recombine two identical *FRT* sites ***(5)***. Although Cre and FLP can in theory be used for chromosomal integration, integration events mediated by these resolvases are quickly reversed by deletion, because the integrated DNA is flanked by *loxP* or *FRT* recognition sites. Therefore, these enzymes are most widely used for making precise deletions.

A different type of site-specific recombinase, a phage integrase, is best suited for efficient integration into chromosomes. Phage integrases recombine two nonidentical sequences, termed *attB* and *attP*, creating hybrid sites after recombination that are not substrates for the integrase. Therefore, phage integrases are unidirectional and provide a high net integration efficiency. The ϕC31 integrase, derived from a *Streptomyces* phage ***(6–8)***, was the first phage integrase shown to provide efficient integration into mammalian chromosomes ***(9–11)***. The enzyme was demonstrated to mediate efficient integration between its own *attB* and *attP* sites in the mammalian environment ***(10,12)***. ϕC31 integrase also mediated integration between an *attB* site on an incoming plasmid and native sequences with partial sequence identity to *attP*, called pseudo *attP* sites (**Fig. 1**) ***(10)***. The ability of ϕC31 integrase to mediate integration into unmodified genomes is important for gene therapy applications (reviewed in **ref. *13***) and for generating transgenic vertebrates ***(14)***. Furthermore, integration into pseudo sites in cultured cells is a rapid way to generate cell lines with high expression of the transgene ***(10,15,16)***. However, this chapter focuses on use of ϕC31 integrase for reproducible, precise placement of genes at the same chromosomal position, a preintegrated *attP* docking site, in mammalian cell lines.

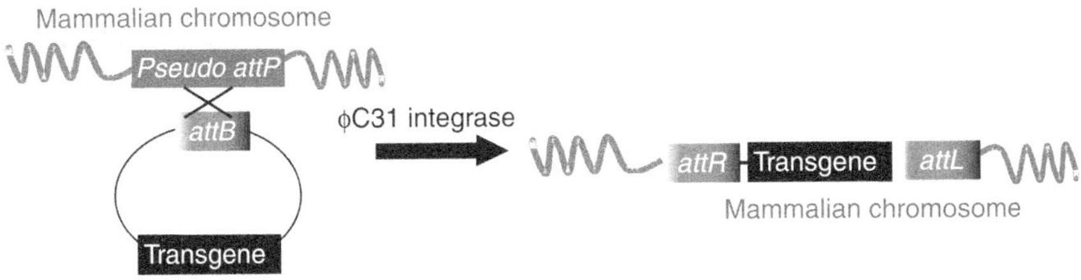

Fig. 1. Integration reaction mediated by φC31 integrase in an unmodified mammalian genome. In the presence of φC31 integrase, an *attB*-bearing donor plasmid carrying a transgene becomes integrated into a pseudo *attP* site in the mammalian genome.

For this purpose, only the precise reaction between an *attB* site on an incoming plasmid and an *attP* site in the genome is desired, and integration into pseudo *attP* sites is unwanted. In small genomes, like that of *Drosophila melanogaster*, a preintegrated *attP* site is the target for nearly all integration by φC31 integrase ***(17–19)***. However, in large, complex genomes, although an integrated *attP* site generally represents a hotspot for integration, many integration events also occur at pseudo *attP* sites ***(10)***. A study of integration in three human cell lines indicated that 19 pseudo *attP* sites received repeated integration events whereas up to several hundred more may receive occasional integrations ***(20)***. This competition between integration at the preintegrated *attP* site vs integration at the set of native pseudo *attP* sites resulted, for example, in 5–15% of integrations occurring at the preintegrated *attP* in human cell lines ***(10,21)***; whereas the remainder occurred at pseudo *attP* sites. To remove this background of integration at pseudo *attP* sites, we set up a selection system in which resistance to an antibiotic is activated only by φC31 integrase-mediated integration of an *attB*-containing plasmid into the desired *attP* site **(Fig. 2)** ***(10)***. This system yields nearly 100% specificity for precise integration at the desired chromosomal location, and its operation is described in this chapter.

New developments with φC31 integrase are likely to create improvements in the utility of the system for creating mammalian cell lines. For example, we have developed mutant forms of φC31 integrase that have higher integration efficiency and higher specificity for integration at *attP*, as opposed to pseudo *attP* sites ***(21)***. In the future, with use of such modified versions of φC31 integrase, screening of a small number of integrants may be adequate for finding those integrated at *attP*, as an alternative to selection.

2. Materials

2.1. Reagents

1. Dulbecco's modified eagle's medium (DMEM) with high glucose and L-glutamine (Gibco/BRL, Bethesda, MD), supplemented with 10% fetal bovine serum ([FBS], Gibco/BRL), and 1% penicillin/streptomycin (Gibco/BRL).

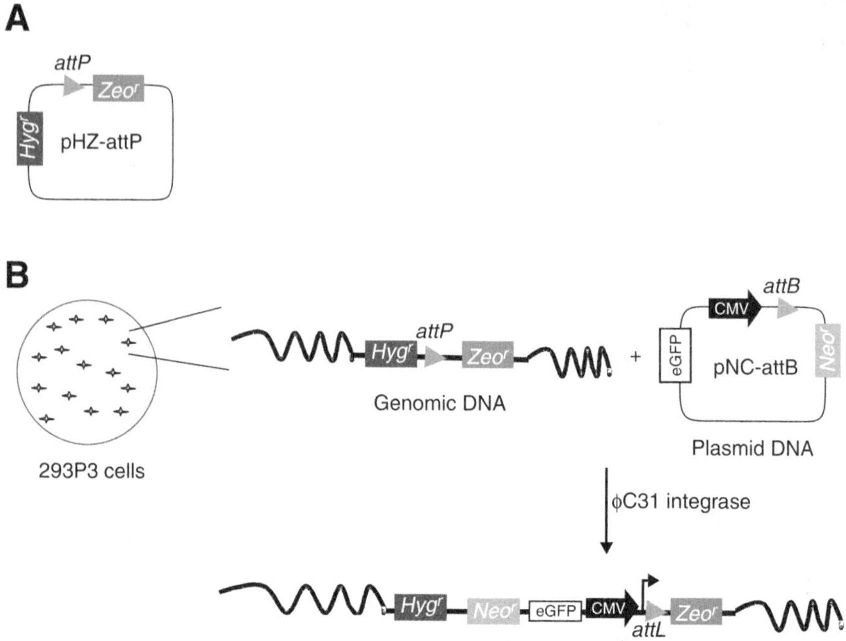

Fig. 2. Chromosomal integration at a preintegrated *attP* site. **(A)** pHZ-*attP* plasmid, used to randomly integrate the φC31 *attP* site into the cellular genome by using the hygromycin-resistance selection marker. **(B)** Schematic diagram describing genomic integration at the preintegrated *attP* site in the genome. The pNC-*attB* plasmid, bearing the neomycin-resistance gene and the cytomegalovirus (CMV) promoter located upstream of the *attB* site, recombines at the chromosomal *attP* site, integrating into the host chromosome (wavy lines). This integration reaction is catalyzed by φC31 integrase. Site-specific integration leads to expression of the zeocin-resistance gene from the CMV promoter introduced on pNC-*attB*.

2. Trypsin (0.05%) and ethylenediamine tetraacetic acid (1 m*M*) (Gibco/BRL).
3. FuGENE 6 transfection reagent (Roche, Palo Alto, CA).
4. Hygromycin B (Calbiochem, San Diego, CA).
5. Geneticin (G418, Gibco/BRL).
6. Zeocin (Gibco/BRL).
7. Agarose (Sigma, St. Louis, MO).
8. DNeasy tissue kit (Qiagen, Valencia, CA).
9. puReTaq ready-to-go polymerase chain reaction (PCR) beads (Amersham Biosciences, Buckinghamshire, England).
10. Restriction enzymes and buffers appropriate for the experiment (New England Biolabs, Ipswich, MA).
11. QIAquick PCR purification kit (Qiagen).

2.2. Plasmids

1. pHZ-*attP*, a plasmid carrying the *attP* site upstream of a promoter-less, *zeocin-resistance* gene, as well as a *hygromycin-resistance* selection marker ***(10)*** (**Fig. 2**).
2. pCMV-Int, plasmid-expressing *wild-type φC31 integrase* ***(9)***.
3. pCSmI, negative control plasmid-expressing *nonfunctional φC31 integrase* ***(22)***.
4. pNC-*attB*, donor plasmid containing the *attB* site downstream of the cytomegalovirus (CMV) promoter, a *neomycin-resistance* gene, and an *enhanced green fluorescent protein* gene (**Fig. 2**) ***(10)***.
5. PCR primers: P3F: 5′-AGGTCTATATAAGCAGAGCTC and P3R: 5′-TGAGCAC-CGGAACGGCACTGG.

2.3. Cell Lines

1. 293 ***(23)***, a transformed human embryonic kidney cell line (ATCC, Manassas, VA) (*see* **Note 1**).

2.4. Equipment

1. Electroporator (Bio-Rad Gene Pulser; Bio-Rad, Hercules, CA).
2. CO_2 incubator set at 37°C.
3. Tissue culture plates (6-well, 60 mm, and 100 mm).
4. Microcentrifuge tubes.
5. Agarose gel electrophoresis equipment.
6. PCR machine.

3. Methods

3.1. Construction of attP Cell Lines

1. Twenty-four hours before electroporation, seed 6–8 × 10^5 293 cells (*see* **Note 1**) per 60-mm dish in 5 mL DMEM supplemented with 10% FBS and 1% penicillin/streptomycin (*see* **Note 2**). The cell number seeded should yield approx 80% confluence on the day of electroporation.
2. Incubate the cells at 37°C in a 5% CO_2 incubator.
3. Concurrently, treat plasmid pHZ-attP with *Hpa*I overnight at 37°C to generate linear molecules.
4. Purify the linearized plasmid DNA using the QIAquick PCR purification kit.
5. Electroporate 5–10 μg of linearized pHZ-attP into 293 cells using the Bio-Rad Gene Pulser according to manufacturer's directions (*see* **Note 3**).
6. Immediately plate the cells on three 100-mm dishes in complete DMEM supplemented with FBS and antibiotics and culture them at 37°C in the CO_2 incubator to allow the cells to recover.
7. Twenty-four hours after electroporation, start selection using complete DMEM containing 200 μg/mL of hygromycin B (*see* **Note 4**). Maintain cells in selective medium under normal growth conditions until well-expanded colonies appear (usually 14–17 d) (*see* **Notes 5** and **6**).
8. Pick single, well-isolated colonies (~6) using trypsin and expand in each well of a 6-well plate using complete DMEM without further selection.

3.2. Selection Assay for Identifying a Cell Line With an attP Site Located at a Favorable Location

1. Plate 1–3 × 10^5 cells of each cell line to be tested (e.g., 293P1–293P6) (*see* **Note 7**) in 2 mL of DMEM with 10% FBS and 1% penicillin/streptomycin in a 6-well plate 24 h before transfection. Incubate the cells at 37°C in a CO_2 incubator. This procedure would achieve 60–80% confluence at time of transfection.
2. On the day of transfection, aliquot 97 μL serum-free DMEM (SFM) into microcentrifuge tubes. Add 3 μL FuGENE 6 (*see* **Note 8**) to each of these tubes and mix gently. Add DNA (980 ng pCMVInt or pCSmI + 20 ng pNC-*attB*) to the tubes, mix gently, and incubate for 15 min at room temperature (*see* **Note 9**).
3. Aspirate medium from cells and replace with 2 mL of complete DMEM per well. Then add 100 μL of the lipid-DNA complex in SFM drop-wise to the appropriate well. Gently rock the plate to distribute evenly over the cells.
4. Incubate overnight at 37°C in a CO_2 incubator.
5. Twenty-four hours posttransfection, harvest the cells using trypsin and divide the cells from each well onto two 100-mm plates.
6. Twenty-four hours after replating, aspirate medium and replace with DMEM containing 350 μg/mL G418 (*see* **Note 4**).
7. Replace the selection medium every 2–3 d, taking care not to dislodge nascent colonies (*see* **Notes 5** and **6**). Maintain the cells under selective pressure for approx 2 wk, and count the number of G418-resistant colonies (*see* **Note 10**).

3.3. Integration Into attP in the 293P3 Cell Line

1. Plate 1–3 × 10^5 293P3 cells in 2 mL of DMEM with 10% FBS and 1% penicillin/streptomycin in a 6-well plate 24 h before transfection. Incubate the cells at 37°C in a CO_2 incubator. This procedure would achieve 60–80% confluence at the time of transfection.
2. On the day of transfection, aliquot 97 μL SFM into microcentrifuge tubes. Add 3 μL FuGENE 6 (*see* **Note 8**) to each of these tubes, and mix gently. Add DNA (980 ng pCMVInt or pCSmI + 20 ng pNC-*attB*) to the tubes (*see* **Note 11**), mix gently, and incubate for 15 min at room temperature (*see* **Note 9**).
3. Aspirate medium from cells and replace with 2 mL of complete DMEM per well. Then add 100 μL of the lipid-DNA complex in SFM drop-wise to the appropriate well. Gently rock the plate to distribute evenly over the cells.
4. Incubate overnight at 37°C in a CO_2 incubator.
5. Twenty-four hours posttransfection, harvest the cells using trypsin and divide the cells from each well onto two 100-mm plates.
6. Twenty-four hours after replating, aspirate medium and replace with DMEM containing 350 μg/mL G418 + 100 μg/mL zeocin (*see* **Note 4**).
7. Replace the selection medium every 2–3 d, taking care not to dislodge nascent colonies, and maintain the cells under selective pressure for approx 2 wk, until well-isolated colonies are formed (*see* **Notes 5** and **6**).
8. Almost all zeocin-resistant colonies will have the plasmid integrated at the chromosomal *attP* site (*see* **Note 12**).

3.4. Supplementary Protocol: PCR Assay to Verify Integration at the Chromosomal attP Site

1. In order to confirm integration of the marker gene into the preintegrated *attP* site in the genome, prepare genomic DNA from each clonal line isolated in **Subheading 3.3.** using a DNeasy tissue kit according to manufacturer's direction.
2. Perform the PCR analysis using the primers P3F (located in the incoming plasmid) and P3R (located in the preintegrated zeocin-resistance gene).
3. Carry out the PCR using the following protocol: 95°C for 30 s; 65°C for 30 s; 72°C for 30 s for 10 cycles, with each cycle decreasing the annealing temperature by 1°C, followed by 25 cycles of 95°C for 30 s; 55°C for 30 s; 72°C for 30 s, followed by 72°C for 7 min.
4. Visualize the amplified PCR products using agarose gel electrophoresis. The presence of a band of the expected size will establish that integration took place at the preintegrated *attP* site.

4. Notes

1. The protocol uses the 293 cell line as an example, but the method is expected to work with any other mammalian cell line.
2. Unless otherwise stated, all solutions should be warmed in a 37°C water bath before use for tissue culture.
3. Electroporation of linearized DNA was used, because it often produces a single-copy plasmid insertion, which was the desired outcome. Copy number of the integrated plasmid can be analyzed by Southern blotting.
4. Before starting any antibiotic selection, it is advisable to do a kill curve for that antibiotic on the cell line to be used for the experiment. Depending on the cell line, the optimal antibiotic concentration required to select cells will vary.
5. Change selection medium every 2–3 d to remove the dead cells and floating debris.
6. 293 cells adhere to the plate loosely. Hence, while changing medium, aspirate and add medium gently along the sides of the plate so the transfected cells will not be lost.
7. Because the *attP*-bearing plasmid becomes integrated into the genome by random integration, some resulting cell lines may have the *attP* site located in an unfavorable region for integration and gene expression. By screening several *attP* cell lines for their ability to generate G418-resistant colonies, one can be assured of choosing an *attP* cell line that has an acceptable level of integration and gene expression at the *attP* site.
8. The transfection method described here utilizes FuGENE 6 as the transfection reagent, which works well with 293 cells. However, the method detailed in this chapter can easily be applied to other cell lines using a variety of different transfection techniques. Plasmid delivery methods generally vary depending on the cell line and application. For cellular uptake of plasmid DNA in culture, such methods include lipid reagents such as FuGENE 6, Lipofectamine (Invitrogen, Carlsbad, CA), and Superfect (Qiagen, Valencia, CA), among others. In addition, standard electroporation ECM 830 (BTX, Holliston, MA) or nucleofection (Amaxa, Cologne, Germany) technologies may also be used.

9. Always dilute FuGENE 6 transfection reagent by pipeting directly into serum-free medium. Do not allow the undiluted reagent to come in contact with any plastic surface.
10. The higher the number of G418-resistant colonies, the better the genomic location of the integrated *attP* site, in terms of the level of integration efficiency and gene expression at the chromosomal location of the *attP* site. Therefore, other things being equal, choose the *attP* cell line that generates the highest number of G418-resistant colonies for use in experiments.
11. The gene of interest to the investigator should be added to pNC-*attB,* for example, in place of enhanced green fluorescent protein if the latter gene is not required.
12. The preintegrated pHZ-*attP* vector contains a promoter-less zeocin-resistance gene that will be transcriptionally activated by integration of the incoming pNC-*attB* plasmid at the *attP* site, by virtue of the CMV promoter adjacent to *attB* (*see* **Fig. 2**).

Acknowledgments

The authors would like to thank Vanessa E. Gabrovsky for her help with editing the chapter. This work was supported by National Institute Health grants HL68112 and EY16702 to M.P.C.

References

1. Ivics, Z., Hackett, P. B., Plasterk, R. H., and Izsvak, Z. (1997) Molecular reconstruction of Sleeping Beauty, a Tc1-like transposon from fish. *Cell* **91,** 501–510.
2. Yant, S. R., Wu, X., Huang, Y., Garrison, B., Burgess, S. M., and Kay, M. A. (2005) High-resolution genome-wide mapping of transposon integration in mammals. *Mol. Cell. Biol.* **25,** 2085–2094.
3. Bushman, F. D. (2003) Targeting survival: integration site selection by retroviruses and LTR-retrotransposons. *Cell* **115,** 135–138.
4. Sauer, B. and Henderson, N. (1988) Site-specific DNA recombination in mammalian cells by the Cre recombinase of bacteriophage P1. *Proc. Natl. Acad. Sci. USA* **85,** 5166–5170.
5. O'Gorman, S., Fox, D. T., and Wahl, G. M. (1991) Recombinase-mediated gene activation and site-specific integration in mammalian cells. *Science* **251,** 1351–1355.
6. Rausch, H. and Lehmann, M. (1991) Structural analysis of the actinophage ΦC31 attachment site. *Nucleic Acids Res.* **19,** 5187–5189.
7. Thorpe, H. M. and Smith, M. C. M. (1998) In vitro site-specific integration of bacteriophage DNA catalyzed by a recombinase of the resolvase/invertase family. *Proc. Natl. Acad. Sci. USA* **95,** 5505–5510.
8. Kuhstoss, S. and Rao, R. N. (1991) Analysis of the integration function of the Streptomycete bacteriophage ΦC31. *J. Mol. Biol.* **222,** 897–908.
9. Groth, A. C., Olivares, E. C., Thyagarajan, B., and Calos, M. P. (2000) A phage integrase directs efficient site-specific integration in human cells. *Proc. Natl. Acad. Sci. USA* **97,** 5995–6000.
10. Thyagarajan, B., Olivares, E. C., Hollis, R. P., Ginsburg, D. S., and Calos, M. P. (2001) Site-specific genomic integration in mammalian cells mediated by phage ϕC31 integrase. *Mol. Cell. Biol.* **21,** 3926–3934.

11. Groth, A. C. and Calos, M. P. (2004) Phage integrases: biology and applications. *J. Mol. Biol.* **335,** 667–678.
12. Belteki, G., Gertsenstin, M., Ow, D. W., and Nagy, A. (2003) Site-specific cassette exchange and germline transmission with mouse ES cells expressing the ϕC31 integrase. *Nat. Biotechnol.* **21,** 321–324.
13. Calos, M. P. (2006) The ϕC31integrase system for gene therapy. *Curr. Gene Ther.* **6,** 633–645.
14. Allen, B. G. and Weeks, D. L. (2005) Transgenic *Xenopus laevis* embryos can be generated using phiC31 integrase. *Nat. Methods* **2,** 975–979.
15. Thyagarajan, B. and Calos, M. P. (2005) Site-specific integration for high-level protein production in mammalian cells, in *Therapeutic Proteins: Methods and Protocols* (Smales, C. M. and James, D. C. eds.), Humana Press, Totowa, NJ, pp. 99–106.
16. Hillman, R. T. and Calos, M. P. (2007) Site-specific integration with phage ϕC31 integrase, in *Gene Transfer: Delivery and Expression of DNA and RNA* (Friedman, T. and Rossi, J. eds.), Chapter 65, Cold Spring Harbor Laboratory Press, Cold Spring Harbor, New York, pp. 653–660.
17. Groth, A. C., Fish, M., Nusse, R., and Calos, M. P. (2004) Creation of transgenic *Drosophila* by using the site-specific integrase from phage ϕC31. *Genetics* **166,** 1775–1782.
18. Bateman, J. R., Lee, A. M., and Wu, C. T. (2006) Site-specific transformation of *Drosophila* via phiC31 integrase-mediated cassette exchange. *Genetics* **173,** 769–777.
19. Venken, K. J. T., He, Y., Hoskins, R. A., and Bellen, H. J. (2006) P[acman]: A BAC transgenic platform for targeted insertion of large DNA fragments in *D. melanogaster. Science* **341,** 1747–1751.
20. Chalberg, T. W., Portlock, J. L., Olivares, E. C., et al. (2006) Integration specificity of phage ϕC31 integrase in the human genome. *J. Mol. Biol.* **357,** 28–48.
21. Keravala, A., Lee, S., Thyagarajan, B., et al. (2007) Mutational derivatives of ϕC31 integrase with enhanced efficiency and specificity (in press).
22. Olivares, E. C., Hollis, R. P., Chalberg, T. W., Meuse, L., Kay, M. A., and Calos, M. P. (2002) Site-specific genomic integration produces therapeutic factor IX levels in mice. *Nat. Biotechnol.* **20,** 1124–1128.
23. Graham, F. L., Smiley, J., Russell, W. C., and Nairn, R. (1977) Characteristics of a human cell line transformed by DNA from human adenovirus type 5. *J. Gen. Virol.* **36,** 59–72.

13

Triplex-Mediated Gene Modification

Erica B. Schleifman, Joanna Y. Chin, and Peter M. Glazer

Summary

Gene targeting with DNA-binding molecules such as triplex-forming oligonucleotides or peptide nucleic acids can be utilized to direct mutagenesis or induce recombination site-specifically. In this chapter, several detailed protocols are described for the design and use of triplex-forming molecules to bind and mediate gene modification at specific chromosomal targets. Target site identification, binding molecule design, as well as various methods to test binding and assess gene modification are described.

Key Words: Antigene; homologous recombination; mutagenesis; peptide nucleic acid (PNA); triplex; triplex-forming oligonucleotide (TFO).

1. Introduction

Triplex formation was first cited in the literature in 1957, when Felsenfeld et al. *(1)* described the ability of polyU and polyA RNA strands to bind in a 2:1 ratio. Triplex-forming oligonucleotides (TFOs) are capable of forming similar structures by binding in the major groove of duplex DNA to polypurine/polypyrimidine runs. These molecules bind to DNA in a sequence-specific manner in either a parallel or antiparallel orientation. In the antiparallel (purine) motif, a polypurine TFO binds through reverse Hoogsteen hydrogen bonds antiparallel to the polypurine strand of the DNA duplex. In the parallel (pyrimidine) motif, a polypyrimidine TFO binds through Hoogsteen base pairing in a parallel orientation to the purine strand of the duplex DNA. Through the manipulation of the cell's own repair machinery, TFOs have been used as tools to effectively inhibit transcription *(2)*, inhibit the binding of proteins to DNA *(3–6)*, inhibit DNA replication *(7,8)*, promote site-specific DNA damage by delivery of a mutagen *(9–12)*, induce

From: *Methods in Molecular Biology, vol. 435: Chromosomal Mutagenesis*
Edited by: G. Davis and K. J. Kayser © Humana Press Inc., Totowa, NJ

mutagenesis ***(13,14)***, and enhance recombination in both chromosomal and episomal targets ***(15–17)***.

Another class of molecules capable of triplex formation is peptide nucleic acids (PNAs). PNAs are oligonucleotide analogs in which the phosphodiester backbone of DNA is replaced with an uncharged polyamide backbone ***(18)***. Because of its hybrid nature, PNA is resistant to both nucleases and proteases, increasing its stability when transfected into cells ***(19)***. PNAs can bind to DNA or RNA through Watson-Crick hydrogen bonds with high thermodynamic stability. Similar to TFOs, PNAs can form triplexes through Hoogsteen base pairing at polypurine:polypyrimidine sites. Triplexes can form when two PNA strands bind to the homopurine strand in a homopurine/homopyrimidine stretch of DNA. Whereas one PNA strand binds through Watson-Crick base pairing in an antiparallel orientation, the second strand binds to the PNA/DNA duplex through Hoogsteen bonding to form the triplex ***(20)***. By linking two PNA molecules through a flexible linker, a *bis*-PNA (clamp) is formed, which can be used to bind sequence specifically to a DNA target with an extremely high melting temperature. PNAs have been shown in vitro to block DNA polymerization, inhibit transcription initiation and elongation, and prevent sequence-specific protein binding. In addition, the formation of the D-loop has been shown to create an artificial transcriptional promoter ***(21–25)***.

There still exist many challenges in the implementation of triplex technology in vivo. TFO and PNA binding can be inhibited by such cellular conditions as potassium and magnesium concentrations. A reduction in binding may also be observed depending on the availability of a target site as a result of its location in chromatin. Efficient delivery of both TFOs and PNAs into cells still remains a challenge. Delivery of PNAs can be improved through the addition of positive lysine residues and conjugation to cell-penetrating peptides such as Antennapedia and trans-activator of transcription (TAT) ***(26)***. Furthermore, conjugation of a nuclear localization signal has been shown to increase targeting of both TFO and PNA molecules to the nucleus ***(27)***.

Modification of TFO backbones and/or bases has been used to increase cellular uptake, improve binding affinity, and prevent degradation on entry into cells. For example, backbone modifications, including replacing the phosphate backbone with cationic phosphoramidate linkages such as N, N-diethylethylene-diamine or N, N-dimethyl-aminopropylamine can increase the binding affinity of TFOs in vitro ***(28)***. The use of N, N-diethylethylene-diamine-modified bases may also eliminate G-quartet formation in G-rich oligonucleotides. To overcome the pH dependence of pyrimidine triplexes, pseudoisocytosine, a base modification which mimics the N3 protonation of cytosine, is readily utilized in both TFOs and PNAs. Other base modifications, 5-methyl-2′-deoxyuridine and 5-methyl-2′-deoxycytidine, as well as the sugar modification 2′-O-(2-aminoethyl)-ribose have all been shown to increase the binding affinity of TFOs in the pyrimidine motif ***(29,30)***.

Triplex formation stimulates mutagenesis by provoking the cell's own DNA repair pathways, primarily the nucleotide excision repair machinery. TFO-induced mutagenesis can lead to heritable and permanent changes in specific genes. TFOs are used to direct site-specific mutagenesis by either delivering a mutagen or directly inactivating a gene of interest. By conjugating a nonspecific mutagen such as psoralen (pso) to a TFO, the mutagen can be directed to a specific target site. Pso-TFOs have been shown to induce mutations in a target site on plasmids in vitro and when transfected in mammalian cells, as well as in chromosomal targets ***(9,13,31)***. Majumdar et al. ***(32)*** showed that by synchronizing cells and treating them in late S-phase targeted mutagenesis could be increased 5.5-fold over those in G_1, and 2.5-fold over cells in early S-phase. By manipulating cell synchronization or transcriptional state, it may be possible to increase the efficiency of triplex-induced mutagenesis.

A reporter system was constructed to detect and quantify TFO-induced mutagenesis. The *supFG* reporter gene, which encodes an amber suppressor tyrosine tRNA and contains a TFO-binding site was cloned into an SV40 vector. In bacteria with an amber mutation in the *lacZ* gene, a functional copy of the *supF* gene produces blue colonies; whereas a nonfunctional (mutated) copy produces white colonies when plated in the presence of 4-chloro-5-bromo-3-indolyl-β-D-galactopyranoside and isopropyl-β-D-1-thiogalactopyranoside. The frequency of mutagenesis can then be calculated by comparing the number of white colonies with blue ***(9,33)***. Using this reporter system, pso-TFOs have been found to induce mutagenesis in mouse cells containing a chromosomally integrated copy of the *supFG* reporter gene ***(34)***. Whereas pso has been shown to induce mutations, triplex formation alone using either TFOs or PNAs has also been shown to be sufficient to stimulate mutagenesis in target genes ***(20,34)***. In transgenic mice containing the same reporter gene integrated in the chromosome, intraperitoneal injection of the TFO, AG30, also led to site-specific mutations ***(13)***.

Another strategy that can be implemented to modify, correct, or replace genes is homologous recombination. Normally in cells, homologous recombination occurs at a low frequency; however, studies have shown that DNA damage at the target site can increase the frequency of homologous recombination ***(35)***. To exploit this idea, several labs have used pso-TFOs to create site-specific damage in order to sensitize a target site for homologous recombination ***(36,37)***. In these experiments it was found that not only could DNA damage by pso increase intrachromosomal recombination ***(38)***, but also a TFO alone was sufficient to induce homologous recombination ***(16,17,39)***. Triplex formation elevated the recombination level at a target site and led to the correction of a specific mutation ***(16)***. Using a plasmid with two tandem *supF* genes, each containing different point mutations, intramolecular recombination was shown to be increased on binding of a TFO, resulting in gene correction of one copy of the gene ***(36)***. When TFOs with or without pso were microinjected into the nuclei of mouse cells c ontaining two mutant copies of the *herpes simplex virus thymidine kinase* gene in tandem, homologous recombination was stimulated at a frequency of 1%, 2000-fold over background ***(17)***. It was hypothesized that by linking a TFO to a short donor fragment, the TFO would act to position the donor over the target area whereas triplex formation would sensitize the site for homologous recombination ***(15)***. The designed donor fragment would be completely homologous to the site adjacent to the triplex-binding site (TBS) except for the mutation or base correction desired in the target gene. Some studies have determined that antisense (homologous to the transcribed strand) donors are preferred over sense (matching the nontranscribed strand) donors at certain target sites ***(39)***. Further studies of this method have shown that the donor does not need to be tethered to the TFO or be adjacent to the TFO-binding site. Homologous recombination has been detected at sites up to 750 bp away from the triplex site ***(40)***. It has also been shown that TFOs can stimulate intermolecular recombination at a single-copy genomic locus in mammalian cells. In this study, both pyrimidine and purine motif TFOs were able to induce site-specific recombination in a dose-dependent manner up to a frequency of 0.11% ***(39)***.

Triplex formation has many applications in chromosomal gene targeting. This chapter will discuss the use of TFOs or PNAs to induce a site-specific mutation in a gene of interest. To use this method of gene targeting, one must first identify potential binding sites (polypurine stretch) within the gene of interest (*see* **Subheading 3.1.**). A TFO or *bis*-PNA, which can bind to the target, must be designed (*see* **Subheading 3.2.**) and evaluated for binding affinity to the target site using **Subheadings 3.3.** and **3.4.** Once a TFO or PNA has been validated by showing a strong binding affinity in vitro and in vivo, experiments can be designed to optimize delivery into the target cells. Gene correction or insertion of a specific mutation can be detected by various means. If the target is a reporter gene, then the relevant phenotypic assay can be used. If not, allele-specific polymerase chain reaction (PCR) can be used to detect a sequence change in the genomic DNA of the target cells.

2. Materials

All solutions and buffers can be stored at room temperature unless otherwise noted.

1. Oligonucleotides: most oligonucleotides discussed here, including pso-conjugated TFOs and single-stranded donors, can be ordered from Midland Certified Reagent Company Inc. (Midland, TX) or other vendors. Single-stranded donors should be modified with the first three (5′ end) and last three (3′ end) base pairs containing phosphorothioate linkages to prevent nuclease degradation. The same protection can be achieved with TFOs with the inclusion of 3′ amine groups. All oligonucleotides should be purified using either high-pressure liquid chromatography or gel purification.
2. PNAs: PNAs (*bis*-PNAs) can be ordered from Bio-Synthesis (Lewisville, TX), or Applied Biosystems (Foster City, CA).

3. Cells: Chinese hamster ovary cells.
4. Vectors: *FLuc*⁺ from pGL3-Basic Vector (Promega, Madison, WI).
5. Cell media—for Chinese hamster ovary cells: Ham's F12 medium with 10% of fetal bovine serum, 2 m*M* of L-glutamine.
6. For TFO-binding assay, **Subheading 3.3.**: triplex-binding buffer: 10 m*M* of Tris-HCl (pH 7.6), 0.1 m*M* of $MgCl_2$ (*see* **Note 1**), 1 m*M* of spermine, and 10% of glycerol (with or without 140 m*M* potassium).
7. For restriction protection assay, **Subheading 3.4.**: lysis buffer: 50 m*M* of Tris-HCl (pH 7.5), 20 m*M* of EDTA, 100 n*M* of NaCl, 0.1% of sodium dodecyl sulfate; TE pH 8.0: 10 m*M* of Tris-HCl (pH 8.0), 1 m*M* of EDTA (pH 8.0).

3. Methods

3.1. Target Site Identification

TFOs and PNAs bind with high affinity to polypurine/polypyrimidine sequences. The target sequence should therefore be a homopurine run (~80%) with few mixed sequences (these decrease binding affinity). By scanning the gene of interest, these homopurine stretches can be identified and TFOs that bind in either a parallel or antiparallel orientation can be designed as discussed in **Subheading 3.2.1. (Fig. 2A)**. TFO-binding sites should be typically 14–30 bp in length, and PNA-binding sites should be at least 8–10 bp in length. It may be necessary to identify several possible target sites as some may be more accessible to triplex formation than others owing to chromatin structure. It is also important to note that if pso is to be conjugated to the TFO (to site-specifically induce crosslinks) then the target site should contain a 5′ TpA site at either the 5′- or 3′-end of the polypurine run. Target sites may include regulatory regions of the gene of interest, which may be effective for modulating gene expression.

3.2. Design and Synthesis

After selecting a target site, several molecules can be designed and synthesized to bind or alter the target. Below are the guidelines that will help to determine which molecule will work best **(Fig. 2B)**.

3.2.1. Triplex-Forming Oligonucleotide (See ***Note 2***)

1. For A-rich target sites, the pyrimidine motif is preferred and requires TFOs containing C and T or their analogs (C^+ will form Hoogsteen bonds with a G in a G:C base pair, and T with the A in an A:T base pair **[Fig. 1]**). These TFOs will bind in an orientation parallel to the purine-rich strand of the target duplex.
2. For G-rich target sites, the purine motif is preferred and requires TFOs containing A (or T) and G or their analogs (G forms Hoogsteen bonds with the G in a G:C base pair and A or T with the A in an A:T base pair). These oligos will bind in an antiparallel orientation, forming reverse Hoogsteen bonds, with the polypurine strand of the duplex.
3. To induce targeted crosslinks at the triplex site, pso can be conjugated on the 5′- or 3′ end of the TFO using pso phosphoramidates.

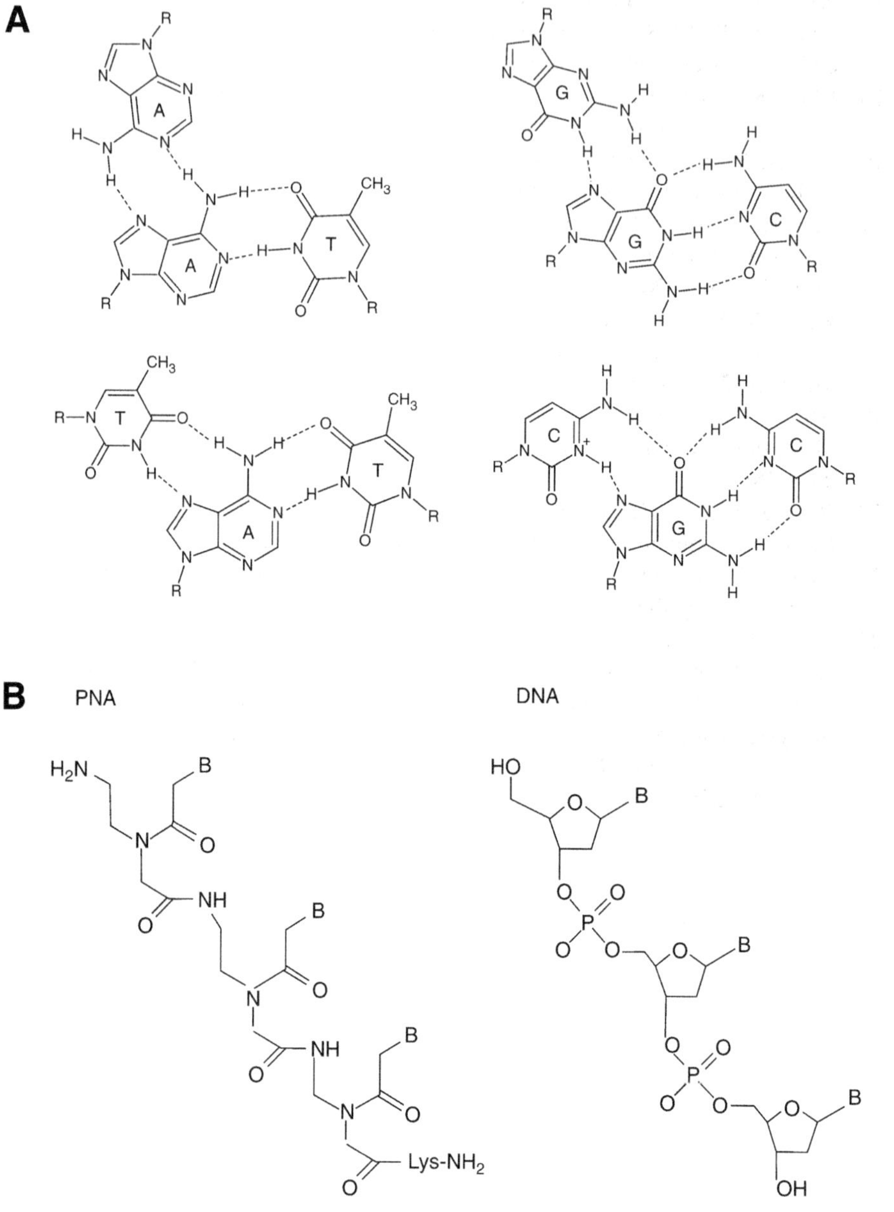

Fig. 1. **(A)** Schematic of triplet base pairing during triplex formation in both the purine (A·A:T and G·G:C) and pyrimidine motif (T·A:T and C^{+}·G:C). **(B)** Chemical structure of PNA vs DNA (reprinted with permission from **ref. *49***).

4. TFOs containing backbone or base modifications as described in **Subheading 1.** can also be synthesized to create a molecule with a high binding affinity to the target site.

3.2.2. Peptide Nucleic Acid

1. Most often a *bis*-PNA clamp containing two pyrimidine PNAs connected by a flexible linker (8-amino-3,6-dioxaoctanoic acid [O]) is used for targeting. This molecule is oriented N-terminus to C-terminus **(Fig. 2B)**.
2. To aid in strand invasion, several lysine residues can be put on the duplex-forming strand of the *bis*-PNA. These positive charges have also been shown to facilitate uptake into cells ***(41)***. Other conjugates such as the cell-penetrating peptides transactivator of transcription, the HIV transactivator of transcription, and antennapedia, the third helix of the *Drosophila* homeodomain transcription factor, have also been shown to increase uptake of PNA ***(42)***.
3. To overcome the pH dependence of N3-protonation of cytosine, pseudoisocytosine (J) can be used in place of cytosine in the Hoogsteen strand. This J base forms two Hoogsteen bonds with a G:C base pair ***(43)***.

3.2.3. Donors

1. Single-stranded donors can be 30–2000 bases in length and can be homologous to any region within 750 bp of the triplex site (*see* **Note 3**) ***(39)***.
2. Antisense (binding to the sense strand of the DNA) or sense (binding to the antisense strand of the DNA) donors can be designed (*see* **Note 4**).
3. Donors can be synthesized by Midland Certified Reagent Company (*see* **Subheading 2., step 1**) or long double-stranded donors can be synthesized by PCR amplification of a plasmid. If donors are ordered from Midland, the first and last three bases should be attached through phosphorothioate linkages to inhibit nuclease degradation.
4. When designing a donor, desired sequence changes should be kept toward the center of the oligo, and a sufficient number of bases must be on either side of the mismatch to allow for homologous recombination within the target **(Fig. 2A,B)**.

3.3. Evaluation of Binding Under Physiological Conditions

3.3.1. TFO-Binding Assay

1. To evaluate the binding of a TFO, a gel mobility shift assay is used. For this, a synthetic duplex containing the potential TFO-binding site should be designed. To do this, complementary oligomers containing the target sequence can be annealed to form duplexes.
2. The duplex DNA can then be 5′ end-labeled using T4 Polynucleotide kinase and (γ-^{32}P) dATP in the reaction. Typical reactions contain 10^{-6} *M* duplex (final concentration) in a total volume of 20 µL.
3. Next electrophorese the duplex on a 15% polyacrylamide gel to purify. Following electroelution of the duplex, add this purified duplex to a Centricon-3 column (Millipore, Bedford, MA) to concentrate (as per the manufacturer's instructions).

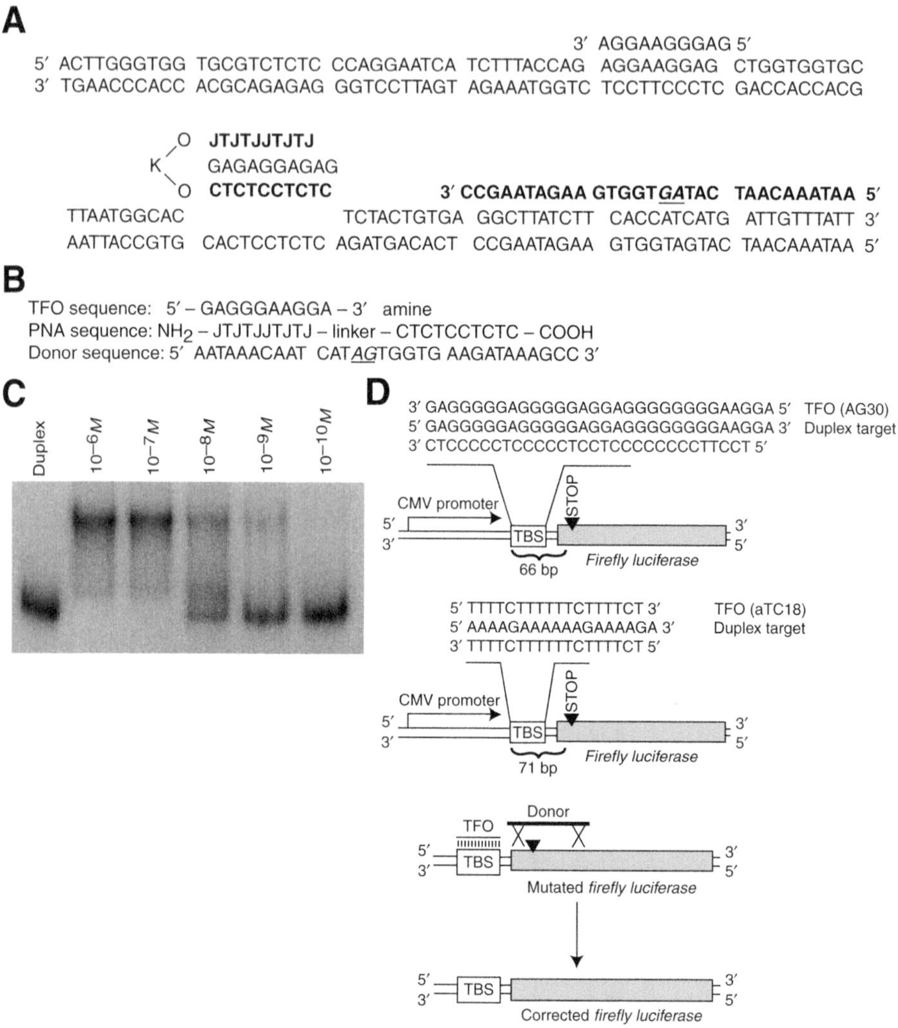

Fig. 2. (**A**) Sample sequence depicting binding of a purine-rich TFO, a *bis*-PNA, and an antisense donor. (**B**) Sequence and orientation of the TFO and *bis*-PNA that would be synthesized to bind to the target sites in A. The single-stranded DNA donor depicted would be used to mutate the underlined site from TC to GA. (**C**) Example of a TFO-binding gel-shift assay. Duplex DNA containing the target site was radiolabeled with ^{32}P and incubated with the indicated amounts of the corresponding TFO. (**D**) Schematic diagram of the luciferase assay. The diagram depicts AG30, a polypurine TFO, and *a*TC18, a polypyrimidine TFO, binding to their respective target sites. When a donor containing the wild-type sequence of the *luciferase* gene is introduced, it can correct the stop codon mutation (as indicated by the inverted triangle [▼]) by triplex-induced homologous recombination (reprinted with permission from **ref. *50***).

4. Incubate the duplex DNA at 37°C for approx 12–24 h in a series of reactions containing increasing concentrations of TFO in triplex-binding buffer. A typical reaction is 20 µL, with 2 µL (10^{-6} *M*) of labeled duplex and 2 µL of 10-fold dilutions of TFO (10^{-12}–10^{-7} *M*).
5. Next, run the binding reactions on a 12% native (19:1 acrylamide:*bis*-acrylamide) gel at 60–70 V in 89 m*M* Tris, 89 m*M* of boric acid with 0.1 m*M* of MgCl to achieve separation of the triplex structures from the duplex.
6. The gel can then be imaged with a PhosphorImager (Amersham Biosciences, Piscataway, NJ) to quantify the band shift. This data can be used to calculate the dissociation constant of the TFO-binding to the target site **(Fig. 2C)**, which represents the concentration at which binding is half-maximal.

3.3.2. PNA-Binding Assay

1. In Eppendorf tubes add the PNA at concentrations ranging from 0 to 1 µ*M*, 2 µg of plasmid DNA containing the target site flanked by known restriction enzymes, KCl to a final concentration of 10 µ*M*, and TE to a final volume of 10 µL.
2. Place these tubes at 37°C overnight.
3. The next day, digest the entire reaction (tube) in a 20 µL digest with the restriction enzymes flanking the binding site (*see* **Note 5**). Allow the reaction to digest for 1–2 h at the enzyme-specific temperature.
4. Add DNA loading dye to each sample and run them on a 10% of polyacrylamide gel until bands are well separated. To visualize the resulting gel shift, stain with silver stain (sodium borohydrate 0.1% silver nitrate [Sigma Aldrich, St. Louis, MO]) (*see* **Note 6**).
5. Expected result: at 0 µ*M*, only a single band representing the duplex fragment containing the target site should be seen. As concentrations of PNA increase, there should be a shift in the band corresponding to binding.

3.4. Evaluation of Intracellular Binding: Restriction Protection Assay/Southern

Several strategies exist to physically evaluate intracellular binding of a TFO or PNA to the target site. If pso is conjugated to the TFO or PNA, a restriction protection assay, a denaturation resistance assay or a PCR-based assay ***(44)*** can be used. In this chapter the first two methods will be described in detail.

1. For this assay, a restriction enzyme (e.g., *Bam*HI) recognition site overlapping the target site is necessary.
2. Cells containing your target site should be used.
3. Design and synthesize a pso-TFO that will bind to the target site and direct a crosslink to a position within the restriction enzyme recognition site. Transfect 2–5 µg of this pso-TFO into cells containing the chromosomal target site (*see* **Note 7**).
4. Incubate cells at 37°C for 2–6 h posttransfection and then irradiate for a total dose of 1.8 J/cm^2 of ultraviolet A (UVA) irradiation (UVA light source centered at 365 nm; Southern New England Ultraviolet, Branford, CT). This will allow pso photoactivation to crosslink the pso-TFO to the duplex DNA.

5. Immediately following UVA irradiation, wash the cells with phosphate-buffered saline and resuspend in heated (60°C) lysis buffer at a concentration of 5×10^6 cells/mL for 15 min.
6. Treat with proteinase K at a concentration of 100 μg/mL overnight at 37°C.
7. Extract the genomic DNA by addition of phenol to the lysates once, followed by two extractions with chloroform/isoamyl alcohol (24:1). Ethanol precipitate the DNA and resuspend in TE pH 8.0 and 100 m*M* of KCl (~100 μL).
8. Incubate samples at 60°C for 2 h to disrupt noncovalent triplexes and remove by filtration through a Centricon column (Millipore) as per manufacturer's instructions (*see* **Note 8**).
9. Digest the genomic DNA with the appropriate restriction enzyme (e.g., *Bam*HI) and two restriction enzymes that will cut out a defined fragment containing the target site.
10. Analyze the DNA by Southern blot using a designed probe to identify the fragment.
11. Expected result: if crosslinking occurred at the target site, the pso adduct in the *Bam*HI site will block cleavage at this site. The product resulting from the restriction protection of the *Bam*HI site can be visualized and quantitated in comparison with the extent of the *Bam*HI digestion products. This can then be used to quantify TFO-directed pso adduct formation in the target.

3.5. Evaluation of Intracellular Binding of Pso-TFOs: Denaturation-Resistance Assay (45)

1. Cells containing the endogenous gene or a chromosomally integrated exogenous gene can be used in the following assay.
2. Design and synthesize a pso-TFO, which binds to the target, and transfect 2–5 μg into cells (*see* **Note 7**).
3. Repeat **steps 4–7** as written in **Subheading 3.4.**
4. Digest 10–20 μg of genomic DNA with flanking restriction enzyme(s) that will cut out the target region.
5. Denature the digested DNA by heating in a solution of 90% formamide (to a final concentration of ~80%) to 80°C for 15 min.
6. Electrophorese the samples in a neutral 1.5% TAE agarose gel.
7. Transfer the DNA from the gel to a nylon filter/membrane by Southern blot and hybridize the membrane with a ^{32}P-labeled probe that contains the target region.
8. Expected results: once denatured, uncrosslinked DNA fails to reanneal after it enters a neutral environment, whereas crosslinked DNA will "snap back" to reform a duplex. Denatured crosslinked DNA when run in a neutral gel will migrate at the same rate as nondenatured DNA, whereas uncrosslinked DNA will migrate faster. The relative amount of the denatured (fast migrating) and the snap back (slow migrating) bands can then be quantified as a measure of target site crosslinking.

3.6. Evaluation of Induced Recombination: Luciferase Assay (40)

A luciferase assay was created to evaluate triplex-induced gene correction by homologous recombination. The *Fluc* reporter gene can be engineered to contain the target site from any gene (TBS) upstream of the coding region of luciferase.

This wild-type construct can then be mutated through site-directed mutagenesis to insert a stop codon mutation within the *Fluc* coding region, which will prevent full translation of the gene. When a TFO or PNA binds to the TBS, it can stimulate recombination between an introduced DNA donor and the *luciferase* gene to correct the mutation. Site-specific recombination can then be evaluated based on the luciferase enzyme activity in the cell lysates when compared with a standard curve, which can be created by mixing cell populations containing the chromosomally integrated wild-type or mutant construct. This reporter system can be used as a model for recombination at any chromosomal gene of interest. By placing the TBS of the target gene upstream of the *luciferase* gene, a construct is created that can be chromosomally integrated to create stable cell lines. These cell lines allow for systematic testing of various DNA-binding molecules for their effects on recombination at chromosomal targets. This assay can be used as an intermediate to evaluate induction of homologous recombination at a chromosomal target by a DNA-binding molecule at a specific target site and allows for quantification of this induction. **Figure 2D** depicts how this reporter is used to measure triplex-stimulated recombination by TFOs.

1. A construct containing the TFO target site upstream of the start site of the wild-type firefly *luciferase* gene, *FLuc*$^+$. A stop codon can then be created at varying distances from the target site (24–750 bp) ***(40)*** using site-directed mutagenesis (QuikChange Site-Directed Mutagenesis Kit, Stratagene, LaJolla, CA). Vector containing the wild-type *Fluc*$^+$ should be used as a control.
   a. Donors are designed containing correct wild-type sequence (~50-mer ***(39)***—*see* **Subheading 3.** for synthesis and design).
   b. TFO designed to bind to target site.
2. Wild-type and mutant cell lines can be stably created using the above constructs.
   a. Determine that the construct is integrated in a single copy by Southern blot.
3. Mix wild-type and mutant cells in fixed ratios to establish standard curves.
   a. Dilute mutant cells to a concentration of 1×10^5 cells/mL and seed 50,000 mutant cells per well into a 12-well plate. Dilute wild-type cells to 5×10^4 cells/mL, serially dilute cells and add to the wells containing the mutant cells.
   b. Grow cells 48 h and then harvest and lyse. Analyze lysates as described in **step 8**.
   c. This standard curve can be used to normalize the experimental data and allow for quantification of recombination frequencies.
4. 24 h before transfection seed 12-well plates with wild-type or mutant cells so they will be approx 50% confluent the day of transfection.
5. The following day cells are transfected using Geneporter 2 (Gene Therapy Systems, San Diego, CA) according to manufacturer's instructions delivering various concentrations of donor and TFO to each well (*see* **Note 9**).
6. Incubate the cells for 24–48 h at 37°C to allow for cellular repair and recombination processes to occur.
7. 48-h posttransfection, rinse the cells two times with phosphate-buffered saline (Gibco BRL, Carlsbad, CA) and then lyse cells with 250 μL of 1X passive lysis buffer (Promega).

8. Assay luciferase activity (gene correction) using the Promega Dual Luciferase Kit (Promega) and a luminometer (Berthold Technologies [Oak Ridge, TN]), as per manufacturer's instructions.
9. Expected result: if TFO-induced recombination between the donor and the gene occurred, correction of the stop codon would have taken place. This would restore tran-slation of a fully functional *Luciferase* gene, resulting in luciferase activity above levels seen in mock transfected cells or cells treated with donor alone.

3.7. Evaluation of Target Gene Correction/Mutagenesis: Genotype Analysis by Allele-Specific PCR (46,47)

After treating cells containing the target gene with TFO, PNA, and/or donor DNA, gene correction or mutation can be detected by allele-specific PCR. Harvest genomic DNA from the treated cells using Promega's Wizard SV Genomic DNA Purification System and dilute it to approx 50 μg/μL. A forward primer can be designed with its 3′ end containing the desired mutation, and the reverse primer should be similar in length and T_m. Add 50 ng of the genomic DNA to a 25 μL PCR reaction and a gradient run to determine the optimal annealing temperature of the primers. Plasmids containing the wild-type and mutant gene can be used as control templates (mutations can be introduced into the wild-type gene with site-directed mutagenesis [QuikChange Site-Directed Mutagenesis Kit, Stratagene]). Expected results: a band should be present on the gel in lanes with mutant template and no bands for wild-type DNA (*see* **Note 10**).

4. Notes

1. These conditions are used to test binding under physiological conditions.
2. All TFOs should be synthesized with a 3′ end cap of amine groups to prevent degradation by 3′ exonucleases.
3. Optimal donor length will vary from site to site often making it necessary to try various lengths *(39)*.
4. Certain targets can be more sensitive to recombination with a sense or an antisense donor. In some cases it may be necessary to test both types of donors for recombination efficiencies.
5. For best results the digest should result in a 150–200-bp band.
6. Multiple bands may be seen in the gel shift owing to the formation of various PNA–DNA triplex structures *(48)*.
7. Any transfection method can be used for this, for example, digitonin permeabilization, electroporation, or cationic lipids. It may be necessary to test several methods to determine which will give the highest efficiency and lowest toxicity in your specific cell type.
8. The Centricon column size will vary depending on the size of the duplex and/or TFO. Please check the manufacturer's specifications to determine which column size is right for you.

9. Control reactions containing only donor DNA should also be run to evaluate the effect of the TFO.
10. When testing for a mutation by allele-specific PCR using genomic DNA, 40 cycles should be used. Alternatively, real-time PCR can be used for a more sensitive measure of PCR products.

Acknowledgments

We gratefully acknowledge members of the Glazer Lab for helpful discussions. Erica Beth Schleifman is supported by a National Institute of Health (NIH) training grant to the Genetics Department, and Joanna Yee King Chin by a grant from the Medical Scientist Training Program. This work was supported by a grant from the NIH (R01CA64186) to Peter Michael Glazer.

References

1. Felsenfeld, G., Davies, D. R., and Rich, A. (1957) Formation of a three-stranded polynucleotide molecule. *J. Am. Chem. Soc.* **79,** 2023–2024.
2. Faria, M., Wood, C. D., Perrouault, L., et al. (2000) Targeted inhibition of transcription elongation in cells mediated by triplex-forming oligonucleotides. *Proc. Natl. Acad. Sci. USA* **97,** 3862–3867.
3. Maher, L. J., III, Wold, B., and Dervan, P. B. (1989) Inhibition of DNA binding proteins by oligonucleotide-directed triple helix formation. *Science* **245,** 725–730.
4. Francois, J. C., Saison-Behmoaras, T., Thuong, N. T., and Helene, C. (1989) Inhibition of restriction endonuclease cleavage via triple helix formation by homopyrimidine oligonucleotides. *Biochemistry* **28,** 9617–9619.
5. Hanvey, J. C., Shimizu, M., and Wells, R. D. (1990) Site-specific inhibition of EcoRI restriction/modification enzymes by a DNA triple helix. *Nucleic Acids Res.* **18,** 157–161.
6. Mayfield, C., Ebbinghaus, S., Gee, J., et al. (1994) Triplex formation by the human Ha-ras promoter inhibits Sp1 binding and in vitro transcription. *J. Biol. Chem.* **269,** 18,232–18,238.
7. Birg, F., Praseuth, D., Zerial, A., et al. (1990) Inhibition of simian virus 40 DNA replication in CV-1 cells by an oligodeoxynucleotide covalently linked to an intercalating agent. *Nucleic Acids Res.* **18,** 2901–2908.
8. Volkmann, S., Jendis, J., Frauendorf, A., and Moelling, K. (1995) Inhibition of HIV-1 reverse transcription by triple-helix forming oligonucleotides with viral RNA. *Nucleic Acids Res.* **23,** 1204–1212.
9. Havre, P. A., Gunther, E. J., Gasparro, F. P., and Glazer, P. M. (1993) Targeted mutagenesis of DNA using triple helix-forming oligonucleotides linked to psoralen. *Proc. Natl. Acad. Sci. USA* **90,** 7879–7883.
10. Takasugi, M., Guendouz, A., Chassignol, M., et al. (1991) Sequence-Specific Photo-Induced Cross-Linking of the Two Strands of Double- Helical DNA by a Psoralen Covalently Linked to a Triple Helix-Forming Oligonucleotide. *PNAS* **88,** 5602–5606.

11. Vasquez, K. M., Wensel, T. G., Hogan, M. E., and Wilson, J. H. (1996) High-efficiency triple-helix-mediated photo-cross-linking at a targeted site within a selectable mammalian gene. *Biochemistry* **35,** 10,712–10,719.
12. Wang, G., Seidman, M. M., and Glazer, P. M. (1996) Mutagenesis in mammalian cells induced by triple helix formation and transcription-coupled repair. *Science* **271,** 802–805.
13. Vasquez, K. M., Narayanan, L., and Glazer, P. M. (2000) Specific mutations induced by triplex-forming oligonucleotides in mice. *Science* **290,** 530–533.
14. Wang, X., Tolstonog, G., Shoeman, R. L., and Traub, P. (1996) Selective binding of specific mouse genomic DNA fragments by mouse vimentin filaments in vitro. *DNA Cell Biol.* **15,** 209–225.
15. Chan, P. P., Lin, M., Faruqi, A. F., Powell, J., Seidman, M. M., and Glazer, P. M. (1999) Targeted correction of an episomal gene in mammalian cells by a short DNA fragment tethered to a triplex-forming oligonucleotide. *J. Biol. Chem.* **274,** 11,541–11,548.
16. Datta, H. J., Chan, P. P., Vasquez, K. M., Gupta, R. C., and Glazer, P. M. (2001) Triplex-induced recombination in human cell-free extracts. Dependence on XPA and HsRad51. *J. Biol. Chem.* **276,** 18,018–18,023.
17. Luo, Z., Macris, M. A., Faruqi, A. F., and Glazer, P. M. (2000) High-frequency intrachromosomal gene conversion induced by triplex-forming oligonucleotides microinjected into mouse cells. *Proc. Natl. Acad. Sci. USA* **97,** 9003–9008.
18. Seidman, M. M. (2004) Oligonucleotide mediated gene targeting in mammalian cells. *Curr. Pharm. Biotechnol.* **5,** 421–430.
19. Demidov, V. V., Potaman, V. N., Frank-Kamenetskii, M. D., et al. (1994) Stability of peptide nucleic acids in human serum and cellular extracts. *Biochem. Pharmacol.* **48,** 1310–1313.
20. Faruqi, A. F., Egholm, M., and Glazer, P. M. (1998) Peptide nucleic acid-targeted mutagenesis of a chromosomal gene in mouse cells. *Proc. Natl. Acad. Sci. USA* **95,** 1398–1403.
21. Hanvey, J. C., Peffer, N. J., Bisi, J. E., et al. (1992) Antisense and antigene properties of peptide nucleic acids. *Science* **258,** 1481–1485.
22. Koppelhus, U., Zachar, V., Nielsen, P. E., Liu, X., Eugen-Olsen, J., and Ebbesen, P. (1997) Efficient in vitro inhibition of HIV-1 gag reverse transcription by peptide nucleic acid (PNA) at minimal ratios of PNA/RNA. *Nucleic Acids Res.* **25,** 2167–2173.
23. Praseuth, D., Grigoriev, M., Guieysse, A. L., et al. (1996) Peptide nucleic acids directed to the promoter of the alpha-chain of the interleukin-2 receptor. *Biochim. Biophys. Acta* **1309,** 226–238.
24. Nielsen, P. E., Egholm, M., Berg, R. H., and Buchardt, O. (1993) Sequence specific inhibition of DNA restriction enzyme cleavage by PNA. *Nucleic Acids Res.* **21,** 197–200.
25. Mollegaard, N. E., Buchardt, O., Egholm, M., and Nielsen, P. E. (1994) Peptide nucleic acid. DNA strand displacement loops as artificial transcription promoters. *Proc. Natl. Acad. Sci. USA* **91,** 3892–3895.

26. Rogers, F. A., Manoharan, M., Rabinovitch, P., Ward, D. C., and Glazer, P. M. (2004) Peptide conjugates for chromosomal gene targeting by triplex-forming oligonucleotides. *Nucleic Acids Res.* **32,** 6595–6604.
27. Branden, L. J., Mohamed, A. J., and Smith, C. I. (1999) A peptide nucleic acid-nuclear localization signal fusion that mediates nuclear transport of DNA. *Nat. Biotechnol.* **17,** 784–787.
28. Vasquez, K. M., Dagle, J. M., Weeks, D. L., and Glazer, P. M. (2001) Chromosome targeting at short polypurine sites by cationic triplex-forming oligonucleotides. *J. Biol. Chem.* **276,** 38,536–38,541.
29. Lacroix, L., Lacoste, J., Reddoch, J. F., et al. (1999) Triplex formation by oligonucleotides containing 5-(1-propynyl)-2′-deoxyuridine: decreased magnesium dependence and improved intracellular gene targeting. *Biochemistry* **38,** 1893–1901.
30. Puri, N., Majumdar, A., Cuenoud, B., et al. (2002) Minimum number of 2′-O-(2-aminoethyl) residues required for gene knockout activity by triple helix forming oligonucleotides. *Biochemistry* **41,** 7716–7724.
31. Wang, G. and Glazer, P. M. (1995) Altered repair of targeted psoralen photoadducts in the context of an oligonucleotide-mediated triple helix. *J. Biol. Chem.* **270,** 22,595–22,601.
32. Majumdar, A., Puri, N., Cuenoud, B., et al. (2003) Cell cycle modulation of gene targeting by a triple helix forming oligonucleotide. *J. Biol. Chem.* **278,** 11,072–11,077.
33. Macris, M. A. and Glazer, P. M. (2003) Transcription dependence of chromosomal gene targeting by triplex-forming oligonucleotides. *J. Biol. Chem.* **278,** 3357–3362.
34. Vasquez, K. M., Wang, G., Havre, P. A., and Glazer, P. M. (1999) Chromosomal mutations induced by triplex-forming oligonucleotides in mammalian cells. *Nucleic Acids Res.* **27,** 1176–1181.
35. Sargent, R. G., Rolig, R. L., Kilburn, A. E., Adair, G. M., Wilson, J. H., and Nairn, R. S. (1997) Recombination-dependent deletion formation in mammalian cells deficient in the nucleotide excision repair gene ERCC1. *Proc. Natl. Acad. Sci. USA* **94,** 13,122–13,127.
36. Faruqi, A. F., Seidman, M. M., Segal, D. J., Carroll, D., and Glazer, P. M. (1996) Recombination induced by triple-helix-targeted DNA damage in mammalian cells. *Mol. Cell. Biol.* **16,** 6820–6828.
37. Sandor, Z. and Bredberg, A. (1995) Triple helix directed psoralen adducts induce a low frequency of recombination in an SV40 shuttle vector. *Biochim. Biophys. Acta* **1263,** 235–240.
38. Faruqi, A. F., Datta, H. J., Carroll, D., Seidman, M. M., and Glazer, P. M. (2000) Triple-helix formation induces recombination in mammalian cells via a nucleotide excision repair-dependent pathway. *Mol. Cell. Biol.* **20,** 990–1000.
39. Knauert, M. P., Kalish, J. M., Hegan, D. C., and Glazer, P. M. (2006) Triplex-Stimulated Intermolecular Recombination at a Single-Copy Genomic Target. *Mol. Ther.* **14,** 392–400.
40. Knauert, M. P., Lloyd, J. A., Rogers, F. A., et al. (2005) Distance and affinity dependence of triplex-induced recombination. *Biochemistry* **44,** 3856–3864.

41. Sazani, P., Kang, S. H., Maier, M. A., et al. (2001) Nuclear antisense effects of neutral, anionic and cationic oligonucleotide analogs. *Nucleic Acids Res.* **29,** 3965–3974.
42. Koppelhus, U., Awasthi, S. K., Zachar, V., Holst, H. U., Ebbesen, P., and Nielsen, P. E. (2002) Cell-dependent differential cellular uptake of PNA, peptides, and PNA-peptide conjugates. *Antisense Nucleic Acid Drug Dev.* **12,** 51–63.
43. Egholm, M., Christensen, L., Dueholm, K. L., Buchardt, O., Coull, J., and Nielsen, P. E. (1995) Efficient pH-independent sequence-specific DNA binding by pseudoisocytosine-containing bis-PNA. *Nucleic Acids Res.* **23,** 217–222.
44. Diviacco, S., Rapozzi, V., Xodo, L., Helene, C., Quadrifoglio, F., and Giovannan-geli, C. (2001) Site-directed inhibition of DNA replication by triple helix formation. *FASEB J.* **15,** 2660–2668.
45. Shahid, K. A., Majumdar, A., Alam, R., et al. (2006) Targeted cross-linking of the human beta-globin gene in living cells mediated by a triple helix forming oligonucleotide. *Biochemistry* **45,** 1970–1978.
46. Orou, A., Fechner, B., Utermann, G., and Menzel, H. J. (1995) Allele-specific competitive blocker PCR: a one-step method with applicability to pool screening. *Hum. Mutat.* **6,** 163–169.
47. Parsons, B. L., McKinzie, P. B., and Heflich, R. H. (2005) Allele-specific competitive blocker-PCR detection of rare base substitution. *Methods Mol. Biol.* **291,** 235–245.
48. Hansen, G. I., Bentin, T., Larsen, H. J., and Nielsen, P. E. (2001) Structural isomers of bis-PNA bound to a target in duplex DNA. *J. Mol. Biol.* **307,** 67–74.
49. Gorman, M. and Glazer, P. M. (2001) Directed gene modification via triple helix formation. *Curr. Mol. Med.* **1,** 391–399; with permission from Bentham Science Publishers Ltd. (Reprinted).
50. Knauert, M. P., Kalish, J. M., Hegan, D. C., and Glazer, P. M. (2006) Triplex-stimulated intermolecular recombination at a single-copy genomic target. *Mol. Ther.* **14,** 392–400; with permission from Elsevier (Reprinted).

14

Allelic Exchange of Unmarked Mutations in *Mycobacterium Tuberculosis*

Martin S. Pavelka, Jr.

Summary

Mycobacterium tuberculosis has been studied since the 19th century, but genetic manipulation of this organism has only become possible within the last decade of the 20th century. One key methodology is the allelic exchange of unmarked, in-frame deletion mutations and point mutations using counter-selectable suicide plasmids. I describe below the challenges of allelic exchange in *M. tuberculosis*, the overall mechanism of a suicide vector system, and how it can be used for efficient allelic exchange in this formally intractable organism.

Key Words: Allelic exchange; counter-selection; electroporation; *Mycobacterium tuberculosis*; mutagenesis; *sacB*; sucrose.

1. Introduction

Experimental manipulation of *Mycobacterium tuberculosis*, the causative agent of the disease tuberculosis, has traditionally been plagued with difficulties. The organism has a slow growth rate of approx 18 h, takes 3–4 wk to form colonies on solid media, has a tendency to aggregate in liquid culture, and must be handled in a biosafety level three facility. These problems slowed the progress of mycobacterial genetic tool development until the 1990s and early 21st century, when an explosion of research yielded new plasmids, electroporation methods, allelic exchange methods, transposon mutagenesis, antibiotic markers, and new mycobacteriophage-based tools to introduce a variety of DNA molecules into mycobacterial cells, reviewed in **refs.** ***1*** and *2*.

There were several obstacles in the development of allelic exchange methods for *M. tuberculosis*. Earlier attempts of allelic exchange in the closely related organism *M. bovis* Bacillus Calmette-Guerin using short, linear, double-stranded DNA molecules were problematic owing to the high degree of illegitimate recombination relative to homologous recombination ***(3)***. Illegitimate recombination is

From: *Methods in Molecular Biology, vol. 435: Chromosomal Mutagenesis*
Edited by: G. Davis and K. J. Kayser © Humana Press Inc., Totowa, NJ

also a problem in *M. tuberculosis*, but in spite of this, the first allelic exchange in this organism was eventually accomplished using long, linear, double-stranded DNA molecules (~50 kb in size) introduced by electroporation *(4)*. The rationale behind the use of long, linear DNA molecules for allelic exchange developed from the idea that recombination in the organism was not efficient, and that providing a larger amount of homologous DNA with free ends could stimulate homologous recombination. This method was laborious and it was soon discovered that the construction of the long recombination substrates was often difficult, if not impossible, for certain genes. Subsequently, it was shown that DNA recombination in this organism is not inherently faulty, and that the primary barrier to allelic exchange in *M. tuberculosis* is probably the entry of DNA into the cell *(5)*.

Methods for allelic exchange in *M. tuberculosis* that used suicide plasmids with counter-selectable markers such as *sacB* (based on sucrose sensitivity) met with great success *(5–7)*, as did a mycobacteriophage-based method to introduce recombination substrates carried within the genome of a temperature-sensitive phage *(8)*. The drawback of the phage method is that the gene of interest must be marked with an antibiotic resistance cassette. The use of suicide plasmids with counter-selectable markers allows one to perform allelic exchange using an in-frame deletion or point mutation allele of the gene under study.

This chapter will describe the rationale and the methodology to construct unmarked mutations in the *M. tuberculosis* genome. The plasmid used for this basic system is shown in **Fig. 1**. The general mechanism for allelic exchange is shown in **Fig. 2**.

2. Materials (*See* Notes 2 and 3)

1. Middlebrook 7H9 liquid medium or 7H10 solid medium (Becton, Dickinson and Company, Franklin Lakes, NJ), each containing 1X albumin-dextrose-saline (ADS) supplement, 0.2% (v/v) glycerol, and 0.05% (v/v) Tween-80 (*see* **Note 4**).
2. 10X ADS: this is prepared as a 10X concentrated stock solution (5% [w/v] bovine serum albumin fraction V [Roche Applied Sciences, Indianapolis, IN], 2% [w/v] dextrose, 8.5% [w/v] NaCl). We typically prepare 1–3 L at a time. The albumin is slowly added until it is dispersed in the water. Take care to not add the albumin all at once, or else it will clump and be difficult to dissolve. Let it stir for several hours. We typically place the beaker in a 4°C cold room and stir overnight. After the albumin has gone into solution, add the dextrose and NaCl and stir until they have dissolved. The stock must then be centrifuged at 1200*g* for 1 h (*see* **Note 5**). Decant the stock solution from the insoluble material and filter through a 0.2 μm filter using a prefilter, as supplied by the manufacturer. The prefilter is important for blocking out any remaining insoluble particulates. Incubate the finished 10X ADS stocks overnight at 37°C to confirm that they are sterile, then store at 4°C. When preparing 7H10 solid media, warm the 10X ADS in the same water bath that is used to temper the media to 50–55°C. Do not add the ADS hot media as the protein will denature. Add a sterile stir bar to the flask and stir well after addition of the ADS.

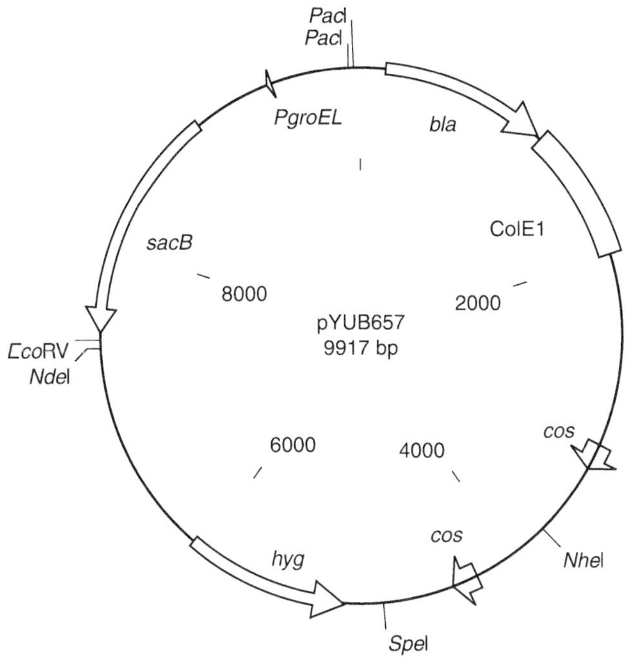

Fig. 1. Counterselectable allelic exchange vector pYUB657 *(5)*. The backbone of this vector is a double λ-*cos* cosmid cloning vector, has a ColE1 origin of replication for *E. coli*, the *bla* gene for selection with ampicillin in *E. coli*, the *hyg* gene, for selection with hygromycin in mycobacteria, and the *sacB* gene driven by the mycobacterial *groEL* promoter, which confers sensitivity to sucrose. The *sacB* gene encodes a secreted levansucrase from *Bacillus subtilis* that cleaves sucrose into glucose and fructose and then polymerizes the fructose. The enzyme is not fully secreted in mycobacteria and thus, fructose polymers are believed to accumulate in the cell envelope with lethal consequences *(11,12)*. Usable cloning sites are shown (*see* **Note 1**).

Do not open the 10X ADS bottle outside of a biosafety cabinet. After the warmed ADS has cooled, a vacuum will form in the airspace above the solution and opening the bottle will allow air to enter, which often brings with it microbial contaminants.

3. 50% Glycerol stock: make a 50% (v/v) stock of glycerol in deionized, distilled water and sterilize by autoclaving. We have noticed no stability problems with glycerol stocks stored at room temperature for several months.
4. 20% Tween-80 stock: made by adding 20 mL of Tween-80 to 80 mL of deionized, distilled water (20% v/v). Mix well with a stir bar and sterilize by filtration with a 0.2 µm filter. It is best to filter this solution into a sterile glass bottle with a phenolic cap, as Tween-80 will eventually weaken the cap on most plastic filtration units, causing it to crack. Contrary to common belief, Tween-80 is not light sensitive. It is stable at room temperature for long periods of time; however, discard the stock if a precipitate appears on the bottom of the bottle.

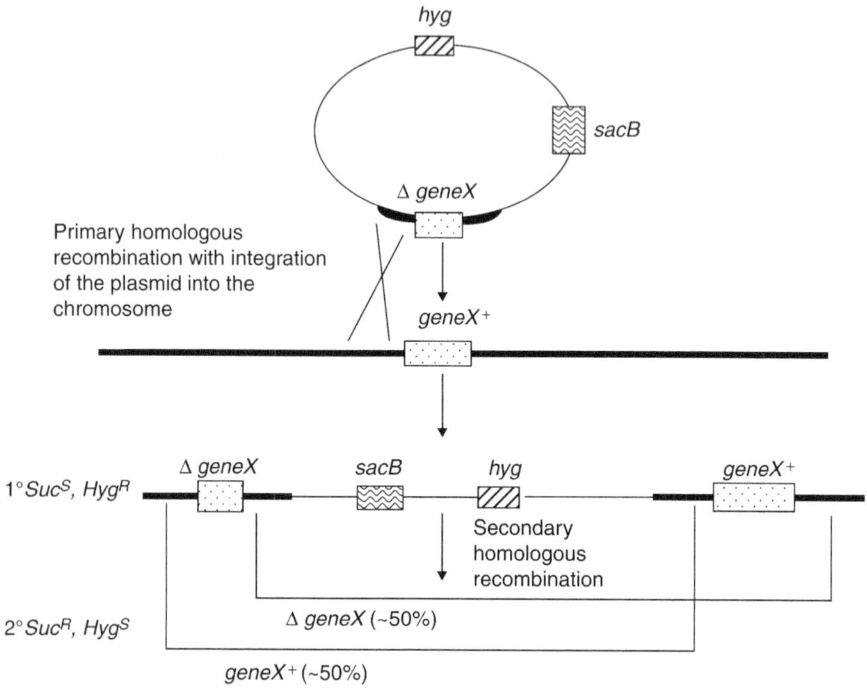

Fig. 2. Allelic exchange with counterselection. This description uses a deletion allele, but the same rationale applies to point mutations. The suicide plasmid based on pYUB657 contains an unmarked deletion allele of *geneX* with approx 1 kb of homologous DNA flanking each side of the gene. The plasmid confers hygromycin resistance and sucrose sensitivity. Introduction of the plasmid into *M. tuberculosis* and selection for hygromycin resistance will yield primary recombinants in which the plasmid has integrated on one side or the other of the wild-type gene in the chromosome. Because the plasmid does not have an origin of replication for mycobacteria, any hygromycin-resistant clones must be the result of a recombination event. The frequency of recombination, based up on comparison with the transformation efficiency with a replication plasmid, is on the order of 10^{-5}. All Hyg^R clones should also be sensitive to sucrose (Suc^S). A primary recombinant is then grown in liquid media lacking antibiotics and sucrose until saturated. During this period, the duplicated homologous regions can undergo a secondary recombinational event in which the intervening plasmid DNA loops out of the genome, taking either the wild-type or the mutated sequences with it. The new plasmid will eventually be lost and the resulting population will be Suc^S and have either the wild-type gene or the mutated allele in the chromosome. The frequency of this secondary recombination is rather low, on the order of 10^{-4} relative to the viable cell count, and so, to select for these clones the population of cells is plated onto media containing sucrose. Those clones that have not undergone a secondary recombinational event will be killed whereas the secondary recombinants that lost the integrated suicide vector and the *sacB* gene will survive. Approximately 50% of the Suc^R population will have the wild-type gene, whereas the remainder will have the mutated allele. These are then identified by phenotype (if possible) and verified by polymerase chain reaction or Southern blot, and in the case of point mutations, DNA sequencings.

5. Middlebrook 7H10/sucrose: made by adding sucrose from a 30% (w/v) stock to a 2% (v/v) final concentration in 7H10 media that has been autoclaved and tempered to 50–55°C. Adjust the volume of water accordingly to compensate for the added volume of sucrose. Make sure that you have a sterile stir bar in the flask and stir well after the addition of sucrose. Failure to do so will result in a gradient of sucrose concentration across the batch of plates.
6. 30% Sucrose stock: make a 30% (w/v) sucrose stock in water and filter sterilize by filtration through a 0.2-μm filter. We have noticed no stability problems with sucrose stocks stored at room temperature.
7. Hygromycin: 50 μg/mL final concentration from a 50 mg/mL commercially prepared stock in phosphate-buffered saline (Roche). We have not noticed any problems with stability of this antibiotic when the stock is stored at 4°C as per the manufacturer's recommendation.
8. 10% Glycerol (v/v) made in deionized, distilled water and filter sterilized. This solution can be autoclaved, but to reduce the use of glass bottles in our BSL-3 facility we filter (0.2 μm) sterilize the solution in plastic. Store at room temperature.

3. Methods

3.1. Bacterial Cell Culture

1. Thaw a frozen 1 mL aliquot of *M. tuberculosis* H37Rv (*see* **Note 6**) stock at room temperature and transfer to a 30-mL media bottle (Nalgene, Rochester, NY) containing 10 mL of Middlebrook 7H9 media. Incubate at 37°C with slow shaking (100 rpm) until saturated (~5 d). Subculture 1:50 into 100 mL of Middlebrook 7H9 in a 490 cm^2 roller bottle (Corning, Corning, NY) and incubate at 37°C on a roller apparatus set for 8–10 rpm for 5–7 d, until the OD_{600} of the culture is between 0.5 and 1.0. Check the optical density of the culture two days before your planned electroporation, to ensure that the cells are doubling (*see* **Note 7**).

3.2. Electroporation and Plating (See Note 8)

1. Transfer cells to two 50-mL conical tubes, pellet cells at 2000*g* for 5 min at room temperature.
2. Gently wash each tube of cells with 40 mL of room temperature 10% sterile glycerol solution by pipeting. Do *not* vortex cells (*see* **Note 9**).
3. Pellet cells 2000*g* for 5 min.
4. Resuspend cell pellets in 10 mL of 10% glycerol for each and combine into a single tube. Add additional glycerol to a final volume of 40 mL.
5. Pellet cells 2000*g* for 5 min.
6. Resuspend cells in 1 mL of 10% glycerol (1/100th the original culture volume).
7. Add the suicide plasmid DNA (at least 1 μg, *see* **Note 10**) to the inside wall of the electroporation cuvet (0.2 cm gap) and then carefully pipet 100 μL of cell suspension into the drop of DNA, letting the liquid run down the inner wall of the cuvet.
8. Cap the cuvet, tap it gently a few times to settle the cells, and then electroporate at room temperature. We use a BioRad Gene Pulser set (Bio-Rad Laborotories, Inc., Hercules, CA) for 2500 mV, 25 μF, and 1000 Ω. The time constant should be around 20 ms (*see* **Note 11**).
9. Add 1 mL of Middlebrook 7H9 media to the cuvet and transfer the cell suspension to a 15-mL conical tube using a 6-in. sterile glass Pasteur pipet (*see* **Note 12**).

10. Repeat the electroporations with the suicide plasmid for a total of 5–10 electroporations (*see* **Note 13**).
11. Perform two control electroporations, one without any DNA added, and the other with a plasmid bearing a mycobacterial origin of replication and the hygromycin resistance marker (*see* **Note 14**).
12. Incubate tubes at 37°C in a 100 rpm shaker overnight.
13. The following day, plate dilutions of the replicating vector, (positive control) out to 10^{-5} on Middlebrook 7H10 media containing hygromycin at 50 μg/mL, and plate dilutions of the DNA negative control out to 10^{-8} on 7H10 media without antibiotic to check the viability of the cells. Plate the remainder of the negative control onto 7H10/hygromycin media to check for spontaneous hygromycin resistance (use 200 μL per plate). Finally, plate out each suicide vector electroporation over five plates, using 200 μL per plate (*see* **Note 15**).
14. Incubate plates, wrapped in foil or encased in metal Petri dish cans for 3–4 wk at 37°C (*see* **Note 16**).

3.3. Screening of Primary Recombinants, Growth of Secondary Recombinants

1. Record the number of transformants obtained with the positive control replicating plasmid, viable counts, and spontaneous hygromycin resistance frequency (*see* **Note 17**).
2. Subculture (by streaking) the Hyg^R primary recombinants onto 7H10 media supplemented with 2% sucrose to test for sucrose sensitivity and streak onto 7H10/Hyg media for working stocks. Incubate plates, wrapped in foil or encased in metal Petri dish cans for 3 wk at 37°C.
3. Simultaneously, subculture each of the Hyg^R primary recombinants into 10 mL of 7H9 broth lacking sucrose and hygromycin in 30-mL media bottles and incubate at 37°C on a 100 rpm shaker until the culture is saturated, which should take about a week (*see* **Note 18**).

3.4. Sucrose Selection and Screening Secondary Recombinants

1. Plate dilutions of the primary recombinant cultures out to 10^{-3} (in duplicate) on 7H10 media containing 2% sucrose. Incubate plates, wrapped in foil or encased in metal Petri dish cans for 3–4 wk at 37°C (*see* **Note 19**).
2. Screen Suc^R clones for hygromycin resistance on 7H10/Hyg^{50}, and rescreen onto 7H10/sucrose to confirm and to identify the Suc^R Hyg^S secondary recombinants (*see* **Note 20**).
3. Screen Suc^R, Hyg^S secondary recombinants (by phenotype if possible), polymerase chain reaction, and Southern blot.

4. Notes

1. It is best to maintain pYUB657 in *Escherichia coli* using media containing ampicillin and hygromycin each at 50 μg/mL. We have found that using ampicillin alone can lead to deletion of the *hyg* gene in a substantial number of clones. This also applies to construction of allelic exchange plasmids in this vector and to subculturing clones for DNA preparation.

2. One of the major determinants in growing *M. tuberculosis* appears to be water quality. Poor quality water often results in cultures that clump prematurely, cells that do not grow well and do not transform well. We have found problems with using house-supplied 18 megohm deionized distilled water for work with *M. tuberculosis*. Using media and solutions made with high-performance liquid chromatography grade water, which typically has a higher conductivity than 18 megohm water but is low in organic contaminants, has solved the problems we have had with our house-supplied water. We hypothesize that this water can become contaminated by organic contaminants leaching out of the filtration system and that *M. tuberculosis* is sensitive to these contaminants.
3. All solution and media preparation should be performed in a biosafety cabinet to prevent contamination with airborne organisms. It is a good standard procedure to have a dedicated set of solutions and reagents to be opened only within the confines of a biosafety cabinet. This precaution is important because of the slow growth rate of *M. tuberculosis*. Pouring plates on an open bench can invite contaminants that will overgrow in the media during the long incubation times.
4. Tween-80 is used in liquid media to prevent clumping of the mycobacteria and is not required for solid media. However, the inclusion of Tween-80 in solid media makes the usually crumbly *M. tuberculosis* colonies softer and easier to pick up with a loop or toothpick. In addition, because mutants arise during growth in liquid media with Tween-80 and are then plated onto solid media, it may be important to have Tween-80 in the solid media as the mutants may be dependent on the detergent for growth. Such was the case for the construction of lysine auxotrophs of *M. tuberculosis* ***(5)***.
5. This is an important step to ensure that any insoluble aggregates are removed from the suspension, for if they remain, they will clog the filter during the sterilization step.
6. Electroporation efficiencies vary between strains of *M. tuberculosis*, with H37Rv having one of the best, with transformation efficiencies of 10^5 colony-forming units per electroporation. We have found that other strains, such as CDC1551 (CSU93) and strain Erdman, do not yield as high a transformation frequency. When using such strains, one can use a counter selectable plasmid system based on a replicating vector with a temperature-sensitive mycobacterial origin of replication ***(6)***. The permissive temperature for this plasmid is 32°C and the nonpermissive temperature is 39°C. Using a conditionally replicating vector relieves the requirement for high transformation efficiency. A drawback to this approach is that selection for the *M. tuberculosis* transformants is done at 32°C, which only yields colonies after 9–10 wk, as this temperature is below the optimal temperature for growth.

   Another way to circumvent low transformation efficiencies is to increase the recombinogenicity of the input plasmid DNA by irradiation of the DNA with ultraviolet light before electroporation ***(9)***.
7. *M. tuberculosis* cultures will occasionally undergo cycles of lysis or clumping, resulting in oscillating readings for optical density. These cultures usually do not transform well. Because of this potential problem, it is prudent to check the optical density of the culture a few days before the culture will be used, to make sure that it is growing and doubling about every day.
8. Most electroporation protocols insist that the cells, solutions, and cuvets be kept on ice. This is not necessary; in fact, doing electroporations at room temperature may aid

the DNA in traversing the lipid portion of the mycobacterial cell envelope by keeping the envelope more fluid than if the cells are ice-cold.

9. Washing the cells should be done gently, by pipeting and not by shaking of vortexing. If the cell suspension is vigorously agitated, the cells will clump and trap air. These clumps will be very difficult to pellet and will both float and coat the sides of the centrifuge tube, resulting in a dramatic decrease in cell pellet yield.
10. We find that transformation of bacteria with a replicating plasmid begins to saturate with about 100–200 ng of DNA, depending on the species and the type of plasmid. We use at least 1 µg of suicide plasmid DNA per electroporation, to ensure that any cell that is capable of taking up DNA through transient pores can take as much as possible. The DNA should be of high quality and in sterile water, preferably prepared using a Qiagen system (Qiagen Inc., Valencia, CA).

    If an allelic exchange is particularly difficult, more DNA can be used up until the quantity begins to effect the electroporation by salt carryover. We think that the additional DNA does not enter additional cells, but that more DNA enters each cell capable of taking up DNA. This would increase the number of plasmid DNA copies per cell and increase the likelihood that a plasmid will integrate into the chromosome. Dispensing the DNA onto the side of the cuvet takes a steady hand; alternatively, one can add the DNA to a microcentrifuge tube and then add the cells, let them sit for a minute, then transfer them to the cuvet. Which technique you use is a matter of personal preference. We prefer the cuvet method because it has a smaller number of steps, and we try to keep manipulations of *M. tuberculosis* to a minimum in our procedures. When using the cuvet method, it is important to add the cells immediately after placing the DNA on the side of the cuvet, as waiting too long will result in the water evaporating from the DNA drop. This is only a concern if you are performing multiple electroporations.
11. The time constant might be lower (15–18 ms) owing to the room temperature electroporation. If the DNA has too much salt or the cells were not well washed, the cuvet may have a visual discharge (arc) and pop. If this happens, discard the electroporation and repeat, adding 50–100 µL of 10% glycerol to a new cell and DNA mixture in a fresh cuvet. It is not clear why certain electroporations arc. Performing 10 electroporations with the same cells and the same DNA, we often find that one to two of the electroporations will arc.
12. When you add the 7H9 media to the cells after electroporation, they will be clumped and grainy but do not despair; after a few hours in fresh media the cells will smooth out again. A sterile glass Pasteur pipet is really the safest and easiest way to remove the cells from the cuvet, as the glass tip is narrow enough to get most of the suspension at the bottom of the cuvet. Trying to tip the cuvet and recover the cells with a pipetor fitted with a 200-µL tip is difficult and is likely to result in a spill.
13. As the best transformation efficiency with a replicating plasmid in *M. tuberculosis* is 10^5 transformants per electroporation, and the frequency of recombination into the chromosome is on the order of 10^{-5} relative to the transformation frequency, one usually finds only a few primary recombinants per electroporation. Thus, it is a good idea to do at least five, and preferably 10, electroporations for each mutation. Some of the primary clones will be illegitimate (nonhomologous) recombinants and doing multiple electroporations will increase the odds that you will obtain a few primary

homologous recombinants. We tend to perform the electroporations in sets of five, and then adding media to each cuvet and transferring the contents to tubes for outgrowth. Adding media to each cuvet immediately after electroporation is not necessary.

14. We typically use the *hyg*-bearing *E. coli*-mycobacterial shuttle vector pMV261.*hyg* as a control ***(10)***.
15. Because the recombination frequency is so low, it is best to plate out the entire 1 mL from each of the transformation mixtures with the suicide plasmid over five plates, 200 μL per plate.
16. Do not use lab tape to package the plates together as this can encourage fungal contaminants to grow on the tape and then spread into the plates during the long incubation. Check the plates about once a week to make sure they are not growing contaminants and to remove any contaminated plates to limit spread to other plates. Any colony taking more than 4 wk to grow is likely slow-growing spontaneous hygromycin-resistant mutants. We do see such mutants occasionally, but they are quite rare and usually do not appear until at after least 4–5 wk of incubation.
17. Expect that there will be approx 10^8–10^9 viable colony-forming units with a transformation efficiency of 10^5 hygromycin-resistant colonies per electroporation.
18. The primary recombinants are streaked out on media to confirm sucrose sensitivity and subcultured to liquid media for outgrowth and development of secondary recombinants at the same time because of the slow growth of *M. tuberculosis*. By the time the liquid cultures are saturated, it is usually possible to score the sucrose phenotype of the plate subcultures. We usually streak out at least 10 putative primary recombinant clones and set up five for broth cultures. The five broth cultures will be plated for secondary recombinants.
19. We plate all five of the broth cultures to ensure that we are working with at least one homologous primary recombinant. Some of the primary recombinants will be illegitimate recombinants that arose from a nonhomologous integration event. These clones do not seem to be able to undergo a secondary recombination event, but can become Suc^R, probably owing to inactivation of the *sacB* gene (*see* **Note 20**, below). However, the frequency of Suc^R from illegitimate primary recombinants is on the order of 10^{-7}, which is far less than the frequency of Suc^R from homologous primary recombinants, which tends to be approx 10^{-4}. So, we disregard those clones that give rise to very low Suc^R frequencies and work up the others.
20. As described in the legend to **Fig. 2**, all of the Suc^R clones in the secondary recombinant pool will be Hyg^S and either mutant or wild-type. This population distribution is a theoretical assumption based on having equal amounts of flanking DNA that is equally recombinogenic. It also assumes that there is no background Suc^R in the population. However, in reality, the population of Suc^R cells also includes clones in which the *sacB* gene is likely to be mutated, either by deletion or interruption by an insertion element.

    We have found that a substantial proportion (~20–70%) of the Suc^R clones have *sacB* inactivated, as revealed by their Hyg^R phenotype, indicating that the integrated plasmid is still present but that the *sacB* gene is not functional or deleted. A faster way to screen for these so-called "*sacB* inactivated" clones is to use a suicide vector with an additional screenable marker such as *lacZ* encoding β-galactosidase (for details, *see* **ref.** *7*). In this scheme, the secondary recombinants are plated on sucrose plates containing the colorometric β-galactosidase substrate 4-chloro-5-bromo-3-indolyl-β-D-galactopyranoside. Primary recombinants are blue on this media and Suc^S. Any secondary clones that lost the integrated plasmid by recombination are Suc^R and white, whereas any undesired "*sacB* inactivated" clones are Suc^R and blue, owing to the continued presence of the *lacZ* gene in the integrated plasmid. Screening for the antibiotic marker carried on the plasmid backbone would be needed to confirm the identity of the clone.

Acknowledgments

The author's research efforts are supported by grant AI068013 from the National Institute of Allergy and Infectious Disease, and the Potts Memorial Foundation. I would like to thank Maria-Magdalena Patru for the graphic in **Fig. 2**, and members of my laboratory for reading this manuscript.

References

1. Jacobs, W. R., Jr. (2000) *Mycobacterium tuberculosis*: a once genetically intractable organism, in *Molecular Genetics of Mycobacteria* (Hatfull, G. F. and Jacobs, W. R., Jr., eds.), ASM Press, Washington, DC, pp. 1–16.
2. Kana, B. D. and Mizrahi, V. (2004) Molecular genetics of Mycobacterium tuberculosis in relation to the discovery of novel drugs and vaccines. *Tuberculosis (Edinb)* **84,** 63–75.
3. Kalpana, G. V., Bloom, B. R., and Jacobs, W. R., Jr. (1991) Insertional mutagenesis and illegitimate recombination in mycobacteria. *Proc. Natl. Acad. Sci. USA* **88,** 5433–5437.
4. Balasubramanian, V., Pavelka, M. S., Jr., Bardarov, S. S., et al. (1996) Allelic exchange in *Mycobacterium tuberculosis* with long linear recombination substrates. *J. Bacteriol.* **178,** 273–279.
5. Pavelka, M. S., Jr. and Jacobs, W. R., Jr. (1999) Comparison of the construction of unmarked deletion mutations in *Mycobacterium smegmatis*, *Mycobacterium bovis* bacillus Calmette-Guerin, and *Mycobacterium tuberculosis* H37Rv by allelic exchange. *J. Bacteriol.* **181,** 4780–4789.
6. Pelicic, V., Jackson, M., Reyrat, J. M., Jacobs, W. R., Jr., Gicquel, B., and Guilhot, C. (1997) Efficient allelic exchange and transposon mutagenesis in *Mycobacterium tuberculosis*. *Proc. Natl. Acad. Sci. USA* **94,** 10,955–10,960.
7. Parish, T., Gordhan, B. G., McAdam, R. A., Duncan, K., Mizrahi, V., and Stoker, N. G. (1999) Production of mutants in amino acid biosynthesis genes of Mycobacterium tuberculosis by homologous recombination. *Microbiology* **145,** 3497–3503.
8. Bardarov, S., Bardarov, S., Jr., Pavelka, M. S., Jr., et al. (2002) Specialized transduction: An efficient method for generating marked and unmarked targeted gene disruptions in *Mycobacterium tuberculosis, M. bovis* BCG, and *M. smegmatis*. *Microbiology* **148,** 3007–3017.
9. Hinds, J., Mahenthiralingam, E., Kempsell, K. E., et al. (1999) Enhanced gene replacement in mycobacteria. *Microbiology* **145,** 519–527.
10. Pavelka, M. S., Jr. and Jacobs, W. R., Jr. (1996) Biosynthesis of diaminopimelate (DAP), the precursor of lysine and a component of the peptidoglycan, is an essential function of *Mycobacterium smegmatis*. *J. Bacteriol.* **178,** 6496–6507.

11. Gay, P., Le Coq, D., Steinmetz, M., Ferrari, E., and Hoch, J. A. (1983) Cloning structural gene *sacB*, which codes for exoenzyme levansucrase of *Bacillus subtilis*: expression of the gene in *Escherichia coli*. *J. Bacteriol.* **153,** 1424–1431.
12. Gay, P., Le Coq, D., Steinmetz, M., Berkelman, T., and Kado, C. I. (1985) Positive selection procedure for entrapment of insertion sequence elements in gram-negative bacteria. *J. Bacteriol.* **164,** 918–921.

15

Mycobacterial Recombineering

Julia C. van Kessel and Graham F. Hatfull

Summary

Although substantial advances have been made in mycobacterial genetics over the past 15 yr, manipulation of mycobacterial genomes and *Mycobacterium tuberculosis* in particular, continues to be hindered by problems of relatively poor DNA uptake, slow growth rate, and high levels of illegitimate recombination. In *Escherichia coli* an effective approach to stimulating recombination frequencies has been developed called "recombineering," in which phage-encoded recombination functions are transiently expressed to promote efficient homologous recombination. Although homologs of these recombination proteins are rare among mycobacteriophages, we have identified one phage, Che9c, encoding relatives of both RecE and RecT of the *E. coli* rac prophage. Expression of the Che9c proteins from an inducible expression system in either slow- or fast-growing mycobacteria provides elevated recombination frequencies and facilitates simple allelic exchange using linear DNA substrates. Mycobacterial recombineering, therefore, offers a simple approach for constructing gene replacement mutants in *M. smegmatis* and *M. tuberculosis*.

Key Words: Allelic exchange; genetic engineering; mutagenesis; mycobacteriophage; *Mycobacterium tuberculosis*; recombineering.

1. Introduction

Genetic manipulation of *Mycobacterial tuberculosis* is complicated by its slow growth rate, tendency of cells to clump, and a relatively high rate of illegitimate recombination compared with homologous recombination events *(1)*. Methods have been developed for constructing mutants by allelic exchange by electroporation of linear DNA molecules, but the combination of inefficient DNA uptake and relatively low host recombination frequencies yields very few desirable recombinants *(2)*. Furthermore, in *M. tuberculosis*, a high proportion (~90%) of the progeny arise through illegitimate recombination elsewhere in the chromosome rather than by gene replacement *(2)*. A variety of approaches have been developed to address these problems including multistep procedures using combinations of selectable and counterselectable markers, and specialized transducing shuttle phasmids *(3,4)*.

From: *Methods in Molecular Biology, vol. 435: Chromosomal Mutagenesis*
Edited by: G. Davis and K. J. Kayser © Humana Press Inc., Totowa, NJ

An alternative approach to contend with these difficulties is to elevate the frequency of homologous recombination in host cells to a level that can yield the desired recombinants even with relatively poor DNA uptake. A model system for this approach has been developed in *Escherichia coli* using phage-encoded recombination proteins such as the λ *red* system or the RecE/RecT system present in the *rac* prophage ***(5)***. Genetic manipulation using this system is known as "recombineering," i.e., genetic engineering using recombination proteins ***(5)***. Transient expression of the recombination proteins leads to high levels of exchange through homologous DNA sequences and has been utilized for targeted gene replacement, gap repair, and creation of point mutations, small insertions, and deletions ***(5)***. It has recently been applied to a variety of other systems, including other Gram-negative bacteria, yeast, and higher organisms such as *Caenorhabditis elegans* ***(6–9)***. Use of the λ *red* system involves the expression of three proteins, Exo (5′–3′ double-stranded DNA exonuclease), β (single-stranded DNA-binding protein), and Gam (an inhibitor of RecBCD), and has the notable property of targeting homologous recombination events through homology lengths as short as 40 bp ***(5)***.

The problems encountered by the relatively high rate of illegitimate recombination in *M. tuberculosis* suggest that recombineering may be an especially important addition to the collection of available tools for mycobacterial genetic manipulation ***(10)***. To our knowledge, there has not yet been a successful exploitation of the λ *red* or other *E. coli*-based systems in the mycobacteria, although there are anecdotal indications that attempts to do so have been unproductive. An alternative approach to address this question is to identify similar recombination proteins in the phages of the mycobacteria (mycobacteriophages) and utilize them for the development of a mycobacterial-specific recombineering system ***(10)***. Whereas it is expected that mycobacteriophages—as with most bacteriophages—are likely to encode recombination enzymes, examination of the proteome of 30 completely sequenced mycobacteriophage genomes (a total of 3345 genes) reveals very few homologs of λ *red* or RecE/RecT proteins ***(11)***. Only two (Che9c and Halo) of the 30 mycobacteriophages encode homologs of RecE, and only one (Che9c) encodes a RecT homolog ***(10)***.

The RecE/RecT versions found in mycobacteriophage Che9c are encoded by genes *60* and *61*. The encoded proteins are distant relatives of RecE and RecT (sharing 28% and 29% amino acid identity in the shared regions), although biochemical characterization demonstrates that Che9c gp60 is indeed a dsDNA-dependent exonuclease and that Che9c gp61 has similar DNA-binding properties to RecT ***(12)***. Furthermore, expression of these proteins in *M. smegmatis* is necessary and sufficient to promote elevated levels of homologous recombination such that gene replacement mutants can be recovered following electroporation of linear DNA molecules containing homologous segments to either chromosomal or plasmid DNA. This mycobacterial recombineering system functions in fast-growing mycobacteria such as *M. smegmatis* as well as in slow-growers such as *M. tuberculosis* ***(10)***.

This mycobacterial recombineering system has many similar features to the λ *red* and RecE/RecT *E. coli* systems. For example, elevated protein expression appears to be deleterious to mycobacteria and thus regulated expression systems must be utilized. Typically, we have utilized the acetamidase expression system on extrachromosomal plasmid vectors and determined expression levels using antigp61 antibodies, although other vector and expression systems could be used ***(10)***. The frequency of recombination is also dependent on the length of the sequence homology in the targeting substrate. Overall levels of DNA uptake are sufficiently poor that substrates with short homologies (i.e., 50 bp) yield only very few colonies, and thus are not yet generally applicable. However, these mycobacterial recombineering methods are still in development, and further advances may reveal additional strategies for further increasing the recovery of recombinants and for extending these methods to the use of oligonucleotide-based, single-stranded DNA substrates.

In this chapter, we present the approaches and technical details for using mycobacterial recombineering methods for performing allelic exchange in slow- and fast-growing mycobacteria. These methods include the growth and preparation of mycobacterial cells in order to optimize DNA uptake frequencies, suitable conditions for expression of the recombineering proteins, preparation of DNA substrates, and recovery and analysis of recombinants.

2. Materials

2.1. Cell Culture

1. Albumin dextrose catalase (ADC): dissolve 20 g dextrose and 8.5 g NaCl in 950 mL dH_2O. Add 50 g albumin (Spectrum Biochem, Spectrum Chemicals and Laboratory products, Gardena, CA) stirring with no heat until dissolved. Filter-sterilize through a 0.22-μm pore membrane and store at 4°C.
2. Hygromycin B (HYG, Sigma, St. Louis, MO) is dissolved at 100 mg/mL in dH_2O, filtered through a 0.22-μm pore membrane, and stored at –20°C in 1-mL aliquots. Use at a final concentration of 150 μg/mL in media for both *E. coli* and *M. smegmatis*; use at a final concentration of 50 μg/mL for *M. tuberculosis*.
3. Kanamycin (KAN, Sigma) is dissolved at 50 mg/mL in dH_2O, filtered through a 0.22-μm pore membrane, and stored at 4°C. Use at a final concentration of 20 μg/mL in media for mycobacteria.
4. Carbenicillin (CB, Sigma, *see* **Note 1**) is dissolved at 50 mg/mL in dH_2O, filtered through a 0.22-μm pore membrane, and stored at 4°C. Use at a final concentration of 50 μg/mL in media.
5. Cycloheximide (CHX, Sigma, *see* **Note 2**) is dissolved at 10 mg/mL in dH_2O, filtered through a 0.22-μm pore membrane, and stored at 4°C. Use at a final concentration of 10 μg/mL in media.
6. Tetracyclin (TET, Sigma) is dissolved at 5 mg/mL in dH_2O, filtered through a 0.22-μm pore membrane, and stored at –20°C. Use at a final concentration of 2.5 μg/mL in mycobacterial media, 6.25 μg/mL in agar plates for *E. coli*, and 10 μg/mL in liquid media for *E. coli*. Tetracyclin is light sensitive—prepare media only at the time needed.

7. Sodium succinate (dibasic hexahydrate, Sigma, *see* **Note 3**) is dissolved at 20% in dH_2O, filtered through a 0.22-μm pore membrane, and stored at 4°C. Use at a final concentration of 0.2% in media.
8. Acetamide (Sigma) is dissolved at 20% in dH_2O, filtered through a 0.22-μm pore membrane, and stored at 4°C. Use at a final concentration of 0.2% in media.
9. Oleic acid (Sigma, *see* **Note 4**) is dissolved at 10 mg/mL in dH_2O by heating ampule in 37°C water bath and stirring into dH_2O with heat until completely dissolved. Add 1 g NaOH pellets and stir until dissolved, filter through a 0.22-μm pore membrane, and store in 10-mL aliquots at –20°C. Use at a final concentration of 50 μg/mL in *M. tuberculosis* media.
10. Tween-80 is dissolved at 20% by heating at 56°C, filtered through a 0.22-μm pore membrane, and stored at 4°C. Use at a final concentration of 0.05% in liquid media.
11. 7H9 broth: Middlebrook 7H9 powder (Difco, BD and Co., Sparks, MD), 900 mL dH_2O, and 5 mL 40% glycerol. Autoclave, and aseptically add 100 mL ADC, 2.5 mL 20% Tween, 1 mL cycloheximide, 1 mL carbenicillin, and other antibiotics as required. For growth of *M. tuberculosis*, add 5 mL oleic acid per liter.
12. 7H9 induction medium: 7H9 powder (Difco), 900 mL dH_2O, 5 mL 40% glycerol. Autoclave, and aseptically add 100 mL dH_2O, 10 mL 20% succinate, 2.5 mL 20% Tween, 1 mL cycloheximide, 1 mL carbenicillin, and 400 μL kanamycin.
13. 7H10 agar: Middlebrook 7H10 powder (Difco), 900 mL dH_2O, and 12.5 mL 40% glycerol. Autoclave, and aseptically add 100 mL ADC, 2.5 mL 20% Tween, 1 mL cycloheximide, 1 mL carbenicillin, and other antibiotics as required. For growth of *M. tuberculosis*, add 5 mL oleic acid per liter.
14. 7H11 agar (*see* **Note 5**): Middlebrook 7H11 powder (Difco), 900 mL dH_2O, and 12.5 mL 40% glycerol. Autoclave, and aseptically add 100 mL ADC, 2.5 mL 20% Tween, 1 mL cycloheximide, 1 mL carbenicillin, 5 mL oleic acid, and other antibiotics as required.

2.2. Transformation Procedures

1. Glycerol is diluted to 10% in dH_2O, filtered through a 0.22-μm pore membrane, and stored at 4°C for *M. smegmatis* and room temperature for *M. tuberculosis*.
2. Electroporation cuvets, BioRad cat. no. 1652086, 0.2 cm.
3. Prepare an aliquot of 7H9 media containing only ADC and Tween (and oleic acid if transforming *M. tuberculosis*).

2.3. Genomic DNA Preparation

1. Glucose-tris-EDTA (GTE) solution: 25 m*M* Tris-HCl, pH 8.0, 10 m*M* EDTA, 50 m*M* glucose. Store at room temperature.
2. Lysozyme solution: dissolve lysozyme at 10 mg/mL in 25 m*M* Tris-HCl pH 8.5.
3. CTAB solution: dissolve 4.1 g sodium chloride in 90 mL dH_2O. While stirring on magnetic stirrer, add 10 g cetrimide. Incubate in 65°C water bath to get cetrimide into solution (cetrimide is hexadecyltrimethlyammonium bromide, Sigma).
4. Sodium dodecyl sulfate is dissolved at 10%, filtered through a 0.22-μm pore membrane, and stored at room temperature.
5. Proteinase K is dissolved at 10 mg/mL in dH_2O and stored in 1-mL aliquots at –20°C.
6. NaCl is dissolved at 5 *M*, autoclaved, and stored at room temperature.
7. A mixture of chloroform:isoamyl alcohol (24:1) should be prepared.

3. Methods

Recombineering in the mycobacteria can be used to target genes in either the chromosome or extrachromosomally replicating plasmids by allelic exchange facilitated by expression of the Che9c recombination proteins ***(10)***. Myco-bacterial recombineering strains are generated using a plasmid (such as pJV53) that contains the Che9c genes *60–61* expressed under control of the inducible acetamidase promoter, origins of replication for *E. coli* and mycobacteria, and a kan^R cassette. The cells are grown to mid-log phase, induced to express the recombination functions, and competent cells are prepared. These strains are transformed with a targeting substrate specific to the locus to be removed that contains 500-bp homology to the locus flanking a hyg^R cassette and γδ res sites. Hygromycin-resistant recombinant colonies are recovered, and the genotype is verified by Southern blot analysis. The strain can be unmarked by transformation with a plasmid expressing γδ resolvase, efficiently removing the hyg^R cassette. Both extrachromosomally replicating plasmids (the resolvase plasmid and the recombineering expression plasmid) are removed—if desired—by growing the cells without selection for *kan*-resistance and *tet*-resistance over two generations ***(13)***. This results in a completely unmarked strain containing a deletion in the locus of choice.

3.1. Preparation of M. smegmatis *Recombineering Strain*

1. Wild-type *M. smegmatis* mc^2155 is inoculated into 3 mL 7H9 broth (containing ADC, CB, CHX, and Tween) in a sterile test tube and grown shaking at 37°C until culture is saturated (~2 d). This is subcultured into 50 mL of medium (7H9 broth as above) in a 250-mL nonbaffled sterile flask to a final OD_{600} = 0.020 and grown shaking at 37°C overnight.
2. Once the cells have reached OD_{600} = 0.800 – 1.000 the following day, electrocompetent cells are prepared (*see* protocol). Electrocompetent cells are transformed with 1–50 ng of plasmid pJV53 (*see* protocol for transformation) and plated on 7H10 agar plates containing KAN (*see* **Note 6**). The plates are incubated at 37°C for 3 d.
3. Colonies (containing plasmid pJV53) are inoculated into 3 mL 7H9 broth (containing ADC, CB, CHX, KAN, and Tween) in a sterile test tube and grown shaking at 37°C until culture is saturated (~2 d). This is subcultured into 50 mL of 7H9 induction medium (containing KAN, *see* **Note 7**) in a 250-mL nonbaffled sterile flask to a final OD_{600} = 0.02 and grown shaking at 37°C overnight (*see* **Note 8**).
4. Once the cells have reached OD_{600} = 0.5 the following day, acetamide is added to a final concentration of 0.2%. The culture is grown shaking at 37°C for 3 h. Electrocompetent cells are prepared as described next.

3.2. Preparation of M. tuberculosis *Recombineering Strain*

1. Wild-type *M. tuberculosis* H37Rv is inoculated into 3 mL 7H9 broth (containing ADC, CB, CHX, oleic acid, and Tween) in a sterile test tube and grown at 37°C (from 10–20 d). This is subcultured into 50 mL of medium (7H9 broth as earlier) and grown at 37°C.

2. Once the cells have reached OD_{600} = 0.800 – 1.000 (~15 d, *see* **Note 9**), electrocompetent cells are prepared (*see* protocol below). Electrocompetent cells are transformed with 1–50 ng of plasmid pJV53 (*see* protocol in **Subheading 3.3.**) and plated on 7H11 agar plates containing KAN (*see* **Note 6**). The plates are incubated at 37°C for 3 wk, or until colonies are of sufficient size for subculturing.
3. Colonies (containing plasmid pJV53) are inoculated into 3 mL 7H9 broth (containing ADC, CB, CHX, KAN, oleic acid, and Tween) in a sterile test tube and grown at 37°C. This is subcultured into 50 mL of 7H9 induction medium (containing KAN, *see* **Note 7**), and grown at 37°C (*see* **Note 8** and **9**).
4. Once the cells have reached OD_{600} = 0.5, acetamide is added to a final concentration of 0.2% and the culture is grown at 37°C overnight (>16 h). Electrocompetent cells are prepared as described next.

3.3. Preparation of Electrocompetent Cells

1. This procedure is used for *M. smegmatis* and *M. tuberculosis* cell preparations. Importantly, *M. smegmatis* cells should be kept cold by using ice-cold glycerol, centrifuging at 4°C, and placing tubes on ice whenever possible. For *M. tuberculosis*, cells and glycerol should be kept warm, and centrifugations should be performed at room temperature (*see* **Note 10**). Any volume of cells can be used; using 50 mL cultures results in approx 20 tubes of competent cells.
2. The cells are transferred from the flask to centrifuge tubes. For *M. smegmatis* only, the tubes are incubated on ice for 30 min to 2 h.
3. The cells are pelleted by centrifuging at 3600*g* for 10 min; the supernatant is carefully discarded into a waste container.
4. The cells are washed with 1/2 vol (25 mL) 10% sterile glycerol by gently pipeting with a 25 mL pipet until the clumps of cells are dissolved.
5. The cells are pelleted as in **step 3**; the cells are washed with 1/4 vol (12.5 mL) 10% sterile glycerol.
6. The cells are pelleted as in **step 3**; the cells are washed with 1/8 vol (5 mL) 10% sterile glycerol.
7. The cells are pelleted as in **step 3**; the cells are washed with 1/10 vol (5 mL) 10% sterile glycerol.
8. The cells are pelleted as in **step 3**; the cells are resuspended in 1/25 vol (2 mL) 10% sterile glycerol. 100 μL of cells are aliquoted in microcentrifuge tubes, flash frozen on dry ice/ethanol, and stored at –80°C, or used immediately (*see* **Note 11**).

3.4. Transformation of Electrocompetent Cells

1. This transformation procedure assumes the use of a BioRad pulse-controller electroporation unit. If cells are being used fresh from preparation for *M. tuberculosis* (*see* **Note 11**), skip to **step 3** and the cells are not incubated on ice after adding DNA.
2. The tubes of cells are placed on ice, one for each transformation, and allowed to thaw for approx 10 min. DNA is pipeted into cells which are mixed gently, and this is incubated for 10 min (on ice if for *M. smegmatis*). The cells are transferred into a cuvet (chilled on ice if for *M. smegmatis*).

3. The cuvet should be wiped off well and place in electroporator holder. Electroporate with the pulse controller set to 2.5 kV, 1000 Ω, 25 μF. If the cells arc, the transformation must be repeated (*see* **Note 12**).
4. For recovery, 1 mL 7H9 broth (containing ADC and Tween, and oleic acid if for *M. tuberculosis*) is added to the cells and pipeted into a test tube. Recover by incubating at 37°C. For plasmid DNAs in *M. smegmatis*, recover for 2 h, and for *M. tuberculosis*, recover overnight. For transformations with linear targeting DNAs, *see* protocols in **Subheading 3.7.**
5. The transformations are plated on selective 7H10 or 7H11 media and incubated at 37°C.

3.5. Preparation of Mycobacterial Genomic DNA

1. 10 mL of culture is transferred to 15-mL conical tube and spun in tabletop centrifuge at 2000*g* for 20 min. The supernatant is discarded and the cell pellet resuspended in 1 mL GTE solution; this is transferred to a microcentrifuge tube and centrifuged for 10 min.
2. The supernatant is discarded and the cell pellet is resuspended in 450 μL GTE solution. To this, 50 μL of a 10 mg/mL lysozyme solution is added and gently mixed; this is incubated at 37°C overnight.
3. After overnight incubation, 100 μL 10% sodium dodecyl sulfate is added and gently mixed. 50 μL 10 mg/mL proteinase K (Sigma) is added and mixed gently; this is incubated at 55°C for 20–40 min.
4. After incubation, 200 μL 5 *M* NaCl is added to the tube and gently mixed. 160 μL of CTAB (preheated at 65°C) is added and mixed gently; this is incubated at 65°C for 10 min.
5. An equal volume (~1 mL) chloroform:isoamyl alcohol (24:1) is added to the tube, mixed gently, and spun in microcentrifuge for 5 min.
6. The aqueous layer (~900 μL) is transferred to a fresh microcentrifuge tube. The extraction is repeated with chloroform:isoamyl alcohol (24:1), mixed gently, and spun in microcentrifuge for 5 min.
7. The aqueous layer (~800 μL) is transferred to a fresh microcentrifuge tube. For BSL3 organisms (*M. tuberculosis*), the tube can be dipped in vesphene to disinfect outer surface, and the supernatant can now be processed in a BSL2 lab.
8. To the aqueous layer, 560 μL (0.7X vol) isopropanol is added; this should be mixed gently by inversion until the DNA has precipitated out of solution. The tube is incubated at room temperature for 5 min., and spun in microcentrifuge for 10 min.
9. The supernatant is aspirated and discarded. 1 mL 70% ethanol is added to wash DNA pellet. This is mixed gently by inversion and spun in microcentrifuge for 5 min.
10. Aspirate supernatant and air-dry pellet for 15 min. Do not overdry. Resuspend in 50 μL TE buffer and incubate at 37°C to dissolve pellet. Store at –20°C.

3.6. Construction and Preparation of the Targeting Substrate

1. Primers are designed to amplify regions of homology flanking the target gene (*see* **Fig. 1** and **Note 13**). Upstream and downstream homologous regions should each be approx 500 bp. It is recommended to include approx 100 bp of sequence on both ends

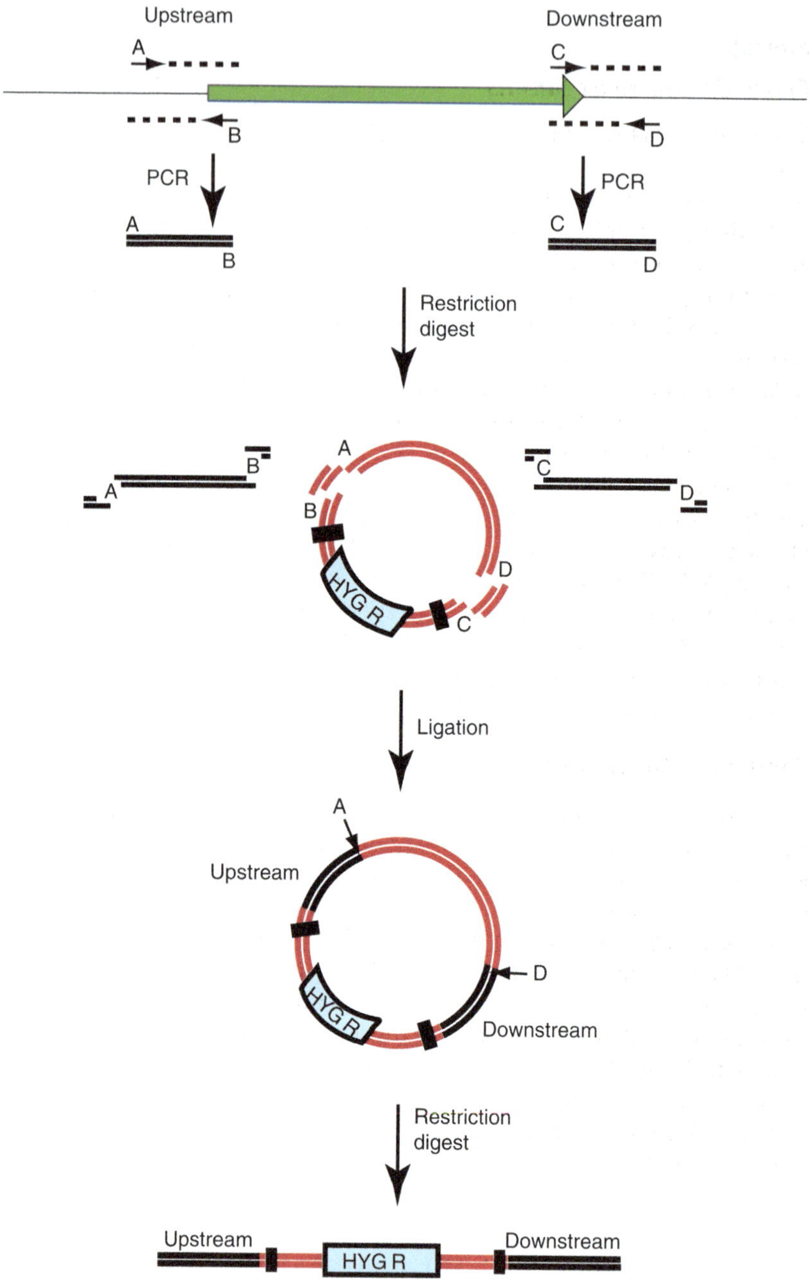

Fig. 1. Construction and preparation of targeting substrate for recombineering. Primers are designed with restriction sites (**A–D**) to amplify upstream and downstream homologous regions flanking a gene. The regions are PCR-amplified, digested, and directionally cloned into a vector flanking a hygromycin-resistance cassette. The plasmid is then linearized by double-digestion with enzymes A and D at the ends of the targeting substrate. Black boxes indicate $\gamma\delta$ res sites that can be used to remove the hygromycin cassette after mutagenesis.

of the gene in order to reduce the chance of causing polar effects. The upstream and downstream homologous regions are polymerase chain reaction (PCR)-amplified using mycobacterial genomic DNA as the template (*see* **Note 14**).

2. The PCR products are cloned into a vector flanking a *hyg*-resistance cassette (*see* **Fig. 1** and **Note 15**). The homologous sequences must be cloned in the same orientation, so that they read from 5′–3′. It is recommended to use plasmid pYUB854 because this plasmid contains γδ res sites flanking the hyg^R cassette, facilitating removal of the cassette after mutagenesis (*see* **Note 16**).
3. The vector (containing the homologous sequences) is linearized by restriction digest with two enzymes, preferably the two enzymes used to clone at the most distal regions of the targeting substrate (**Fig. 1**). Alternatively, the section including the homologous regions can be amplified by PCR. This should yield two fragments; one should contain the HYG-cassette flanked by the two homologous regions.
4. The digest reaction is cleaned up to remove enzyme (no need to gel extract) and DNA is eluted in dH_2O (in order to minimize salt, *see* **Note 12**). It is recommended to use the QIAGEN QIAquick PCR purification protocol for enzymatic reaction cleanup. The linear DNA containing the homology is quantified by agarose gel electrophoresis or ultraviolet spectrometry.

3.7. Transformation of the Recombineering Strains With a Targeting Substrate

1. Electrocompetent cells of strains containing plasmid pJV53 are transformed with 100 ng targeting substrate DNA (*see* **Note 17**) as described in the transformation protocol. The transformations are recovered by incubating at 37°C in 7H9 broth (containing only ADC and Tween, and oleic acid if for *M. tuberculosis*). For *M. smegmatis*, recover for 4 h (*see* **Note 18**); for *M. tuberculosis*, recover for 72 h.
2. The entire reaction is plated on 7H10 or 7H11 agar plates (containing KAN and HYG, and oleic acid for *M. tuberculosis*), and incubated at 37°C until colonies are of sufficient size for subculturing. Typically, between 50 and 200 recombinant colonies are recovered (*see* **Table 1**). This can vary between batches of competent cells (*see* **Note 19**). A hyg^R replicating plasmid can be used to test the efficiency of transformation to compare cell batches.

3.8. Verification of Recombinant Strain Genotype

1. Colonies recovered from transformations with the targeting substrate are inoculated into 7H9 broth (~10 mL, containing ADC, Tween, KAN, and HYG, and oleic acid for *M. tuberculosis*). The cultures are incubated at 37°C for 3 d (*M. smegmatis*) or 10 d (*M. tuberculosis*) until culture has a substantial amount of visible growth. It is recommended to screen 4 or more colonies. In addition, the pJV53 strain is inoculated in media (without HYG) as a control. The cells are collected and genomic DNA is prepared as described in **Subheading 3.5.**
2. Southern blot analysis is performed on each recombinant strain to be tested, using the pJV53 strain as a control. It is suggested that probes are used that hybridize to either the upstream or downstream homologous region in order to distinguish between different band sizes expected. Alternatively, a probe to the hyg^R cassette will determine the location of the targeting substrate.

Table 1
Example Results From a Recombineering Experiment Targeting the *groEL1* Gene in *M. smegmatis*

	Recombinants[a]	Cell competency[b] (cfu/μg DNA)	Recombineering frequency[c]
mc²155:pJV53	154	2.0×10^6	7.7×10^{-4}
Control strain[d] (*mc²155:pLAM12*)	0	1.5×10^6	0

[a]Number of colonies recovered following transformation with 100 ng linear targeting DNA substrate (described in **Subheadings 3.6.** and **3.7.**) and plating on media containing hygromycin.

[b]Cell competency is determined by transformation with 50 ng of integration–proficient vector DNA and is expressed as colony-forming units per μg DNA (cfu/μg).

[c]The recombineering frequency is expressed as the number of allelic exchange recombinants per μg/cell competency.

[d]The control strain contains an empty vector as a control for background recombination events.

3.9. *Unmarking Mutant Strains in* M. smegmatis

1. Unmarking of *M. smegmatis* strains is accomplished using a plasmid expressing γδ resolvase (*see* **Note 20**). Following transformation with this plasmid, the hyg^R cassette is removed, and both the recombineering plasmid and resolvase plasmid are removed by serially diluting 1:10,000 and growing in the absence of selective antibiotics; this procedure is repeated twice (*see* **Note 21**) ***(13)***. This protocol describes the process for *M. smegmatis*, but would likely also be applicable to *M. tuberculosis*.
2. The mutant strain is inoculated in 7H9 media (containing HYG, not KAN) and incubated shaking at 37°C until culture is saturated (~2 d). This is subcultured and cells are grown to prepare electrocompetent cells (*see* protocols in **Subheadings 3.1.** and **3.3.**).
3. Electrocompetent cells are transformed with 50 ng of plasmid pGH542 (*see* **Note 20**, *see* protocol for transformation) and plated on 7H10 agar plates containing TET (*see* **Note 6**). The plates are incubated at 37°C for 3–4 d until colonies are of sufficient size for subculturing.
4. The recovered colonies are patched on multiple selective 7H10 agar plates containing antibiotics in the order as follows: (1) HYG, (2) TET, and (3) CB/CHX only. The plates are incubated at 37°C for 3–4 d.
5. Colonies that are *hyg*-sensitive and *tet*-resistant are subcultured into 10 mL cultures of 7H9 media (containing CB and CHX only) and incubated shaking at 37°C until culture is saturated (~2 d). This culture is subcultured into another 10 mL culture exactly as above using 1 μL of the culture (1:10,000). Incubate while shaking at 37°C until culture is saturated (~2 d). Dilutions (10^{-4}–10^{-7}) are plated on 7H10 agar plates (containing CB and CHX only).
6. The recovered colonies are patched on multiple selective 7H10 agar plates containing antibiotics in the order as follows: (1) HYG, (2) TET, and (3) KAN, 4) CB/CHX only. The plates are incubated at 37°C for 3–4 d. Colonies that are now *tet*-sensitive, *kan*-sensitive, and *hyg*-sensitive can be further verified by Southern blot analysis as the correct unmarked mutant strain.

4. Notes

1. *M. smegmatis* and *M. tuberculosis* are naturally resistant to ampicillin; carbenicillin is a more stable form and is recommended for laboratory use.
2. Addition of cycloheximide to media can prevent growth of molds in media, but does not inhibit mycobacterial growth.
3. Sodium succinate dibasic hexahydrate is also known as succinic acid disodium salt.
4. It is critical to add oleic acid to any media for growing *M. tuberculosis*, but is not required for *M. smegmatis*.
5. Growing *M. tuberculosis* on 7H11 plates (instead of 7H10) can increase the growth rate.
6. When transforming with plasmid DNA, it is recommended to plate one or more serial dilutions of the transformation reaction in order to obtain isolated single colonies.
7. It is critical that the growth of pJV53 cells to prepare electrocompetent cells is performed in the 7H9 induction media containing succinate, as described in **Subheading 2.** Growth in media containing ADC will not facilitate proper expression of the recombination functions.
8. It will be observed that growth in succinate induction medium will result in a slightly slower growth rate than in ADC medium.
9. Growth of *M. tuberculosis* strains may take longer than 10 d, up to 20 d. Optical density should be monitored routinely, as the growth rate can vary depending on growth conditions.
10. It is critical to treat *M. smegmatis* cells with cold glycerol and centrifugation at 4°C. *M. tuberculosis* cells should be at room temperature, and glycerol can be warmed to 37°C when not in use. This improves transformation efficiency for both species ***(14)***.
11. Using *M. tuberculosis* cells fresh after preparation without freezing will improve efficiency significantly.
12. Recurrent problems with arcing can be avoided by ensuring thorough washing of the cells during electrocompetent cell preparation to remove salts from the media. Additionally, elution buffer for DNA should contain little to no salt, as this also can cause arcing
13. Primers should be between 22 and 27 nt in length, with 50–60% GC content, and the melting temperatures work best above 60°C. It is also recommended to synthesize restriction sites in the primers to facilitate directional cloning into the vector.
14. Addition of dimethyl sulfoxide (1–5%) in the reaction can sometimes reduce production of nonspecific bands, and can increase product quantity.
15. Any plasmid containing a hyg^R cassette and sites for cloning adjacent to the cassette can be used. Conventional cloning techniques should be applied, and plasmids should be propagated in *E. coli*. If it is desired to unmark the strain, use a plasmid containing γδ res sites (*see* **Note 16**).
16. If using a plasmid containing γδ res sites, such as pYUB854, the plasmid must be propagated in *E. coli* HB101 strains that are deleted for the resolvase function.
17. Transformation with 100 ng targeting substrate is critical as this will yield optimal numbers of recombinants. If possible, do not transform with more or less DNA, as both can limit recovery of recombinants.
18. Recovery for 4 h after transformation of the recombineering strain with the targeting substrate is vital to recovery of recombinants, as compared with the 2 h recovery required during canonical transformations with plasmid DNA.

19. Optimizing the number of recombinants is achieved by making competent cells with good transformation efficiencies, and not by increasing the amount of DNA transformed (*see* **Note 17**). Typically, between 10^5 and 10^6 cells cfu/μg are recovered when testing cells with a replicating plasmid.
20. Plasmid pGH542 expresses γδ resolvase and contains a *tet*^{R} cassette. Transformation with this plasmid efficiently promotes recombination between the res sites flanking the *hyg*^{R} cassette to remove the cassette, generating an unmarked mutation.
21. Growing the strain by diluting serially twice, 1:10,000, in media lacking selective antibiotics effectively causes loss of the instable replicating plasmids pJV53 and pGH542.

Acknowledgments

The authors would like to thank J. Flynn and her colleagues for help with *M. tuberculosis* manipulations and W. Jacobs, Jr. for helpful discussions.

References

1. Hatfull, G. F. (1996) The molecular genetics of *Mycobacterium tuberculosis. Curr. Top Microbiol. Immunol.* **215,** 29–47.
2. Kalpana, G. V., Bloom, B. R., and Jacobs, W. R., Jr. (1991) Insertional mutagenesis and illegitimate recombination in mycobacteria. *Proc. Natl. Acad. Sci. USA* **88,** 5433–5437.
3. Bardarov, S., Bardarov, S., Jr., Pavelka, M. S., Jr., et al. (2002) Specialized transduction: an efficient method for generating marked and unmarked targeted gene disruptions in *Mycobacterium tuberculosis, M. bovis* BCG and *M. smegmatis. Microbiology* **148,** 3007–3017.
4. Pelicic, V., Jackson, M., Reyrat, J. M., Jacobs, W. R., Jr., Gicquel, B., and Guilhot, C. (1997) Efficient allelic exchange and transposon mutagenesis in *Mycobacterium tuberculosis. Proc. Natl. Acad. Sci. USA* **94,** 10,955–10,960.
5. Court, D. L., Sawitzke, J. A., and Thomason, L. C. (2002) Genetic engineering using homologous recombination. *Annu. Rev. Genet.* **36,** 361–388.
6. Datta, S., Costantino, N., and Court, D. L. (2006) A set of recombineering plasmids for gram-negative bacteria. *Gene* **379,** 109–115.
7. Ranallo, R. T., Barnoy, S., Thakkar, S., Urick, T., and Venkatesan, M. M. (2006) Developing live Shigella vaccines using lambda Red recombineering. *FEMS Immunol. Med. Microbiol.* **47,** 462–469.
8. Sarov, M., Schneider, S., Pozniakovski, A., et al. (2006) A recombineering pipeline for functional genomics applied to *Caenorhabditis elegans. Nat. Methods* **3,** 839–844.
9. Erler, A., Maresca, M., Fu, J., and Stewart, A. F. (2006) Recombineering reagents for improved inducible expression and selection marker re-use in *Schizosaccharomyces pombe. Yeast* **23,** 813–823.
10. van Kessel, J. C. and Hatfull, G. F. (2007) Recombineering in *Mycobacterium tuberculosis. Nat. Methods* **4,** 147–152.
11. Hatfull, G. F., Pedulla, M. L., Jacobs-Sera, D., et al. (2006) Exploring the mycobacteriophage metaproteome: phage genomics as an educational platform. *PLoS Genetics* **2,** E92.

12. Gottesman, M. M., Gottesman, M. E., Gottesman, S., and Gellert, M. (1974) Characterization of bacteriophage lambda reverse as an *Escherichia coli* phage carrying a unique set of host-derived recombination functions. *J. Mol. Biol.* **88,** 471–487.
13. Lee, M. H., Pascopella, L., Jacobs, W. R., Jr., and Hatfull, G. F. (1991) Site-specific integration of mycobacteriophage L5: integration-proficient vectors for *Mycobacterium smegmatis, Mycobacterium tuberculosis,* and bacille Calmette-Guerin. *Proc. Natl. Acad. Sci. USA* **88,** 3111–3115.
14. Wards, B. J. and Collins, D. M. (1996) Electroporation at elevated temperatures substantially improves transformation efficiency of slow-growing mycobacteria. *FEMS Microbiol. Lett.* **145,** 101–105.

16

Chromosomal Engineering of *Clostridium Perfringens* Using Group II Introns

Phalguni Gupta and Yue Chen

Summary

Clostridium perfringens is a major natural pathogen of human and domestic animals owing to the production of multiple toxins. Defined clostridial mutants are essential for studying the role of toxins in disease pathogenesis. However, it has been very difficult to introduce mutations into *C. perfringens*. We recently developed a clostridia-modified targetron that can specifically and efficiently inactivate *C. perfringens* genes. The usefulness of this system has now been demonstrated by specifically inactivating four different *C. perfringens* toxin genes.

Key Words: α-Toxin; chromosomal mutagenesis; *Clostridium perfringens*; gene knockout; targetron; group II intron.

1. Introduction

Clostridium perfringens, a major natural pathogen of human and domestic animals, is one of the most prolific bacterial toxin producers *(1)*. Site-specific mutations on the chromosome or plasmids of *C. perfringens* are essential for studying the pathogenesis of toxins during disease. Unfortunately it has been very difficult to introduce mutations into *C. perfringens* by classical allelic exchange approaches. A major problem in constructing knockout mutants is an extremely low recombination frequency.

Commercially available TargeTron™ technology has been used to insertionally knockout genes in *Escherichia coli* and other bacteria. TargeTron methodology is based on mobile group II introns, which are site-specific retroelements that use a retrohoming mechanism to directly insert the excised intron lariat RNA into a specific DNA target site and reverse-transcribe into DNA that inactivates the disrupted gene. As the DNA target site is recognized primarily by base pairing of intron RNA and a few intron-encoded protein (IEP) recognition spots, the intron targeting sequences can be modified to insert into a specific DNA target *(2)*. But,

From: *Methods in Molecular Biology, vol. 435: Chromosomal Mutagenesis*
Edited by: G. Davis and K. J. Kayser © Humana Press Inc., Totowa, NJ

the existing TargeTron technology, based on *E. coli T7lac* promoter and lacking a *C. perfringens* origin of replication, would not work in clostridia. We have constructed a *C. perfringens*-specific targetron based on the Sigma TargeTron vector by inserting *C. perfringens*-specific promoter and introducing the targetron cassette into the plasmid pJIR750, an *E. coli–C. perfringens* shuttle vector. The constructed plasmid has been used to successfully inactivate multiple toxin genes in *C. perfringens* plasmids as well as its chromosome ***(3)***. In this chapter, the methods of constructing a *Clostridium*-specific targetron donor plasmid and inactivating a chromosome gene in *C. perfringens* with the plasmid are outlined.

2. Materials

2.1. Bacteria and Bacterial Culture

1. *E. coli* DH5α, SOC medium (Catalog 15544–034, Invitrogen, 1600 Faraday Avenue PO Box 6482, Carlsbad, CA).
2. Luria Bertani (LB) medium: 10 g tryptone, 5 g yeast extract, and 10 g NaCl, 15 g bacto-agar, dissolved in 1 L H_2O, and autoclaved.
3. *C. perfringens* type A ATCC3624.
4. Fluid thioglycolate broth (FTG).
5. Modified Duncan-strong medium: 15 g peptone, 4 g yeast extract, 1 g sodium thioglycollate, 10 g $Na_2HPO_4{\cdot}7H_2O$, dissolved in 950 mL H_2O. 4 g soluble starch dissolved in 50 mL H_2O and boiled for 3 min and added to the 950 mL mix and autoclaved. Cool the solution and add theophylline (200 μg/mL final) (*see* **Note 1**).
6. Egg yolk brain heart infusion (BHI) agar plates: 52 g brain heart infusion agar (Difco Brain Heart Infusion Agar. Ref No 241830, Becton, Dickinson and Company, Sparks, MD 21152) dissolved in 900 mL H_2O and autoclaved. Cool the solution to about 50°C, add two egg yolks suspended in 100 mL normal saline, mix by swirling, and pour plates.

2.2. Construction of the Targetron Donor Plasmid and Bacterial Transformation

1. Plasmids, pACD3 (Sigma-Aldrich), pDF(B2h1)v751 ***(4)***, and pJIR750 (ATCC87015).
2. Primers (*see* **Table 1**).
3. *Taq* polymerase, 10X *Taq* polymerase buffer, dNTPs (Promega, Madison, WI).
4. AccuPrime *Pfx* DNA polymerase, 10X reaction mix (Invitrogen, 1600 Faraday Avenue PO Box 6482, Carlsbad, CA).
5. 1X TAE (40 m*M* Tris-acetate, 1 m*M* EDTA).
6. Agarose (Bio-Rad, Hercules, CA).
7. HyperLadder I (Bioline, London, UK).
8. Gel electrophoresis equipment.
9. Restriction enzymes: *Spe*I, *Xba*I, *Bsr*GI, *Hin*dIII, and *Pvu*I (NEB New England Biolabs, Ipswich, MA).
10. UltraClean DNA purification kit (Mo Bio laboratories, Carlsbad, CA).
11. Alkaline phosphatase, shrimp (Roche, Grenzacherstrasse 124, CH–4070 Basel, Switzerland).
12. T4 DNA ligase (Promega).

Table 1
Oligonucleotide Primers

Primer sequence (5′–3′)
IBS1/IBS2 AAAAAAGCTTATAATTATCCTTACCAGCCCATAGGGTGCGCCCAGATAGGGTG
LtrBAsEBS2 CGAAATTAGAAACTTGCGTTCAGTAAAC
EBS1/delta CAGATTGTACAAATGTGGTGATAACAGATAAGTCCATAGGCTTAACTTACCTTTCTTTGT
EBS2 TGAACGCAAGTTTCTAATTTCGGTTGCTGGTCGATAGAGGAAAGTGTCT
Alpha toxin F GTGAGGTTATGTTAATTATATGGTATAATTTCAATGC
Alpha toxin 100R AGTTACAATCATAGCATGAGTTCCTGTTCC
B2PF ACTAGTGAATTCCATTATTAACTGAT
B2PR AAGCTTTGGTTTCCCCCCTGAATTTTT
Biotin-labeled targetron specific probe GTGTTTACTGAACGCAAGTTTCTAATTTCGGT
Plc targetron cassette F CGATCGGAATTCCATTATTAACATGATAAATCCTATAACCCA
Plc targetron cassette R CGATCGCACTTGTGTTTATGAATCACGTGACGATGA
4931F TCAATAAGGTCAAAAATCTTAAAGGCAAAGAAAA
PJIR750R CAGATGCGTAAGGAGAAAATACCGC

13. LB-kanamycin or chloramphenicol plates: 10 g tryptone, 5 g yeast extract, and 10 g NaCl, 15 g bacto-agar, dissolved in 1 L H_2O, and autoclaved. Cool the solution to about 50°C, and add kanamycin (50 μg/mL final) or chloramphenicol (15 μg/mL), mix by swirling, and pour plates.
14. Plasmid minipreps kit (Promega).
15. Gene pulser (Bio-Rad).
16. 0.4-cm Cuvet (BTX).
17. TGY broth: 30 g trypticase, 10 g yeast extract, and 20 g glucose, 1 g L-cysteine, dissolved in 1 L H_2O, and autoclave.
18. SMP buffer: 270 m*M* sucrose, 1 m*M* $MgCl_2$, 7 m*M* sodium phosphate buffer, pH 7.3.
19. Brain-heart infusion chloramphenicol plates: 52 g brain-heart infusion agar dissolved in 1 L H_2O and autoclaved. Cool the solution to about 50°C, and add chloramphenicol (15 μg/mL final), mix by swirling, and pour plates.
20. GasPak (BD).

2.3. Southern Hybridization

1. MasterPure Gram-positive DNA purification kit (Epicentre, Madison, WI).
2. Restriction enzyme *Eco*RI (NEB).
3. Biotin-labeled targetron-specific probe (*see* **Table 1**).
4. TE 22 tank transfer unit.
5. TAE: 40 m*M* tris-acetate, 1 m*M* EDTA, pH 8.0.
6. Nylon membrane (Pierce, Rockford, IL).
7. Ultraviolet (UV) stratalinker 1800 (Stratagene, La Jolla, CA).
8. North2South chemiluminescent hybridization and detection kit (Pierce).
9. X-ray film (Kodak, Rochester, NY).

3. Methods

3.1. *Insertion of a* C. perfringens-*Specific Promoter in Front of the Intron*

In order to insert the intron into a specific gene in *C. perfringens* plasmid or chromosomal DNA, the intron DNA must be transcribed in *C. perfringens* to allow its RNA to perform its retrohoming function. The commercially available intron donor plasmid used is pACD3, a chloramphenicol-resistant plasmid containing the *Ll.LtrB* intron expressed from a *T7lac* promoter. Because the yue chen *T7lac* promoter would not function in *C. perfringens*, we inserted the promoter region of the β-2 toxin gene (*cpb2*) from a *C. perfringens* type A isolate into the pACD3 plasmid upstream of the intron sequence to drive intron transcription.

3.1.1. Amplification of cpb2 *Promoter Region by Polymerase Chain Reaction (PCR)*

The following components are mixed together:

1. 1 μL 400 ng/μL temperate DNA, plasmid pDF(B2h1)v751 containing β-2 toxin gene cassette (final amount = 400 ng).
2. 2 μL of 20 μ*M* primer B2PF containing *Spe*I site as a linker at 5′ end (final concentration of each primer = 800 n*M*).
3. 2 μL of 20 μ*M* primer B2PR containing *Hin*dIII site as a linker at 5′ end (final concentration of each primer = 800 n*M*).
4. 5 μL of 10X PCR buffer (final concentration = 1X).
5. 3 μL of 3.8 m*M* dNTPs (final concentration = 228 μ*M*).
6. 0.5 μL of 5 U/μL *Taq* polymerase (final amount = 2.5 U).
7. 36.5 μL of H_2O.

The PCR cycling program: initial denaturation, 94°C, 10 min. Amplification, 94°C, 1 min; 50°C, 1 min; 72°C, 1 min for 35 cycles. Final extension, 72°C for 7 min.

3.1.2. Cloning of Amplified PCR Products Into pACD3 Upstream of Intron Sequence

1. Run the PCR products on a 1.8% agarose gel containing ethidium bromide by electrophoresis. A HyperLadder I is used as a molecular marker.
2. Visualize the 200-bp PCR products under UV light, and excise the DNA band from agarose gel by a scalpel.
3. Extract DNA fragments from gel by UltraClean DNA purification kit according to manufacturer's instruction.
4. Digest PCR products with *Spe*I and *Hin*dIII and vector pACD3 with *Xba*I and *Hin*dIII. Incubate at 37°C for 4 h.

	Digestion 1	Digestion 2
DNA	10 μL (5 μg) purified PCR product	5 μL (5 μg) pACD3
10X buffer	2 μL	2 μL
*Spe*I	1 μL	–
*Xba*I	–	1 μL
*Hind*III	1 μL	1 μL
H_2O	6 μL	11 μL

5. Separate digested DNA fragments by agarose gel electrophoresis. Excise and purify the DNA fragments from the gel.
6. Ligate the digested pACD3 and PCR product by T4 ligase.
   a. Setup ligation by adding the following components:
      i. 2 μL (2 μg) digested pACD3 DNA.
      ii. 2 μL (2 μg) PCR product.
      iii. 5 μL of 2X ligation buffer.
      iv. 1 μL (4 U) T4 ligase.
   b. Incubate at 4°C overnight.
7. Transform ligation mix into *E. coli* DH5*a* competent cells according to manufacturer's instruction. Spread 100 μL or 900 μL of the transformation mixture onto LB agar plates containing 15 μg/mL chloramphenicol and incubate the plates overnight at 37°C.
8. Purify and sequence the plasmid DNA. Inoculate single bacterial colony grown from the plates into 5 mL of LB medium containing 15 μg/mL chloramphenicol. After overnight culture with shaking at 37°C, isolate plasmid DNA, pACD3wb2p, from the culture by minipreps kit. Sequence the joint regions of the insertion point in the vector and entire inserted fragment.

3.2. Modification of the **L1.LtrB** *Group II Intron for Targeting to* **C. perfringens** *α-Toxin Gene* **(plc)**

To target the intron to *plc*, the *L1.LtrB* intron sequence in pACD3wb2p is modified based on the sequences of predicted insertion sites in the *plc* gene using the Sigma website (www.sigma-aldrich.com/targetron). The program predicts 10 intron insertion sites across the 1197-bp *plc* gene. For optimal gene interruption and stable insertion, the insertion site in the antisense strand at position 50/51 from the initial ATG is chosen for intron modification. There are three short sequence elements involved in the base pairing interaction between the DNA target site (*IBS1, IBS2*, and *δ'*) and intron RNA (*EBS1, EBS2*, and *δ*). Modifications of intron RNA sequences (*EBS1, EBS2*, and *δ*) to base pair with the *plc* target site sequences are introduced through PCR.

3.2.1. *PCR Using a* IBS1/IBS2 *and* LtrBAsEBS2 *Primer Pairs to Generate α 257-bp Amplicon*

Setup PCR reaction as following:

1. 1 μL 400 ng/μL temperate DNA, plasmid pACD3wb2p (final amount = 400 ng).
2. 2 μL of 20 μ*M* primer *IBS1/IBS2* (final concentration of each primer = 800 n*M*).
3. 2 μL of 20 μ*M* primer *LtrBAsEBS2* (final concentration of each primer = 800 n*M*).
4. 5 μL of 10X PCR buffer (final concentration = 1X).
5. 3 μL of 3.8 m*M* dNTPs (final concentration = 228 μ*M*).
6. 0.5 μL of 5 U/μL *Taq* polymerase (final amount = 2.5 U).
7. 36.5 μL of H_2O.

Perform PCR with initial denaturation, 94°C, 10 min. Amplification, 94°C, 1 min; 55°C, 1 min; 72°C, 40 s for 35 cycles. Final extension, 72°C for 7 min. Run PCR product by agarose gel electrophoresis. Excise and purify the 257-bp PCR product from the gel.

3.2.2. *PCR Using* EBS1/Δ *and* EBS2 *Primer Pairs to Generate an 117-bp Amplicon*

Add following components together:

1. 1 μL 400 ng/μL template DNA, plasmid pACD3wb2p (final amount = 400 ng).
2. 2 μL of 20 μ*M* primer *EBS1*/Δ (final concentration of each primer = 800 n*M*).
3. 2 μL of 20 μ*M* primer *EBS2* (final concentration of each primer = 800 n*M*).
4. 5 μL of 10X PCR buffer (final concentration = 1X).
5. 3 μL of 3.8 m*M* dNTPs (final concentration = 228 μ*M*).
6. 0.5 μL of 5 U/μL *Taq* polymerase (final amount = 2.5 U).
7. 36.5 μL of H_2O.

Perform PCR with initial denaturation, 94°C, 10 min. Amplification, 94°C, 1 min; 55°C, 1 min; 72°C, 30 s for 35 cycles. Final extension, 72°C for 7 min. Run PCR product by agarose gel electrophoresis. Excise and purify the 117-bp PCR product from the gel.

3.2.3. *PCR Using* IBS1/IBS2 *and* EBS1/Δ *Primers to Generate a 353-bp* plc *Targetron*

The 257-bp and 117-bp amplicons, having 21 overlapping nucleotides, are used as templates for PCR with *IBS1/IBS2* and *EBS1*/Δ primers to generate a 353-bp *plc* targetron.

Add following components together:

1. 1 μL 200 ng/μL 257-bp fragment (final amount = 200 ng).
2. 1 μL 200 ng/μL 117-bp fragment (final amount = 200 ng).
3. 2 μL of 20 μ*M* primer *EBS1*/Δ (final concentration of each primer = 800 n*M*).
4. 2 μL of 20 μ*M* primer *IBS1/IBS2* (final concentration of each primer = 800 n*M*).
5. 5 μL of 10X PCR buffer (final concentration = 1X).
6. 3 μL of 3.8 m*M* dNTPs (final concentration = 228 μ*M*).

7. 0.5 μL of 5 U/μL *Taq* polymerase (final amount = 2.5 U).
8. 35.5 μL of H_2O.

Perform PCR with initial denaturation, 94°C, 10 min. Amplification, 94°C, 1 min; 55°C, 1 min; 72°C, 1 min for 35 cycles. Final extension, 72°C for 7 min. Run PCR product by agarose gel electrophoresis. Excise and purify the 353-bp PCR product from the gel.

3.2.4. Cloning of plc *Targetron Into pACD3wb2p*

The gel-purified 353-bp fragments and pACDwb2p vector are digested with *Hin*dIII and *Bsr*GI as following:

	Digestion 1	Digestion 2
DNA	10 μL (5 μg) 353-bp fragment	5 μL (5 μg) pACD3wb2p
10X buffer	2 μL	2 μL
*Bsr*GI	1 μL	1 μL
*Hin*dIII	1 μL	1 μL
H_2O	6 μL	11 μL

Incubate at 37°C for 4 h. Run the digested mixtures in an agarose gel and purify the digested fragments from the gel. The gel purified 350-bp insert and 6-kb vector are ligated by T4 ligase. Transform ligation mix into *E. coli* DH5α-competent cells. Plate the transformed bacteria on LB agar plates containing 15 μg/mL chloramphenicol and incubate overnight at 37°C. Pick up single colony from the transformation plates and grow overnight in LB medium containing 15 μg/mL chloramphenicol. Purify plasmid DNA, pACD3wb2pai, from the culture by minipreps and sequence the joint regions of the insertion point in the vector and the entire inserted fragment.

3.3. Cloning of plc *Targetron Gene Cassette From pACD3wb2pai Into an* E. coli-C. perfringens *Shuttle Vector, pJIR750*

As pACD3 lacks an origin of replication in *C. perfringens*, the targetron would not be maintained in *C. perfringens* long enough to obtain efficient targetron insertion into a *C. perfringens* target gene. In response, the whole *plc* targetron gene cassette (including *β*-2-promoter, *plc* targetron, and downstream *LtrA* ORF) is PCR amplified from pACD3wb2pai and ligated into pJIR750, creating pJIR750ai.

3.3.1. PCR Amplification of Whole Targetron Gene Cassette

PCR is performed on plasmid pACD3wb2pai using sequence specific primers, *plc* targetron cassette F and R, which amplify the whole 3.7-kb targetron gene cassette and have *Pvu*I linker sites at 5′ ends of both primers (*see* **Note 2**).

The PCR mixture is setup as follows:

1. 1 μL of 200 ng/μL pACD3wb2pai (final amount = 200 ng).
2. 2 μL of 20 μ*M* primer *plc* targetron cassette F (final concentration = 800 n*M*).
3. 2 μL of 20 μ*M* primer *plc* targetron cassette R (final concentration = 800 n*M*).
4. 5 μL of 10X AccuPrime reaction buffer (final concentration = 1X).
5. 1 μL of AccuPrime *Pfx* DNA polymerase (final amount = 2.5 U).
6. 39 μL of H_2O (final volume: 50 μL).

Perform PCR with initial denaturation, 95°C, 10 min. Amplification, 95°C, 15 s; 56°C, 30 s; 68°C, 4 min for 35 cycles. Final extension, 68°C for 7 min. Run PCR product by agarose gel electrophoresis. Excise and purify the 3.7-kb PCR product from the gel.

3.3.2. TOPO Cloning of the Amplified plc Targetron Cassette

1. Setup TOPO Zero Blunt TOPO PCR Cloning Kit, Invitrogen, Carlsbad, CA. cloning and transformation according to manufacturer's instruction (*see* **Note 3**). Plate the transformed bacteria on LB agar plates containing 50 μg/mL kanamycin. Incubate the plates at 37°C overnight.
2. Inoculate individual colony from the plates into LB broth containing 50 μg/mL kanamycin. Incubate overnight at 37°C.
3. Purify plasmid DNA, pCR-Blunt II-TOPO-3.7-kb, from the culture by Minipreps and sequence the entire inserted fragment in the plasmid.

3.3.3. Cloning of plc *Targetron Cassette Into a pJIR750 Vector*

1. Digest pJIR750 and pCR-Blunt II-TOPO-3.7-kb with *Pvu*I. Separate digested DNA fragments by agarose gel electrophoresis. Excise and purify the 6.5-kb fragment from digested pJIR750 vector and 3.7-kb DNA from digested pCR-Blunt II-TOPO-3.7-kb from the gel.
2. Dephosphate the 6.5-kb vector DNA by alkaline phosphatase according to manufacture's instruction. Run the mixture in an agarose gel and purify the DNA from the gel.
3. Ligate the 3.7-kb DNA fragment containing the whole *plc* targetron cassette to the digested and dephosphated pJIR750 vector DNA by T4 ligase. Transform the ligation mixture into *E. coli* DH5α competent cells. Plate the transformed cells onto LB agar plates containing 15 μg/mL chloramphenicol, isopropyl-β-D-1-thiogalactopyranoside, and 4-chloro-5-bromo-3-indolyl-β-D-galactopyranoside. Incubate the plates at 37°C overnight.
4. Inoculate white colonies into LB culture medium containing 15 μg/mL chloramphenicol, incubate at 37°C overnight. Purify plasmids, pJIR750ai, from the culture by miniprep. Sequence the 3.7-kb *plc* targetron cassette in pJIR750ai by primer walking.

3.4. Creation of a plc *Gene Knockout Mutant of* C. perfringens *by the Constructed* plc *Targetron*

C. perfringens is one of the prolific bacterial toxin producers. According to the profile of toxins produced, *C. perfringens* have been classified into types A–E.

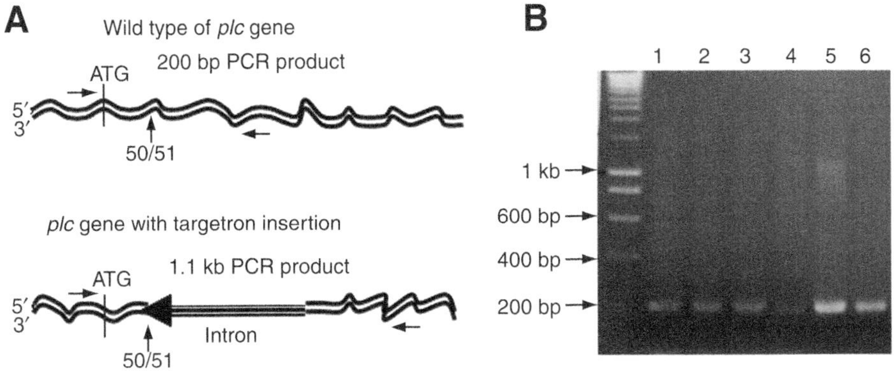

Fig. 1. Screening of *plc* gene insertion. **(A)** Diagram of the *plc* gene with and without insertion of the targetron. → forward primer, ←reverse primer. **(B)** Screening of *C. perfringens* colonies for targetron insertion by PCR. The numbers at the top are bacterial colony numbers. Colony 5 shows both 1.1- and 200-bp bands.

However, all types of *C. perfringens* possess *plc* gene on their chromosomes and produce α-toxin, the most important toxin for development of gas gangrene. Type A *C. perfringens* ATCC3624 is used for the chromosome α-toxin gene knockout.

3.4.1. Preparation of ATCC3624 for Electroporation and Electroporation

1. Inoculate 0.2 mL *C. perfringens* ATCC3624 spores stock into the bottom of the tube containing 10 mL FTG, heat-shock the spores by putting the tube in 70°C water bath for 20 min, incubate the tube with lid tightly closed at 37°C overnight (*see* **Note 4**).
2. Inoculate 0.2 mL of an overnight FTG culture into 10 mL TGY broth and incubate overnight.
3. Harvest the bacterial cells and wash them twice in SMP buffer. Resuspend the cells in 400 μL SMP buffer and put them into 0.2 cm cuvet with 5 μg pJIR750ai DNA (*see* **Note 5**).
4. Electroporate the bacteria with a single pulse of 2.5 kV at 25 μF capacitance and 200 Ω resistance. Remove the electroporated cells to 3 mL prewarmed TGY broth. Rinse the cuvet twice with medium to recover any cells adhering to the electrodes, and add to the broth. Incubate the electroporated cell in TGY medium at 37°C for 3 h.
5. Plate the bacteria on BHI agar plates containing 15 μg/mL chloramphenicol. Incubate the plates in an anaerobic jar with a GasPack overnight at 37°C.

3.4.2. Screening of Bacterial Colonies With plc *Targetron Insertion Into Chromosome*

1. Colony PCR is performed with two primers hybridizing to either side of the insertion point in chromosome α-toxin gene, so that 1.1-kb PCR product is produced in the bacteria with the intron insertion in predicted position and orientation, while 200-bp PCR product is produced in wild-type bacteria (**Fig. 1**). When the *plc* targetron in the plasmid is transcribed and excised out and the IEP is produced in the transformants, the IEP and excised intron lariate RNA carry out the targetron insertion into the *plc* gene.

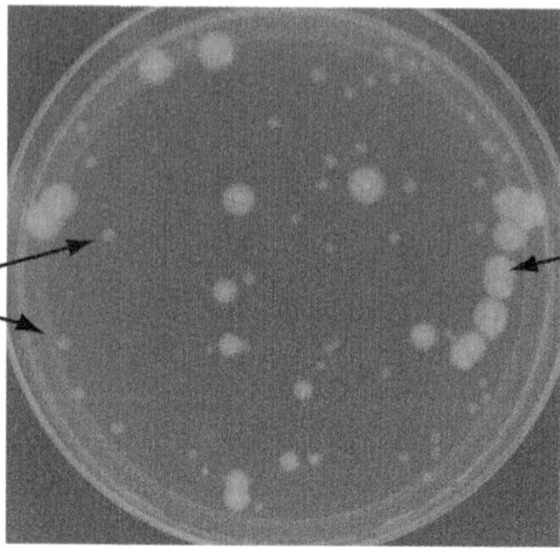

Fig. 2. Growth of wild-type and *plc* mutants on egg yolk BHI agar plates after 24 h of incubation in anaerobic conditions.

When a single transformed bacterium replicates to form a colony, the targetron insertion occurs in some of the progeny bacteria but not all of them. As a result, a single colony coming from one transformed cell may contain some bacteria with the targetron-inserted *plc* genes and some bacteria with wild-type *plc* genes. PCR on such a mixture colony with primers described, produce both 200-bp as well as 1.1-kb PCR products.

Setup colony PCR mixture: pick up the bacteria from a colony with a pipet tip, rinse the bacteria into 5 μL of FTG buffer in a tube. Add following components together:

a. 2.5 μL of the bacterial mixture.
b. 1 μL of 20 μ*M* primer α-toxin F (final concentration = 800 n*M*).
c. 1 μL of 20 μ*M* primer α-toxin 100R (final concentration = 800 n*M*).
d. 2.5 μL of 10X PCR buffer (final concentration = 1X).
e. 1.5 μL of 3.8 m*M* dNTPs (final concentration = 228 μ*M*).
f. 0.25 μL of 5 U/μL *Taq* polymerase (final amount = 2.5 U).
g. 16.25 μL of H_2O.

Perform thermo cycling with initial denaturation, 94°C, 10 min. Amplification, 94°C, 1 min; 54°C, 1 min; 72°C, 2 min for 35 cycles. Final extension, 72°C for 7 min. Run PCR product by agarose gel electrophoresis. Visualize PCR product under UV light.

2. Monitoring α-toxin activity by observing the white halo around colonies on BHI agar containing egg yolk. Because *C. perfringens* α-toxin breaks down lecithin (a normal component of egg yolk) to insoluble diglycerides, an α-toxin producing colony when grown on egg yolk agar plate gives rise to an opaque halo surrounding the colony (**Fig. 2**).

Pick up the PCR-confirmed mixture colony that show two PCR products with a pipet tip, put them into 50 μL of FTG medium (Fluid thioglycolate broth). Make serially 10-fold dilution of the bacteria for eight times with FTG medium. Vortex each dilution vigorously for 30 s. Plate each dilution on egg yolk BHI agar plates. Incubate the plates in the anaerobic jars with GasPaks overnight at 37°C. Pick up the bacterial colonies without surrounding white halo, culture them in FTG medium and incubate overnight at 37°C.

3.5. Confirmation of plc *Gene Inactivation*

To detect the precise targetron insertion point on the chromosome, sequence analysis is performed at joint regions. To monitor whether the targetron inserts only into *plc* gene, Southern blot analysis is performed on the bacterial chromosome DNA using targetron sequence-specific probe.

3.5.1. Analysis of the Joint Regions Between plc *Targetron and Chromosome DNA by Sequencing*

Perform PCR with 5 μL of bacterial culture from colonies without a halo using primer α-toxin F or α-toxin 100R hybridizing to each side of flanking sequences of the insertion site in α-toxin gene. Run PCR amplification products in a 1% agarose gel. Excise the 1.1-kb DNA band and extract the DNA from the gel as described above (*see* **Section 3.1.2**). Sequence the 1.1-kb PCR product using primer α-toxin F or α-toxin 100R. Analyze the sequences using Vector NTI suite (Informax, Oxford, UK).

3.5.2. Southern Blot Analysis on Bacterial Chromosome DNA

3.5.2.1. Purification of *C. perfringens* DNA

Culture *C. perfringens* in BHI medium overnight. Purify *C. perfringens* DNA from the culture using masterpure Gram-positive DNA purification kit according to manufacturer's instruction.

3.5.2.2. Enzymatic Digestion of *C. perfringens* DNA and Southern Blot

1. Mix following components together: 14 μL of purified DNA (containing 10 μg DNA), 2 μL of 10X *Eco*RI enzyme buffer, 2 μL of H_2O, 2 μL of 20 U/μL *Eco*RI enzyme and incubate overnight at 37°C.
2. Separate digested DNA by agarose gel electrophoresis and transfer DNA onto nylon membrane according to manufacture's instruction (*see* **Note 6**). After the transferring, remove the nylon membrane from the cassette; cross-link the membrane twice using a UV Stratalinker 1800.
3. Perform the hybridization and detection according to manufacturer's instruction (*see* **Note 7**).

3.6. Cure of the Targetron Donor Plasmid pJIR750ai From Mutant Bacteria

Although targetron insertional mutagenesis requires the targetron donor plasmid to initiate the targetron integration, the donor plasmid is not required for maintaining the integration. Removing the donor plasmid is necessary for future gene manipulation in the same bacterial cell.

Culture *plc* mutants in FTG lacking chloramphenicol for 5 d with daily subculture. Plate the day 5 subculture on BHI agar plates lacking chloramphenicol, incubate the plates in anaerobic jar with GasPaks overnight at 37°C.

Perform colony PCR with the colonies grown on the plates after overnight incubation using plasmid-specific primers 4931F and pJIR750R. Subculture the colonies with PCR-negative results.

4. Notes

1. When making Modified Duncan-strong medium, make the stock solution of theophylline to 20 mg/mL in ddH_2O. Theophylline is not dissolvable in H_2O with neutral pH. But increasing pH by adding a little 10 *N* NaOH in the solution helps dissolving theophylline in H_2O. Filter theophylline stock solution before use.
2. We have found that *Taq* DNA polymerase introduced some mutations when amplifying the 3.7-kb DNA fragment. It is desirable to use high-fidelity DNA polymerase *pfx* to amplify the 3.7-kb DNA fragment.
3. *C. perfringens* is an anaerobic bacterium. It is preferable to use a narrow culture tube to hold the culture medium to reduce the medium surface area exposed to air. When inoculating *C. perfringens* into a fresh culture medium, put the bacteria near the bottom of the tube, incubate the tube with lid tightly closed in a nondisturbed place at 37°C.
4. Instructions for Zero Blunt TOPO PCR cloning kit from Invitrogen.
5. Keep the cuvet on ice and make sure there is no bubble in the bacterial mixture.
6. Technical Bulletin no. 121 (Amersham Biosciences).
7. Instructions for North2South chemiluminescent hybridization and detection Kit, Pierce.

References

1. Rood, J. I. (1998). Virulence genes of *Clostridium perfringens*. *Annu. Rev. Microbiol.* **52,** 333–360.
2. Lambowitz, A. M. and Zimmerly, S. (2004). Mobile group II introns. *Annu. Rev. Genet.* **38,** 1–35.
3. Chen, Y., McClane, B. A., Fisher, D. J., Rood, J. I., and Gupta, P. (2005) Construction of an α-toxin gene knockout mutant of *Clostridium perfringens* type A by use of a mobile group II intron. *Appl. Environ. Microbiol.* **71,** 7542–7547.
4. Fisher, D. J., Miyamoto, K., Harrison, B., Akimoto, S., Sarker, M. R., and McClane, B. A. (2005) Association of β-2 toxin production with *Clostridium perfringens* type A human gastrointestinal disease isolates carrying a plasmid enterotoxin gene. *Mol. Microbiol.* **56,** 747–762.

Index

GPSR Compliance

The European Union's (EU) General Product Safety Regulation (GPSR) is a set of rules that requires consumer products to be safe and our obligations to ensure this.

If you have any concerns about our products, you can contact us on ProductSafety@springernature.com

In case Publisher is established outside the EU, the EU authorized representative is:

Springer Nature Customer Service Center GmbH
Europaplatz 3
69115 Heidelberg, Germany

Batch number: 10614999

Printed by Printforce, the Netherlands